T0405838

Heat and Mass Transfer

Series Editors: *D. Mewes and F. Mayinger*

Henning Bockhorn · Dieter Mewes ·
Wolfgang Peukert · Hans-Joachim Warnecke

Micro and Macro Mixing

Analysis, Simulation and Numerical Calculation

With 202 Figures

 Springer

Series Editors

Prof. Dr.-Ing. Dr.h.c. Dieter Mewes
Universität Hannover
Inst. Mehrphasenprozesse
Callinstr. 36
30167 Hannover, Germany

Prof. Dr.-Ing. Dr.E.h.mult. Franz Mayinger
Technische Universität München
Lehrstuhl für Thermodynamik
85747 Garching, Germany

Editors

Prof. Dr. Henning Bockhorn
Universität Karlsruhe
Inst. Technische Chemie und
Polymerchemie
Kaiserstr. 12
76131 Karlsruhe, Germany

Prof. Dr.-Ing. Wolfgang Peukert
Universität Erlangen-Nürnberg
LS für Feststoff- und
Grenzflächenverfahrenstechnik
Cauerstr. 4
91058 Erlangen, Germany

Prof. Dr.-Ing. Dr.h.c. Dieter Mewes
Universität Hannover
Inst. Mehrphasenprozesse
Callinstr. 36
30167 Hannover, Germany

Prof. Dr.-Ing. Hans-Joachim Warnecke
Universität Paderborn
Fak. Naturwissenschaften
FG Technische Chemie
Warburger Str. 100
33098 Paderborn, Germany

ISBN 978-3-642-04548-6   e-ISBN 978-3-642-04549-3

DOI 10.1007/978-3-642-04549-3

Heat and Mass Transfer   ISSN 1860-4846

Library of Congress Control Number: 2009937158

Typeset design: Scientific Publishing Services Pvt. Ltd., Chennai, India.

Cover design: Deblik, Berlin

Printed in acid-free paper

9 8 7 6 5 4 3 2 1

springer.com

Preface

The homogenization of single phase gases or liquids with chemical reactive components by mixing belongs to one of the oldest basic operations applied in chemical engineering. The mixing process is used as an essential step in nearly all processes of the chemical industry as well as the pharmaceutical and food industries. Recent experimentally and theoretically based results from research work lead to a fairly good prediction of the velocity fields in differend kinds of mixers, where as predictions of simultaneously proceeding homogeneous chemical reactions, are still not reliable in a similar way. Therefore the design of equipment for mixing processes is still derived from measurements of the so called "mixing time" which is related to the applied methods of measurement and the special design of the test equipment itself.

The cooperation of 17 research groups was stimulated by improved modern methods for experimental research and visualization, for simulations and numerical calculations of mixing and chemical reactions in micro and macro scale of time and local coordinates. The research work was financed for a six years period within the recently finished Priority Program of the German Research Foundation (DFG) named "Analysis, modeling and numerical prediction of flow-mixig with and without chemical reactions (SPP 1141)". The objective of the investigations was to improve the prediction of efficiencies and selectivities of chemical reactions on macroscopic scale. The results should give an understanding of the influence of the design of different mixing equipment on to the momentum, heat and masss transfer as well as reaction processes running on microcopic scales of time and local coordinates.

The careful analysis of the previous existing results for mixing processes in equipment with different velocity fields revealed

- a lack in measurement methods for three dimensional concentration and temperature fields giving a high accuracy in time and spacial coordinates,
- a very limited set of closure equations derived from experiments inter relating velocity- and concentration profiles with simultaneously running homogeneous chemical reactions,
- large deficiencies in extended knowledge for the derivation of models predicting relations between the different scales for mixing processes in macroscopic single phase flows with simultaneous mass transport and chemical reactions.

Based on these deficiencies in our basic knowledge in micro and macro-mixing, the scientific objectives for the Priority Program were derived.

The result was a joined effort of cooperating scientists proposing new experimentally and theoretically based research projects in order to develop new insights into the physics and reaction kinetics of mixing processes in single phase liquid mixtures of varying concentrations and rheological behavior with and without chemical reactions. On one hand side momentum-, heat- and mass transport on different spacial and time scales were considered, on the other hand side simultaneously running chemical reactions were applied. The latter were studied locally on molecular scale with different reaction kinetics according to their time scales.

As results newly developed methods of measurement are adjusted to the scales of the selected special transport and conversion processes. They allow a more detailed modelling of the mixing processes by the formulation of an appropriate set of momentum-, heat- and masss balance equations as well as boundary conditions in time and spacial coordinates together with constitutive equations and reaction kinetics equations as closure laws for numerical and analytical calculations. The latter were empirically derived in the past and therefore of limited reliability only.

The improved and more detailed modelling leads to a major progress in predicting mixing processes on the different scales adjusted to transport and reaction processes in molecular, micro- and macro dimensions.

As a consequence improved numerical calculations are performed on the basis of newly derived experimental, measurement and modelling methods which are the basis for the prediction of mixing time as well as conversion rates and selectivities of chemical reactions during the mixing process. The research efforts are focused onto the design of the technical equipment for flow mixing processes. Mixing is performed inside velocity fields leading to deformation gradients from free or wall induced boundary layers. The different kinds of process equipment are jet mixer, static mixer and mixing vessels equipped with rotating stirrers. Especially in micro mixing newly developed constructions are investigated permitting the scale up from laboratory to technical dimensions.

The results from the Priority Program are presented in three parts named by the working groups:

- Experimental methods for visualization and measurements in macro- and micro-scale dimensions of mixing.
- Theoretical methods for modelling and numerical calculations of mixing processes.
- Macro- and micro-mixing in micro channel flow.

The three groups were established, in order to enhance the cooperation between the involved researcers, whose background ranged from technical chemistry and chemical engineering to applied mathematics and micro processing.

The entire financial support of about 6.5 million Euro granted by the Deutsche Forschungsgemeinschaft was greatly appreciated by the involved researchers and authors of the following chapters of this book. Special thanks goes to the group of referees: Dr. Appel, Dr. Benfer, Prof. Ebert, Prof. Janicka, Prof. Schmidt-Traub,

Prof. Schubert and Prof. Sommer, who evaluated the proposals for three consecutive two years periods and to Dr. Giernoth, who administered the program.

August 2009   Henning Bockhorn
                                                        Dieter Mewes
                                                     Wolfgang Peukert
                                              Hans-Joachim Warnecke

Contents

Part 1: Quality of Mixing

Part 2: Experimental Methods for Visualization and Measurements in Macro- and Micro-Scale Dimensions of Mixing

Part 3: Theoretical Methods for Modelling and Numerical Calculations of Mixing Processes

Part 4: Macro- and Micro-Mixing in Micro Channel Flow

Part 1: Quality of Mixing

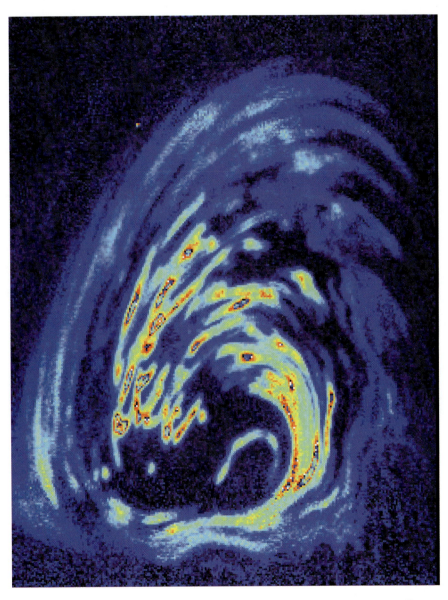

Local concentrations of an inert dye and a reacting dye inside the vortex adjacent to a rotating impeller (from Kling, Mewes see page …..)

The Variance as Measured Variable for the Evaluation of a Mixing Process or for the Comparison of Mixtures and Mixers

Karl Sommer

Lehrstuhl für Verfahrenstechnik disperser Systeme
Technische Universität München
Am Forum 2, 85354 Freising
k.sommer@bl.tum.de

1 Introduction

In the DFG program: *"Analysis, modelling and calculation of flow mixers with and without chemical reactions (SPP 1441)"* there are used different mixing machines (reactors), different objectives and different dimensions. All research programs, however, have in common, that they have to evaluate the mixture concerning its "quality". There are totally different measured variables, measuring methods and methods of taking samples employed. Together with different "mixedness definitions", it is difficult, if not impossible, to compare the results with regard to their "mixing efficiency". The publication presented tries to express the fundamental requirements and dependencies of the definition "mixedness" to avoid these problems.

2 Total Mixture

From the theory of fluid flow it is well known that a turbulent flow can be divided into macroscopic convection and a microscopic vortex-like velocity field which is dominated by the diffusion. Caused by this mechanistic imagination the mixing process is often divided into the "macro- and micro-mixing". Such a division may make sense for a description of a model and thus for the time development of a mixing procedure, but not for the definition of a "mixedness" as description for the condition of a "mixture". Later it will be shown that for instance a "macro-mixedness" cannot be compared with a "micro-mixedness" or cannot even be considered as a competitor (for example "the macro-mixture is better than the micro-mixture or vice versa"). According to this description of the actual condition first of all there has to be defined the mass being the "total mix". With batch mixing it is mostly the total contents of the vessel. When mixing continuously the interval has to be defined, for which the emerging mass flow is measured as "total mix". A "continuous" mixedness is not reasonable and can be defined as quality of cross-mixture only for a specific period Δt (even though it may be a very short, but measurable time period), which represents then the "total mix".

3 Mixedness

3.1 Measurable Properties

Aim of the procedure of mixing is the "homogeneity" of the product. Lexically, homogeneity means that something is "consistently composed", "spacially equal", "steady", "all of a piece". All these expressions refer to measurable values of two or more products which are to be mixed or of one product with different physical properties (for example temperature) or both. A product is homogeneous, if every part of mass (or space) of a mix has the same measurable property. A product is inhomogeneous, if the properties can differ from point to point, even if an ideal, accurate measurement method would exist. In the procedure of batch mixing there is a functional relationship between the measurable properties and the location in the mix which is to be examined. In case of a continuous mixing process there is a functional relationship with the time (for example at a fixed point of measure at the outlet). These functions are local (resp. temporary) property functions $\xi(r)$ resp. $\xi(t)$. The choice of the measured variable for defining the properties which "are to be mixed" is eligible. If possible, it should be identical to the relevant application properties. In many cases, for practical reasons the measured variable is different property. Then there has to be a strict correlation between the application property ξ_A and the measure property ξ_M.

$$\xi_A = f(\xi_M) \tag{3.1}$$

The most common property for mixing analysis is the concentration. It is assumed that the concentration rules all other application properties. In practice, however, the concentration is not easy to measure. Therefore, other measurement properties (for example optical ones) are applied. In order to compare the measurement results with the results of other scientists, there should be always made a reference to the concentration, for example by calibration.

3.2 Definition of Mixedness

The mixedness is an own quality characteristic, which does not judge the property itself, but only its homogeneity. The term "mixing" is defined as the "distribution of different masses in a given volume", where the masses differ at least in one property, for example in colour, density, viscosity etc. The aim of the mixing is the homogeneity of the components and thus, of the afore-mentioned properties. A degree of mixedness has to accommodate the definition of the mixing as well as the aim of the mixing process. The description of homogeneity can be made qualitatively or quantitatively. The qualitative methods judge from appearance. This, for example, can be made by spreading a sample on a board and evaluating the parts of inhomogeneity. Frequently, there is taken a standard with points or marks and one assigns the mixed product to the system. The collection and analysis of data for example is technically feasible for frozen or photographed samples with help of the automatic picture-analysis. Generally, as with molecular-disperse systems one limits the information on temporal and local concentration functions. When analysing the data quantitatively it is a precondition that the aforementioned properties present measurable factors, both as property of the "pure" components and of all mixing conditions. A full description of the mix - with respect to the properties to be looked at – is given if one knows the properties on every spot r (or at any time t) in the volume given. They indicate the maximum information on the inhomogeneity of the measured variable. Local, or rather time property functions would give a full degree of mixedness. The local (time) property functions, however, have a some big practical disadvantage:

1. The whole mix has to be analysed, that means measuring of properties on every spot.
2. There are infinitely possible property functions.
3. Non-identical property functions are hardly comparable amongst themselves.

In practice the property functions are not or at least hardly comparable. In most of the cases it is not necessary. One confines oneself (of course under loss of information) to the investigation and the comparison of mean variables.

a) The expected variable E (ξ) is calculated out of the local property functions $\xi(r)$:

$$E(\xi) = \frac{1}{R} \cdot \int_0^R \xi(r) \, dr \qquad (3.2)$$

With homogeneous goods, this expected value is identical to the "real" value. A comparison of different property functions makes only sense – with some exceptions – for identical reference values (real values).

b) Theoretical variance with infinite small sample size
Infinite small sample size in the used sense means the property is known in every point of location within the volume of the mixing product. $\sigma_{syst}^2(\xi)$ is calculated as follows:

$$\sigma_{syst}^2 = \frac{1}{R} \cdot \int_0^R (E(\xi) - \xi(r))^2 \, dr \qquad (3.3)$$

$\sigma_{syst}^2(\xi)$ is identical to the variance known from statistics as the mean square deviation of the reference value. It is here unimportant, if the deviation is positive or negative. The variance is in accordance with our understanding of identifying a mixture. **The theoretical variance, therefore, is a good measure for mixedness.** (The value is called "theoretical", because of the calculation of the variance with the full distribution of properties in every spot). For such exact values, in this publication there are always used Greek letters. The variance stresses huge deviations considerably more (square of deviation) than smaller deviations. Only the user can recognize, if this is reasonable for the properties of a mixed product. Due to the

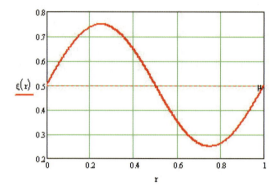

$$\sigma_{syst}^{2} = \frac{\Delta p^{2}}{2} = 0.03$$

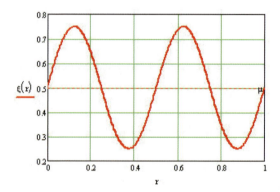

$$\sigma_{syst}^{2} = \frac{\Delta p^{2}}{2} = 0.03$$

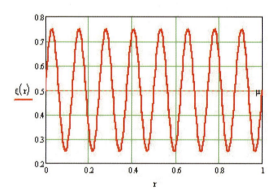

$$\sigma_{syst}^{2} = \frac{\Delta p^{2}}{2} = 0.03$$

Fig. 3.1. Spacial concentration frequency to calculate the variance σ_{syst}^{2}

calculus provided by the statistics, the variance has established itself as the degree for mixedness. Many other mixing qualities are deducted from the variance. The different mixedness degrees shall either "standardise" the empirical standard deviation or gain an increased independency against sample size or sample concentration. Frequently, variances as degree for mixedness are classified overall "too global". The reason for this is that the user observes a big amount of the mixing product whereas emotionally he has a much smaller amount of mixing product in mind. If the variance is classified "too global", it has to be checked, if the amount of the mixing product taken for the measurement corresponds to the application. The same has to be considered for the scale of segregation, which describes the structure of the property distribution "in the neighbourhood" of given points of location. Thus, reference is made to parts of the mixing goods, instead of making a global consideration as assumed before. Under consideration of this fact, the variance is an ideal degree of mixedness. For "sinusoidal" concentration distributions with the constant amplitude Δp, we get the variances:

$$\sigma_{syst}^{2} = \frac{\Delta p^{2}}{2} \tag{3.4}$$

The implementation of variances according to equation (3.3) as degree of mixedness means a loss of information to such an extent, that also different local property distributions with higher frequency have the same variance (Figure 3.1). This does not seem to be logical, because the higher frequent distribution of concentration seems to be mixed better. The results differ when the sample sizes get infinite.

4 Choice of the Sample Size (Theoretical Variance σ^{2}_{N} with Finite Sample Size)

In practice it is not possible to measure the property in infinite small volumes, but it is necessary to take finite sample sizes. This means, it is possible to divide the total mix in N samples of a finite size. The variance

$$\sigma_{N}^{2} = \frac{1}{N} \cdot \sum_{i} \left(E(\xi) - \xi(r_{i}) \right)^{2} \tag{4.1}$$

is also called theoretical variance, because it contains all possible samples and, thus, presumes knowledge about the whole mixed product. The value of σ^{2}_{N} is different to σ_{syst}^{2} dependent on the sample size. For the sinusoidal concentration distribution the ratio of the theoretical variance $\sigma^{2}_{N}/\sigma_{syst}^{2}$ is dependent on the sample size, expressed as part of a period (Figure 4.1).

One recognizes that for sample sizes of 1/10 of the period the variances are nearly identical. For bigger samples the variance σ^{2}_{N} is getting smaller, that means the "mixedness" improves. Some information about the distribution of properties

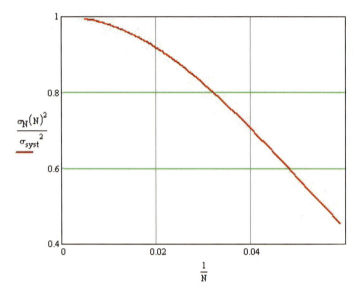

Fig. 4.1. The theoretical variance $\sigma^2_N/\sigma_{syst}^2$ dependent on the sample size, expressed as part of a period

within a sample is "lost", that means, there is a better measured mixedness. This becomes even more apparent, the nearer the sample size comes to the "range of fluctuation" (for example periods). For Figure 4.2, with increasing sample size, first the small variations (micro-mixings) and then the bigger variations (macro-mixings) would be "smoothened" by sampling. Whether the sample size is a reasonable degree of mixedness the application has to show. The sample size has to be chosen in such a way, that probable "inhomogeneities" within the sample do not have any effect on the subsequent application. The choice of sample size has therefore to be given by the user, after critical consideration. If the sample size is too large (often the method of measurement determines the sample size), the mixing quality measurement is too uncritical, that means, it has not enough significance for the appliance. If the sample size is chosen too small, the requirements towards homogeneity are bigger than necessary and the mixing process will me made difficult or impossible. An extreme case: If the sample size according to the molecular size is chosen, the mixing always seems to be demixed even if the molecules would be perfectly arranged, like on a chess-board. <u>Mixing qualities can only be compared with equal sample sizes.</u>

The determination of the measured variable ξ is only possible with a finite accuracy. This measuring accuracy is normally only to be defined by preliminary tests. The variance of the measuring accuracy has to be estimated, for example, by means of a known, preset property with sufficient measurements of the chosen sample size. It must be taken care that the measurement technique corresponds closely to reality. Furthermore, one has to check how strong the measuring

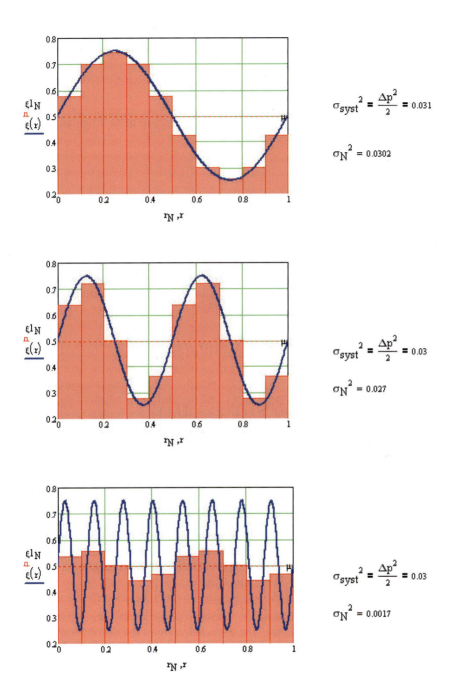

Fig. 4.2. Spacial concentration frequency to calculate the variance σ_N^2

accuracy is affected by the property itself. The variance of the measuring accuracy is very important for the calculation of the "Confidence interval" of the real value of a mixing product. Independency assumed, it is imperative:

$$\sigma_{ges}^2 = \sigma_{syst}^2 + \sigma_M^2 \tag{4.2}$$

The variances due to measurement inaccuracies are not to be to differentiate from the variances of the mixing conditions.

5 Empirical Variances

5.1 Definition of the Empirical Variance s^2

In praxis, for the investigation of the mixedness, neither the analysis of an infinite number of samples (as with σ_{syst}^2) nor the analysis of all samples N of a mixing is possible. In fact, there has to be taken a finite number k of samples for the analysis of the mixedness. Consequently, according to the distribution data for the expected value μ and variances σ^2 it is imperative:

$$\xi_m = \frac{1}{k} \cdot \sum_i \xi(r_i) \tag{5.1}$$

$$s_{syst}^2 = \frac{1}{k} \cdot \sum_i \left(E(\xi) - \xi(r_i) \right)^2 \tag{5.2}$$

$$s_{syst}^2 = \frac{1}{k-1} \cdot \sum_i \left(\xi_m - \xi(r_i) \right)^2 \tag{5.3}$$

Equation (5.2) is taken, if the expected value (reference value) $E(\xi)$ of the mixing is known with the so-called degree of freedom $f = k$. Equation (5.3) is taken, if it has to be estimated from the average value ξ_m, with the degree of freedom $f = (k-1)$. Equation (5.2) and 5.3) are equivalent.

Also when having identical mixing conditions, one receives different values of s^2 if one makes repeated analysis of s^2. These s^2, consequently, are statistical values forming an empirical variance. The expected value of the distribution is in accordance with the theoretical mixedness σ^2:

$$E\left(s_{syst}^2 \right) = \sigma_{syst}^2 \tag{5.4}$$

$$E\left(s_N^2\right) = \sigma_N^2 \tag{5.5}$$

s^2_{syst}, s^2_N therefore are estimated values for the mixedness, they are fluctuating around their expected value σ^2_{syst}, σ^2_N. With help of the empirical variance it is not possible to draw safe conclusions, however, one can draw conclusions, which are "safe" with a certain simulated probability.

5.2 Chi-Square Distribution

As mentioned before, the estimated values s^2 usually vary from the expected value σ^2. In order to make safe probability conclusions concerning the extent of the variance (safe conclusions are not possible in mathematical statistics), one has to know the distribution function of s^2. In the following it is assumed that the distribution of the property values can be described sufficiently accurate by a Gaussian distribution (normal distribution). Except for an initial phase, when both components are still largely separated, this assumption will be generally fulfilled for the usual sample sizes. It can also be explained theoretically with help of the central limit theorem of statistics. Then, s^2 form a Chi-Square distribution.(Figure 5.1). The degree of freedom is either (k-1) or the sample number k.

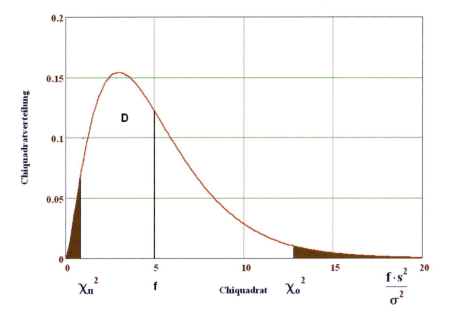

Fig. 5.1. Chi-square-Distribution

6   Confidence Intervals

When determining the mixedness (empirical variance), the real mixedness (theoretical variance) is unknown. The Chi-square-distribution shows, that with probability D and the degree of freedom it is:

$$\chi_o^2 \geq f \cdot \frac{s^2}{\sigma^2} > \chi_u^2 \tag{6.1}$$

This conclusion remains also valid, if the expected distribution value, the theoretical variance, is unknown. This conclusion remains also valid, if equation (6.1) is changed as follows:

$$s^2 \cdot \frac{f}{\chi_o^2} < \sigma^2 < s^2 \cdot \frac{f}{\chi_u^2} \tag{6.2}$$

The conclusion now is: With probability D the theoretical variance lies between the given limits (confidence interval). Misleadingly, with this result one often implies, that there is a distribution of the theoretical variance. That is not the case! The theoretical variance is a fixed quality value (mixedness) around which the empirical standard deviations are still fluctuating.

7   Choice of the Number of Samples

To compare measured mixing qualities only make sense with the calculated confidence intervals as mentioned above. If the two confidential intervals do not overlap, one can draw the conclusion – with the chosen probability - that the mixedness differs, independent of the number of samples. Do the confidence intervals overlap, the statistical conclusion is: "It can – with the given probability - *not be excluded* that both mixing products are equal". The conclusion is not, that with the given probability both mixing products *are identical*, moreover, the confidence intervals at small numbers of samples will overlap nevertheless, although the mixedness in real will be different (producer's risk). In order to minimise this risk, the amount of samples can be increased.(Figure 7.1) The amount of samples is only crucial for the producer's risk and can be chosen freely by the experimenter under these preconditions. In practice, for financial reasons it is often reasonable, to choose the samples successively and to go on analysing them, until the quality of the mixes differ. If the confidence intervals are also overlapping when having a huge number of samples, then this is an indication that the mixing qualities are very similar or alike (Figure 7.2). A total equality cannot be confirmed with a finite number of samples.

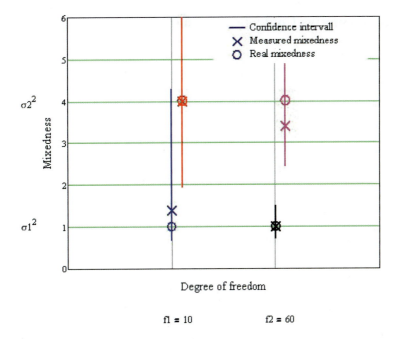

Fig. 7.1. Confidence intervals with different mixedness (variances)

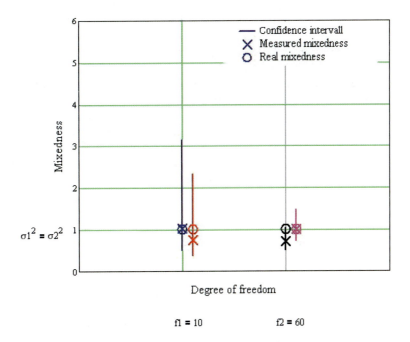

Fig. 7.2. Confidence intervals with equal mixedness (variances)

8 Representative Controlled Sampling

In order to define the mixedness one has to use measuring values out of a number of samples of finite size. An assumption, which has to be made in this connection, is the representativeness of the samples taken. The general public understands "representative" as being "identical to". When evaluating mixing products, the samples shall allow that one can draw a conclusion on the mixedness in form of the theoretical variance. The empirical variance which is to be calculated out of a finite number of samples has – therefore – to be an unbiased estimated value of the theoretical variance. This is possible in two ways:

a) Regular sampling
The theoretical variance follows from the local (resp. temporal) property function. A representative theoretical variance can be calculated if the measured properties display this profile sufficiently. This way of sampling is frequently realised by "continuous measuring" of the properties at the end of a continuous mixing. However, at such a systematic measuring it has to be guaranteed that the temporal (local) progression of the measuring points can collect the frequencies of the property functions adequately correct. For the sample size the contents of chapter 3 applies.

b) Stochastic sampling
If a sufficient measuring of the concentration profile is not possible, a conclusion about the theoretical variance has to be made possible with a lesser number of samples. Under utilization of the central limit theorem it is possible – for any distributions – to approximate the property functions by a normal distribution with the same expected value and with the same variance. Then by stochastic sampling one can estimate the real value as well as the theoretical variance (mixedness) with a certain probability (confidence intervals). Stochastic sampling: After the choice of the sample size the total content of the mixer, the mixing product or the charge is divided into parts of the same size and every part is numbered. After choosing the necessary number of samples, with a stochastic generator the indices of samples will be chosen. The average value ξ_m of these samples is an estimated value for the real value (expected value) $E(\xi)$ and the empirical variance s^2 is an estimated value for the theoretical variance σ^2.

9 Conclusion

For analysing a mixing procedure, for the evaluation and the comparison of mixing products, the following rules are to be obeyed:

1. The theoretical variance of property functions is an optimal degree of mixedness
2. The amount of the mixing product, for which the analysed mixedness shall be valid, has to be defined.

3. The measured variable has to be in an obvious functional connection to the relevant properties. When comparing different measuring methods, the properties have to be clearly convertable towards each other (calibration).
4. The choice of the sampling size is made according to application-technological principles. A comparison between different kinds of measures of mixedness is only possible if identical sample sizes are taken.
5. Sampling is made systematically or stochastically. The latter requires, that the whole mixing product is virtually divided into sample-size-units. With the help of a stochastic generator the samples are chosen and measured.
6. The scientific success in comparing the mixing qualities can only be well founded with the respective confidence intervals.

Evaluating the Quality of a Mixture: Degree of Homogeneity and Scale of Segregation

Dedicated to Hans-Joachim Warnecke

Dieter Bothe

Center of Smart Interfaces, TU Darmstadt
Petersenstraße 32, 64287 Darmstadt

Abstract. The evaluation of mixing requires measures for the quality of mixtures. Integral measures like the intensity of segregation are advantageous for practical purposes and such variance-based measures are shown to be especially meaningful in the context of reactive mixing. But the variance is insensitive to the scales on which segregation occurs, hence it is complemented by an integral scale of segregation. The latter is closely related to the concept of contact area and can be efficiently computed using a formula from geometric measure theory. This integral scale of segregation is also characterized by the rate at which the minimizing-total-variation-flow from image processing would alter the intensity of segregation. Applied to secondary flow mixing in a T-shaped micro-mixer, the dependence of the integral scale of segregation shows a new type of dependence on the Schmidt number, while the smallest length scales display the kind of dependence known from the Batchelor's length scale.

Keywords: Specific contact area, intensity of mixing, integral length scales, Schmidt number dependence, decay of variance.

1 Introduction

The term *mixing* refers to the reduction of inhomogeneities concerning a certain property like color, density, composition or temperature which are initially present in the material to be mixed. Depending on the kind of application, a decrease of the length scales on which these variations are present or reductions of their amplitudes – or both – are desired. While it may be sufficient to diminish the scale at which segregation persists below some moderate value to obtain the desired product quality in case of the production of blends, mixing on the molecular level is necessary for any chemical reaction to occur. The latter can only be achieved by molecular diffusion, i.e. by the chaotic thermal motion of particles, and, hence, only a statistical homogeneity can be reached. Since molecular diffusion can only affect the macroscopic distribution if gradients in the respective property are present, the aim of convective macro- or mesoscopic mixing processes is to generate gradients, or to shrink the length scale at which these gradients occur, although the final goal is to diminish them. The rate of many reactive processes is in fact controlled by mixing and not by diffusion – hence a deeper understanding of mixing is highly demanded.

Optimization of mixing devices or processes requires quantitative means to evaluate mixing. Since mixing requires relative motion within the material to be mixed, the underlying kinematics is of utmost importance. Hence, in particular in the case of fluid mixing, a thorough analysis of the velocity field can yield valuable information about convective mixing. Therefore, methods from the mathematical theory of dynamical systems like Poincaré sections, rates of length- and area-stretch as well as Lyapunov exponents are employed to assess the mixing efficiency; cf. the monograph of Ottino (1989). Since Lyapunov exponents quantify the sensitivity of the flow trajectories to initial conditions, they can prove the appearance of so-called chaotic advection (Aref, 1984). A fundamental quantity for this kind of analysis is the velocity gradient $\nabla \mathbf{v}$, respectively the rate of strain tensor

$$\mathbf{D} = \left(\nabla \mathbf{v} + \nabla \mathbf{v}^{\mathsf{T}}\right), \tag{1}$$

since $\mathbf{D}$ constitutes that part of $\nabla \mathbf{v}$ which corresponds to relative motion. In viscous fluids, the latter leads to dissipation of energy at the specific rate

$$\varepsilon = 2v\, \mathbf{D} : \nabla \mathbf{v}, \tag{2}$$

where v denotes the kinematical viscosity. Therefore, the energy dissipation rate ε is often employed as a measure of mixing efficiency, especially for turbulent mixing; cf. Finn et al. (2004), or one of the monographs Frisch (1995), Baldyga and Bourne (1999), Pope (2000) or Fox (2003) as well as the contributions on turbulent mixing in the present volume. In fact, a true measure of the efficiency of mixing should relate the achieved degree of mixing to the energy required. In this respect, it is shown in Raynal and Gence (1997) that chaotic laminar mixing can be more energy-efficient than turbulent mixing. Let us also note that (2) is appropriate for evaluation of isotropic mixing, while modifications of (2) are required in case of continuous processes like mixing in pipes or ducts with axial main flow but desired mixing only in cross directions; see Bothe et al. (2006).

Besides local measures for the rate of progress of a mixing process, integral and even global quantities for the evaluation of mixing are needed. Such condensed information, for example a mixing time, is especially important in case of industrial applications. While there are many different definitions of mixing time, any such definition evidently requires a measure of the integral or global quality of mixing. The present paper focuses on this aspect; i.e., given a scalar field $c(\mathbf{x})$ which can be considered as the (molar) concentration of a chemical species, what mixing condition does the scalar field possess? In the following, main emphasis is given on the intensity and scale of segregation, their intrinsic meaning in the context of chemical reaction, possible extensions and efficient ways for their computation. There is a bulk of related work: Reviews concerning the evaluation of mixing quality mainly by statistical means are provided in Fan et al. (1970) for mixing of solids and in Boss (1986) and in Rielly et al. (1994) for liquid and solid mixing. Below, we only touch upon the issue of reactive mixing. For mixing with chemical reaction see Villermaux (1982), Ottino (1994), Baldyga and Bourne (1999) and Fox (2003). Further references to related work are given at the appropriate places later on.

2 Homogeneity of a Mixture

Since mixing of educts is necessary for their chemical reaction, the rate at which a chemical reaction occurs depends in particular on the homogeneity of the mixture and, hence, can be used to measure the degree of homogeneity. In the typical case of a bimolecular reaction of the type

$$A + B \xrightarrow{k} P , \tag{3}$$

this leads to the quantity

$$\langle k\, c_A\, c_B \rangle = \frac{1}{|V|} \int_V k\, c_A\, c_B\, dV \tag{4}$$

as a measure of the degree of mixing inside the domain V, where $|V|$ denotes the volume of the spatial region V. Recall that $k\, c_A\, c_B$ is the reaction rate according to mass action kinetics, where c_A and c_B denote the molar concentrations of A and B, respectively.

Evidently, the quantity in (4) depends on the rate constant k and this would result in arbitrary values. Therefore, instead of (4) the ratio

$$\frac{\langle k\, c_A\, c_B \rangle}{k \langle c_A \rangle \langle c_B \rangle} = \frac{\text{integral rate of reaction at given mixing condition}}{\text{rate of reaction at homogeneous mixing condition}} \tag{5}$$

yields a meaningful measure of the degree of homogeneity. Since k cancels, the quantity in (5) is independent of the particular rate constant. The latter is important for experimental realizations of this approach, since it allows for the use of extremely fast (instantaneous) reactions in which case the (finitely fast) transport processes do not change this quantity on the reaction time scale. In fact, as reported in Hiby (1981), it has already been noticed in Toor (1962) that the degree of conversion of a neutralization reaction coincides with the appropriately defined degree of mixing. This leads to

$$\frac{\langle c_A\, c_B \rangle}{\langle c_A \rangle \langle c_B \rangle} = 1 + \frac{\langle c_A\, c_B \rangle - \langle c_A \rangle \langle c_B \rangle}{\langle c_A \rangle \langle c_B \rangle} = 1 + \frac{\mathrm{Cov}[c_A, c_B]}{\langle c_A \rangle \langle c_B \rangle} , \tag{6}$$

where $\mathrm{Cov}[X, Y]$ denotes the covariance of square-integrable functions X and Y. The ratio $\langle c_A\, c_B \rangle / \langle c_A \rangle \langle c_B \rangle$ in (6) is called segregation coefficient in Warhaft (2000) in the context of the modeling of turbulent reactive mixing; this quantity is nothing but the closure term that is needed for a scale-reduced model based on concentration averages.

In the special case $A = B$, i.e. in case of a single educt A which reacts according to $A + A \xrightarrow{k} P$, the quantity in (6) reduces to

$$1 + \frac{\langle c_A^2 \rangle - \langle c_A \rangle^2}{\langle c_A \rangle^2} = 1 + \frac{\sigma^2}{\mu^2}. \tag{7}$$

This value is closely related to the variation coefficient σ/μ from statistics, where σ denotes the standard deviation and μ the expectation. For given experimental or numerical data in the form of discrete values

$$c_i, \quad i = 1, \ldots, N, \tag{8}$$

belonging to N spatial cells of equal volume, the expectation μ and the variance σ^2 are estimated by the sample mean

$$\bar{c} = \frac{1}{N} \sum_{i=1}^{N} c_i \tag{9}$$

and the sample variance

$$s^2 = \frac{1}{N-1} \sum_{i=1}^{N} (c_i - \bar{c})^2, \tag{10}$$

respectively.

Alternatively, the concept of entropy can be applied to evaluate the state of a mixture. Indeed, a so-called quality of mixedness which is based on entropy of mixing is defined in Ogawa and Ito (1975). Despite the fact that both mixing and entropy are intuitively related to the amount of disorder in a system, this approach is rarely employed. But, recently, entropic measures of mixing received new interest in the context of so-called Renyi entropies; see Wang et al. (2003).

Let us note in passing that the segregation coefficient from (6) is related to the so-called spatial mixing deficiency,

$$\text{SMD} = \frac{\sigma_{\text{space}}[\phi(\mathbf{x})]}{\langle \phi(\mathbf{x}) \rangle_{\text{space}}} \quad \text{with} \quad \phi(\mathbf{x}) = \langle c(t, \mathbf{x}) \rangle_{\text{time}}, \tag{11}$$

which is used in Prière et al. (2004) to evaluate the mixing of a scalar quantity in turbulent flows. Note that SMD is nothing but the variation coefficient of the previously time-averaged scalar field, i.e. the time and space dependent concentration is averaged with respect to time before the spatial homogeneity of this averaged field is computed. Implications concerning the rate of chemical reactions cannot be drawn from the value of SMD, since a time-dependent completely segregated quantity can appear as homogeneous after time-averaging.

In case of a transient species distribution $c(t,\mathbf{x})$, the standard deviation $\sigma = \sigma(t)$ and, hence, the variation coefficient will also depend on time, although the mean value stays constant if no flux of tracer across the domain boundaries occurs. The evolution of $\sigma(t)$ indicates how the quality of mixing changes in the course of time and can be used to define a mixing time. But, while the segregation coefficient σ/μ allows for an evaluation of the homogeneity of a scalar field at a given time, it cannot be directly employed for the definition of a mixing time or a mixing length. This is due to the fact that it is not normalized to the range $0\ldots1$, say, but can attain any value between zero and infinity. Hence, while an ideally homogeneous mixture has a segregation coefficient of zero, it is not clear which threshold should be reached during a mixing process. An appropriate normalization can be derived more easily in case of non-reactive mixing and for an isolated mixing region. As mentioned above, the expectation μ stays constant then, hence only the variance σ^2 needs to be normalized. This leads to

$$I = \frac{\sigma^2}{\sigma_{max}^2} , \tag{12}$$

where σ_{max}^2 denotes the maximum variance. Unfortunately, the latter is not defined in general. In case of a binary system, consideration of the molar fraction

$$x_A = \frac{c_A}{c_{total}} \tag{13}$$

instead of the molar concentration yields

$$\sigma_{max}^2 = \mu\left(1-\mu\right) \tag{14}$$

as the maximum possible variance which is attained for any totally segregated distribution of the same amount of species A. These considerations result in Danckwerts' intensity of segregation (Danckwerts, 1952), defined by

$$I_S = \frac{\sigma^2}{\mu\left(1-\mu\right)} . \tag{15}$$

Based on Danckwerts' intensity of segregation, measures for the intensity of mixing can be derived and a large number of related measures which are similar in the sense that they depend on statistical parameters are reviewed in Boss (1986). A typical example, which is recommended there, is

$$I_M = 1 - \sqrt{I_S} = 1 - \frac{\sigma}{\sigma_{max}} . \tag{16}$$

Since I_S is normalized, this measure is zero for completely segregated mixtures and attains the value one in the homogeneously mixed case.

A general definition of σ_{max}^2 is problematic, since a universal maximum value for the molar concentration does not exist. One possible adaptation to molar concentrations or, more generally, intensities is to let σ_{max}^2 be the variance of a totally segregated distribution composed of the same amount of substance or the same total intensity and the same maximum value c_{max}. Again, this variance is independent of the concrete structure of the segregated field and leads to $\sigma_{\max}^2 = \mu\,(c_{\max} - \mu)$. This reference value allows for a normalization of the variance of the distribution based on the scalar field itself, i.e. without need for an additional external value. This is useful for assessing a species distribution by itself, but, since $c_{\max}$ will usually change with time or space it cannot be employed for defining a mixing time or a mixing length. On the other hand, if a mixing time or length is to be defined, a mixing process instead of a single mixture is to be evaluated. In this case, mixing evolves in the course of time or along a spatial direction and, hence, the variance σ_0^2 at the initial time or the entrance of the mixing channel can be used instead of $\sigma_{\max}^2$. For instance, a mixing time t_{mix} is obtained as the first time t_0 such that

$$s(t) \leq \varepsilon\, s(0) \qquad \text{for all } t \geq t_0, \tag{17}$$

where ε (with $0 < \varepsilon \ll 1$) is a given fraction; typical values are $\varepsilon = 0.05$ or $\varepsilon = 0.01$. This criterion corresponds to the condition $I_M \geq 1 - \varepsilon$ with I_M from (16).

The determination of mixing times from numerical tracer experiments with the help of CFD simulations can be difficult, since a value of $I_M = 0.99$, say, will often be only reached after long and expensive calculations. This is especially true for highly resolved simulations which need to be used in case of high Schmidt numbers. On the other hand, there are several references – apparently starting with Lacey (1954, p. 261) – in which an exponential asymptotic behavior of the type

$$\sigma^2(t) \sim \exp(-t/\tau_{\mathrm{mix}}^\infty) \quad \text{for large } t \tag{18}$$

with an asymptotic mixing time $\tau_{\mathrm{mix}}^\infty$ is assumed. In case of a continuous mixing process, this type of behavior – with time replaced by axial position – is employed in Godfrey (1992) to define a mixing length. While (18) is often used without justification, there is increasing mathematical evidence for (18) to be valid in several situations; cf. Liu and Haller (2004) and Thiffeault (2004). In situations where (18) holds, this can of course be used to reduce the computational effort needed for numerical mixing time calculation. Indeed, it then suffices to run a numerical tracer experiment until a linear dependence of $\log \sigma$ versus time is reached and the mixing time can then be obtained by extrapolation.

While numerical tracer experiments are of increasing importance as a means to analyze mixing processes, there is a principle problem which becomes a severe obstacle in case of highly viscous liquid mixing or, more generally, in case of high tracer Schmidt numbers. In this situation, a sufficiently accurate solution of the species equation can be spoiled by the effect of so-called numerical diffusion, i.e. the artificially generated smoothing of a tracer profile due to errors from the discretization, and the numerical method can be much stronger than the true physical diffusion. In fact, almost all of the mixing that is observed in a numerical simulation may then be artificial. A way to avoid this problem is to replace the continuous tracer concentration by a number concentration obtained from Lagrangian (i.e. inertia free) particles that are tracked during the simulation. This approach does not suffer from artificial diffusion, since the position of tracer particles can be re-solved with sub-grid-scale accuracy and the velocity field at these particle posi-tions can be obtained by interpolation from the grid values. But notice that random displacements according to a Wiener process need to be added in order to incorpo-rate the Brownian motion of diffusive particles within this approach. Additional information on Lagrangian particle tracking based mixing calculations can be found in the recent paper of Phelps and Tucker (2006).

3 Scale of Segregation

The intensity of segregation from (14), like the intensity of mixing defined in (15), is not sensitive to the length scales on which segregation occurs. For a meaningful characterization of mixing quality, the intensity of segregation therefore has to be supplemented with a measure for the scale of segregation. This has already been noted in Danckwerts (1952) and the scale of segregation given there is based on autocorrelations of the concentration field, but only refers to one-dimensional cases. Unfortunately, an extension of this concept to 3D induces large numerical efforts and – perhaps even more important – has no obvious intuitive meaning.

There are several other ways to define such a scale of segregation. First of all, given any measure for the intensity of segregation, a measure for the scale of seg-regation can be related to it. The simple idea behind is that the scale on which in-homogeneities occur correlates with the rate at which the intensity of segregation will change under the action of diffusion. More precisely, let $I(t_0)$ be the intensity at time t_0 of a concentration profile $c_0(\mathbf{x})$ and $c(t,\mathbf{x})$ be the solution of the diffusion equation

$$\frac{\partial c}{\partial t} = \Delta c, \quad t > t_0, \mathbf{x} \in V, \quad \frac{\partial c}{\partial \mathbf{n}} = 0 \text{ on } \partial V, \quad c(t_0, \mathbf{x}) = c_0(\mathbf{x}) \text{ in } V. \quad (19)$$

Then the instantaneous rate of change

$$-\frac{dI}{dt}(t_0) \quad (20)$$

provides an integral measure of the length scale on which $c_0(\mathbf{x})$ is segregated. In case of mixing in a batch process, i.e. mixing with conservation of species mass due to impermeable walls, the rate in (20) is determined by the rate at which the variance is decaying. If μ denotes the (constant) mean of $c(t,\mathbf{x})$, the latter satisfies

$$\frac{d}{dt} \frac{\sigma^2}{2} |V| = \int_V (c - \mu) \frac{\partial c}{\partial t} dV = \int_{\partial V} (c - \mu) \frac{\partial c}{\partial \mathbf{n}} dA - \int_V \nabla(c - \mu) \cdot \nabla c \, dV . \quad (21)$$

Hence,

$$-\frac{d}{dt} \frac{1}{2} \sigma^2 = \frac{1}{|V|} \int_V \|\nabla c\|^2 \, dV , \quad (22)$$

which is nothing but the mean value of the scalar dissipation rate known from the theory of turbulent flow; cf. Frisch (1995), Pope (2000) or Fox (2003). The corresponding quantity for the momentum balance equation is directly related to a simple physical quantity, namely the total rate of viscous heating given by

$$\int_V \mathbf{S} : \nabla \mathbf{v} \, dV = \eta \int_V \|\nabla \mathbf{v}\|^2 \, dV \qquad (23)$$

which is a measure for the total internal friction due to relative motion and, hence, a measure for convective mixing. It is in turn proportional to the integral mass-specific power input; cf. (2). The scalar dissipation rate from (22) does not have such an immediate physical meaning, especially not as a measure for the scale of segregation.

In order to define a meaningful scale of segregation for a scalar quantity, it is necessary to know the structure of its distribution. In case of convection-dominated mixing, the small scale structures are often of a lamellar type. Such lamellar structures occur for instance if the contact area between regions of higher and lower concentration is rolled up in a secondary vortex-like velocity field. A typical example, showing the distribution of a fluorescence tracer in the mixing zone of a T-shaped micro-mixer is displayed in Figure 1; cf. Section 4 below. Somewhat surprisingly, highly resolved experimental measurements, e.g. in Dahm et al. (1991), show that for turbulent flows the finest-scale structures of passive scalars are also of a layered, lamellar type in case of high Schmidt numbers. Let us note in passing that for lamellar structures simplified mechanistic models of reduced complexity can be developed to describe reactive mixing, exploiting the fact that a lamellar structure is basically one-dimensional; cf. Szalai et al. (2003).

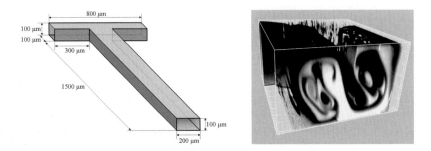

Fig. 1. T-micromixer (left), tracer distribution inside the mixing zone (right)

Given a (locally) lamellar tracer distribution, the striation thickness s as introduced in Mohr et al. (1957), i.e. s is half the mean distance between two successive lamellas, is the relevant length scale to characterize the finest scales with respect to mixing. Notice that the so-called micro-mixing acting on the finest scales is solely due to diffusion. Hence, at least for isotropic (Fickian, say) diffusion, the one-dimensional distance in the direction of the concentration gradient is the length scale relevant for diffusive fluxes. Locally, this direction is perpendicular to the iso-surfaces of the tracer concentration and, hence, to the contact area. Furthermore, the striation width s is related to the specific contact area a by means of

$$a = \frac{|A|}{|V|} = \frac{1}{s}, \tag{24}$$

and the contact area has already been considered in Spencer and Wiley (1951) as an important quantity to characterize mixing. The relation (24) is true for a completely lamellar structure, and it holds locally if the fine-scale structure is locally of such type. In the context of mixing and its characterization, several quantities have their roots in the concept of specific contact area. Prominent examples are the (rate of) stretching of interfacial area (cf. Ottino, 1989), Lyapunov exponents (see, e.g., Ottino et al. 1979) and the area tensor (see Wetzel and Tucker, 1999 and Galaktionov et al., 2002).

Due to (24), the striation width or, more generally, the mean distance between layers of high and low tracer concentrations can be computed from the contact (or interfacial) area. But, unfortunately, an explicit tracking of this interface during a numerical simulation of convective mixing is extremely expensive as noted in Clifford et al. (1999) and other references therein. On the other hand, given a tracer distribution $c(\mathbf{x})$, the interface between regions of high and low concentration is given by

$$\Gamma = \left\{ \mathbf{x} : c(\mathbf{x}) = \frac{1}{2} c_{max} \right\}, \tag{25}$$

say, i.e. as the iso-surface on which $f = c/c_{max}$ attains the value ½. In case of a completely segregated distribution, the surface Γ separates the regions $\{\mathbf{x} : f(\mathbf{x}) = 0\}$ and $\{\mathbf{x} : f(\mathbf{x}) = 1\}$. In this case, geometric measure theory provides the formula

$$|\Gamma| = \int_V \| \nabla f \| dV, \tag{26}$$

where $|\Gamma|$ denotes the area content of the surface Γ and $\| \nabla f \|$ is – in a formal sense – the Euclidean length of the gradient of f. The mathematically rigorous meaning of the right-hand side of (26) is that of the total variation of the Radon measure ∇f; cf. Giusti (1984) or Evans and Gariepy (1992). The integral in (26) can also be defined as the limit of the corresponding integral for a sequence of approximating smooth functions f_k. For the latter, the simple interpretation mentioned above is valid. To understand (26), replace f by such a smooth approximation (again denoted by f), the gradient of which vanishes outside a sufficiently small layer around Γ. Then

$$\int_V \| \nabla f \| dV = \int_{\Gamma_\varepsilon} \nabla f \cdot \frac{\nabla f}{\| \nabla f \|} dV = \int_\Gamma \int_{-\varepsilon}^{\varepsilon} \nabla f \cdot \mathbf{n} \, ds \, dA = \int_\Gamma \int_{-\varepsilon}^{\varepsilon} \frac{\partial f}{\partial \mathbf{n}} \, ds \, dA, \tag{27}$$

where $\mathbf{n}$ denotes the unit normal vector on the respective iso-surface of f, pointing into the region $\{\mathbf{x} : f(\mathbf{x}) = 1\}$, and

$$\Gamma_\varepsilon = \{\mathbf{x} + s\mathbf{n}_\Gamma : \mathbf{x} \in \Gamma, -\varepsilon \le s \le \varepsilon\}. \tag{28}$$

Since $\mathbf{n}$ is close to $\mathbf{n}_\Gamma$, we approximately obtain

$$\int_V \| \nabla f \| dV \doteq \int_\Gamma (f(\mathbf{x} + \varepsilon \mathbf{n}_\Gamma) - f(\mathbf{x} - \varepsilon \mathbf{n}_\Gamma)) dA = \int_\Gamma (1 - 0) dA = |\Gamma|. \tag{29}$$

For the following, it is more important to observe that the integration in (26) is done over V and not over Γ. Consequently, Γ need not be identified in order to compute its area!

According to (24) and (26), we consider the quantity

$$\Phi = \frac{1}{|V|} \int_V \| \nabla f \| dV \qquad \text{with} \qquad f = c/c_{max} \tag{30}$$

as a measure for the contact area between regions of high and low concentration. As explained above, the quantity Φ exactly gives the specific contact area (resp., the specific contact length in 2D) in case of completely segregated species distributions. The reciprocal of Φ, which has the dimension of a length, can be interpreted as an average distance between regions of high and low species concentration. This interpretation is exact in case of a lamellar structure, hence Φ^{-1} is the mean striation thickness then. The quantity Φ keeps these meanings for almost segregated concentration fields, while these interpretations loose their significance for close to homogeneous fields. For arbitrary tracer distributions, Φ is a measure for the total driving force within the concentration field for diffusive dissipation of concentration gradients. Below, the quantity Φ will therefore be called the *potential for diffusive mixing* and Φ^{-1} will be termed *integral scale of segregation*.

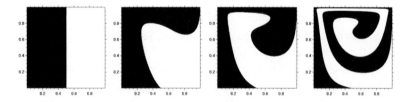

Fig. 2. Rollup of contact area in a Driven Cavity Flow

The fact that Φ is not sensitive against moderate smearing of the interface is illustrated by means of a Driven Cavity Flow. Figure 2 shows the rollup of an initially vertical contact area by a low Reynolds (Re = 10) Driven Cavity Flow. To avoid numerical diffusion, the contact area has been computed using the Volume of Fluid (VOF) method. Therefore, black and white regions in Figure 2 correspond to two different non miscible fluids of the same physical properties and with negligible small interfacial tension. If diffusion is acting as well, the contact line between zero and maximum concentration is smeared; cf. Figure 3.

The corresponding specific contact areas, computed by means of (30), are compared in Figure 4 where the Driven Cavity Flow refers to a Reynolds Number of 10 and the diffusivity corresponds to a Schmidt number of 500. The scale of the abscissa is such that the upper left point on the driven lid needs $\Delta t = 10$ (dimensionless time units) to move to the right end of the top boundary.

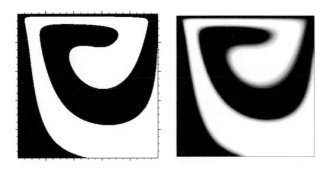

Fig. 3. Convective vs. convective/diffusive transport in a Driven Cavity Flow

Figure 3 shows the mixing states at $t = 65$ and Figure 4 displays the corresponding values of Φ. While the diffusive smearing is clearly visible, these values are still very close to each other, i.e. Φ is barely influenced by diffusive smearing up to this point. This is due to the fact that Φ is the *total variation* of the quantity $f = c/c_{max}$; in 1D, the latter equals the sum of the amplitudes of all fluctuations and, hence, will not change due to smearing as long as the local maxima and minima remain unchanged.

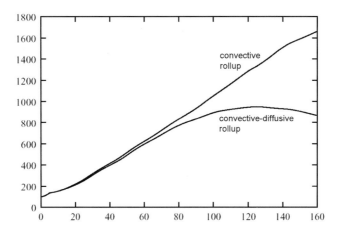

Fig. 4. Evolution of Φ during rollup in Driven Cavity Flow

Since Φ is a total variation, its definition can also be related to the general concept mentioned at the beginning of this section. Indeed, if the linear diffusion equation (19) is replaced by the so-called *minimizing total variation flow*, i.e.

$$\frac{\partial c}{\partial t} = \nabla \cdot \frac{\nabla c}{\|\nabla c\|} \quad t > t_0, \mathbf{x} \in V, \quad \frac{\partial c}{\partial \mathbf{n}} = 0 \text{ on } \partial V, \quad c(t_0, \mathbf{x}) = c_0(\mathbf{x}) \text{ in } V, \quad (31)$$

then

$$-\frac{d}{dt} \frac{1}{2} \sigma^2 = \frac{1}{|V|} \int_V \|\nabla c\| dV, \quad (32)$$

which resembles (30) except for the normalizing factor $c_{\max}$. The nonlinear diffusion equation (31) plays an important role in the context of image processing by minimization of the total variation; cf. Osher et al. (2003) and Andreu et al. (2004).

In case of kinematical mixing, i.e. mixing without or with negligible diffusive effects, the timely evolution of the contact area is of present interest; cf., e.g., Wetzel and Tucker (1999) and Dopazo et al. (1999). In contrast to (30) above, surface integrals over the contact area instead of a volume integral over the whole mixing domain are employed there. For numerical purposes, this requires a representation of the interface which causes additional computational effort.

4 An Example: Mixing in a T-Shaped Micro-mixer

While diffusive transport in micro-devices, in relation to the time-scale of chemical kinetics, is faster than in conventional mixers, mixing by diffusion alone is usually not sufficient to perform fast or highly exothermic reactions.

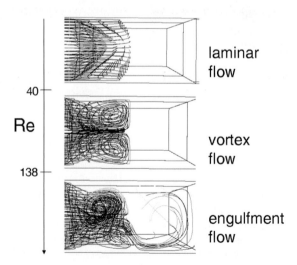

Fig. 5. Streamlines in a T-shaped element for different Reynolds numbers

Since, on the other hand, turbulent flow conditions are too expensive in terms of pressure drop, secondary flows provide an important means to promote mixing. Combined with the concept of chaotic advection (Aref, 1984) this can lead to very efficient mixing; cf., e.g., Stroock et al. (2002) and Ottino and Wiggins (2004).

To perform chemical reactions in a continuously operated reactor requires the mixing of initially separated feed streams. In microreactors, junctions like T- or Y-shaped elements are used for this purpose. Therefore, a deeper understanding of the mixing behavior of convective-diffusive transport in such junction elements is required. The flow and the mixing behavior of a prototype T-shaped micro-mixer with rectangular cross sections (see Figure 1) has been theoretically investigated on the basis of numerical simulations in Bothe et al. (2006), (2008) and by experimental techniques in Hoffmann et al. (2006); cf. also the subsequent contributions within this volume. Depending on the mean velocity, three different stationary flow regimes can be observed (see Figure 5) up to a Reynolds number of about 240, where instationary, periodic flow phenomena set in. At low Reynolds numbers, strictly laminar flow behavior occurs where both inlet streams run parallel through the mixing channel, without formation of vortices. Above a critical Reynolds number a secondary flow in form of a double vortex pair is build due to centrifugal forces. In this regime, symmetry concerning a plane perpendicular to the inlet channels is still maintained:

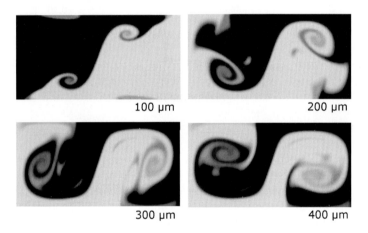

100 µm   200 µm

300 µm   400 µm

Fig. 6. Numerically computed cross sectional tracer distribution inside the mixing zone at various distances from the entrance

Therefore, mixing across this symmetry plane can only occur due to diffusion and, hence, the intensity of mixing is approximately zero. With increasing velocity this flow symmetry is destroyed and fluid elements reach the opposite side of the mixing channel, leading to tracer distributions as shown in Figure 1. This so-called engulfment flow regime promotes mixing, since the resulting intertwinement of both fluid streams generates additional contact area and, hence, raises the potential for diffusive mixing. This is illustrated by Figure 6, which shows the tracer distribution on cross sections inside the mixing zone at distances of 100 µm to 400 µm from the entrance into the mixing channel. The roll up of the contact area causes a decrease of the integral scale of segregation and thereby intensifies diffusive transport. This leads to an increase of the intensity of mixing. In fact, as can be seen in Figure 7, both quantities show a jump at the critical mean velocity. Above a mean velocity of approx. 1.1 m/s (Re=146) the integral scale of segregation, computed from Φ in (30) for the cross section at a distance of 300 µm from the mixing channel entrance, falls to about 50 µm and further decreases to 30 µm. As mentioned above, these values can be interpreted as a mean distance between regions of high and low tracer concentration which is also illustrated by Figure 6.

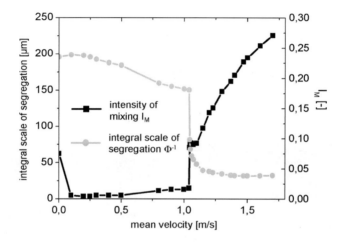

Fig. 7. Intensity of mixing (black) and scale of segregation (grey) versus mean velocity

How much potential for diffusive mixing is generated in a given flow depends, in particular, on the Schmidt number. In the engulfment flow (cf. Figure 5 and Figure 7), very fine structures are generated near the center of both vortices which is especially true in the case of the high Schmidt number of approx. 3600 of the fluorescence tracer used in the experimental work cited above. This is accompanied by a locally high specific contact area. In numerical simulations these fine structures are to some extend smeared by numerical diffusion. By means of locally refined grids of up to 18 million cubic grid cells with local cell sizes down to 0.6 µm and massively parallel numerical simulations using up to 36 parallel processes it has been possible to compute grid independent (i.e. independent against further refinement of the regions containing the small scale tracer structures) tracer distributions at relatively large Schmidt numbers of up to 500. In Bothe et al. (2006), these numerical simulations have been performed for different diffusivities. Computation of the specific contact area by means of (30) for a subregion containing only the thinnest striation then allows for an easy extraction of the size of the smallest structures. Concerning their dependence on Sc, it turns out that the finest structures λ_{conc} of the tracer distribution scale as

$$\lambda_{conc} = \frac{\lambda_{vel}}{Sc^{\alpha}}, \tag{33}$$

where the reference scale λ_{vel} depends in particular on Re, with α approximately equal to 0.5. Now note that (33) means that the finest length scale is given by the Batchelor scale, although the flow field is even stationary! Since the finest structures are quickly dissipated by diffusion, they can only persist if the convective transport to generate those acts on the same time-scale as diffusive smearing. If the convective time-scale is assumed to be given by the characteristic time for diffusion of momentum, this yields

$$\frac{\lambda_{vel}^2}{\nu} = \frac{\lambda_{conc}^2}{D}, \tag{34}$$

which suggests a simple explanation for the Batchelor length scale, i.e. (33) with $\alpha = 0.5$.

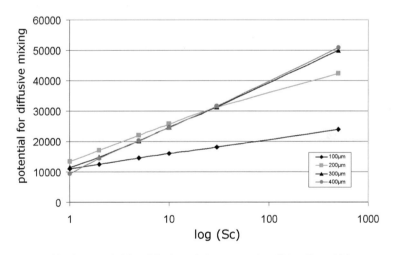

Fig. 8. Potential for diffusive mixing versus log (Sc) at Re = 186

On the other hand, since the tracer distributions are far from being statistically homogeneous, the integral length scale is more important than the smallest one. To investigate how the latter depends on the Schmidt number, the potential for diffusive mixing Φ has been computed on various cross sections and for different Reynolds numbers. Surprisingly, in the engulfment flow regime, the observed dependence of Φ on the logarithm of Sc is always linear within very small error bounds. Figure 8 displays this dependence for a Reynolds number of 186; the depicted linear regressions show a correlation coefficient better than 0.99. Hence the relation

$$\Phi = a_0 + a_1 \log(\mathrm{Sc}) \tag{35}$$

holds with high accuracy. Now let $\langle \lambda_{\mathrm{conc}} \rangle := \Phi^{-1}$ be the integral scale of segregation. Then this length scale depends on the Schmidt number according to

$$\langle \lambda_{\mathrm{conc}} \rangle = \frac{\langle \lambda_{\mathrm{vel}} \rangle}{1 + a \log(\mathrm{Sc})} \tag{36}$$

with an integral length scale $\langle \lambda_{\mathrm{vel}} \rangle$ of the flow field – which will attain different numerical values than the velocity scale from (33) – and with a certain coefficient that depends especially on the flow, i.e. $a = a(\mathrm{Re})$. The numerical values of $\langle \lambda_{\mathrm{vel}} \rangle$ corresponding to the linear regressions displayed in Figure 8 lie in the range of 70 µm to 140 µm and increase along the axial direction which is reasonable because of the redeveloping flow.

The relevance of a dependence of the integral scale according to (36) – in contrast to (33) – is twofold. Firstly, conversion of a chemical reaction might be completely insensitive against the smallest length scales, given that these are only realized in small subregions, but will strongly depend on the integral length scale. Secondly, at least if locally refined meshes are employed, the computational effort to resolve the tracer distribution increases significantly slower with increasing Sc. Indeed, since most of the grid cells are needed to cover the contact area with sufficient resolution, the number of grid cells needed behaves roughly like Φ, hence like log(Sc) if (35) is valid. For comparison, recall that for turbulent tracer transport (in homogeneous, isotropic turbulence) the Batchelor scale applies globally, hence at fixed Re the number of degrees of freedom increases according to Sc^3, leading to unaffordable numerical effort in case of realistic Schmidt numbers of about 1000, say, in the case of aqueous systems. In fact, it is only due to the scaling from (36) that the secondary flow mixing described above can be computed numerically with sufficient resolution using today's computer facilities.

5 Concluding Remarks

Since mixing is a fundamental process step applied in widely different overall processes with different purposes, it is obvious that meaningful evaluations of mixing – both the mixing processes as well as the mixtures – will require different approaches depending on the respective context. While this might call for a deep local analysis of velocity and scalar fields, in practical applications there often is a demand to quantify mixing by one or two, say, plain numbers. One possible choice then is the intensity of mixing, supplemented by the integral scale of segregation. Since mixing is a multi-scale process by its very nature, an enormous compression of the local data is needed in order to obtain such a few characteristic numbers. There are recent attempts to define "multi-scale measures" of mixedness which somehow incorporate all the scales present; cf. Mathew et al. (2005). This condensation into a single quantity obviously has to be guided by the actual mixing purpose. If, for example, reactive mixing is considered, there usually is a key chemical reaction which is to be performed. Given the characteristic time τ_R of this reaction, mixing by convection needs to reduce the scale of concentration gradients to a length l_{micro} such that mixing by diffusion (micro-mixing) will homogenize the mixture in a time much smaller than τ_R, in order to avoid the disadvantages of masking by mixing (Rys, 1992). Hence one should have

$$\tau_D = \varepsilon \tau_R \quad \text{for} \quad \tau_D := \frac{l_{micro}^2}{D} \qquad \text{with some } \varepsilon \ll 1. \qquad (37)$$

This in turn means that segregation of the concentration field on scales smaller than l_{micro} from (37) is not relevant for the chemical process. Even if the species is completely segregated but on a smaller scale, the distribution would appear as if completely mixed although its intensity of mixing is zero. In this case, samples of radius l_{micro} should be considered as homogeneous, i.e. the averaged field

$$\langle c \rangle_r(\mathbf{x}) = \frac{1}{|B_r(\mathbf{x})|} \int_{B_r(\mathbf{x})} c(\mathbf{y}) \, dV \qquad (38)$$

with $r = l_{micro}$ instead of the true scalar field $c(\mathbf{x})$ should be analyzed. The interesting point now is the variance of the random variable $\langle c \rangle_r$; the smallest scale on which fluctuations can occur is clear by its definition. The dependence of the variance on the sample size as a way to measure the scale of segregation has already been employed in Missiaen and Thomas (1995). More information about the scale-dependence of the intensity of mixing is of course contained in the map $r \to \sigma^2(\langle c \rangle_r)$; this map has recently been studied in Tucker and Peters (2003).

   In our opinion, especially in the case of reactive mixing, more emphasis has to be put on integral measures of the scales produced by mixing processes that have a clear meaning for ongoing chemical reactions. Here, it seems worth to further bridge the gap between experimental techniques, like chemical probes (cf., e.g., Villermaux, 1982), and theoretical, simulation-based approaches. Moreover, many of the known techniques to assess mixing processes are designed for the evaluation of batch processes. These need to be adapted to the case of continuous flow processes, where a distinguished axial direction exists which causes the need for anisotropic mixing. Finally, especially in the context of microfluidic applications, there is a strong demand to model and assess reactive secondary-flow mixing.

Acknowledgment

The author gratefully acknowledges financial support by the DFG within the scope of the Priority Programme 1141. Special thanks go to the group of N. Räbiger for providing the visualized μ–LIF data shown in Figure 1.

References

Ottino, J.M.: The kinematics of mixing: stretching, chaos and transport. Cambridge University Press, Cambridge (1989)

Aref, H.: Stirring by chaotic advection. J. Fluid Mech. 143, 1–21 (1984)

Finn, M.D., Cox, S.M., Byrne, H.M.: Mixing measures for a two-dimensional chaotic Stokes flow. J. Eng. Math. 48, 129–155 (2004)

Frisch, U.: Turbulence. In: The Legacy of A. N. Kolmogorov. Cambridge University Press, Cambridge (1995)

Baldyga, J., Bourne, J.R.: Turbulent mixing and chemical reactions. John Wiley & Sons, New York (1999)

Pope, S.: Turbulent Flows. Cambridge University Press, Cambridge (2000)

Fox, R.O.: Computational Models for Turbulent Reacting Flows. Cambridge University Press, Cambridge (2003)

Raynal, F., Gence, J.N.: Energy saving in laminar mixing. Int. J. Heat Mass Transfer 40(14), 3267–3273 (1997)

Bothe, D., Stemich, C., Warnecke, H.J.: Fluid mixing in a T-shaped micro-mixer. Chem. Eng. Sci. 61(9), 2950–2958 (2006)

Fan, L.T., Chen, S.J., Watson, C.A.: Solids mixing. Ind. & Eng. Chem. 62(7), 53–69 (1970)

Boss, J.: Evaluation of the homogeneity degree of a mixture. Bulk Solids Handling 6(6), 1207–1215 (1986)

Rielly, C.D., Smith, D.L.O., Lindley, J.A., Niranjan, K., Phillips, V.R.: Mixing process for agricultural and food materials: part 4, assessment and monitoring of mixing systems. J. Agric. Engng. Res. 59, 1–18 (1994)

Villermaux, J.: Mixing in chemical reactors. In: Wei, J., Georgakis, C. (eds.) Chemical Reaction Engineering – Plenary Lectures. ACS Symposium Series, vol. 226, pp. 135–186. American Chemical Society, Washington (1982)

Ottino, J.M.: Mixing and chemical reaction. A tutorial. Chem. Eng. Sci. 49(24A), 4005–4027 (1994)

Hiby, J.W.: Definition and measurement of the degree of mixing in liquid mixtures. Int. Chem. Eng. 21(2), 197–204 (1981)

Toor, H.L.: Mass transfer in dilute turbulent and non-turbulent systems with rapid irreversible reactions and equal diffusivities. AIChE Journal 8, 70–78 (1962)

Warhaft, Z.: Passive scalars in turbulent flows. Annu. Rev. Fluid Mech. 32, 203–240 (2000)

Ogawa, K., Ito, S.: A definition of quality of mixedness. J. Chem. Eng. Japan 8(2), 148–151 (1975)

Wang, W., Manas-Zloczower, I., Kaufman, M.: Entropic characterization of distributive mixing in polymer processing equipment. AIChE Journal 49, 1637–1644 (2003)

Prière, C., Gicquel, L.Y.M., Kaufmann, P., Krebs, W., Poinsot, T.: Large eddy simulation predictions of mixing enhancement for jets in cross-flows. J. Turbulence 5, 005 (2004)

Danckwerts, P.V.: The definition and measurement of some characteristics of mixtures. Appl. Sci. Res. A3, 279–296 (1952)

Lacey, P.M.C.: Developments in the theory of particle mixing. J. Appl. Chem. 4, 257–268 (1954)

Godfrey, J.C.: Static mixers. In: Harnby, N., Edwards, M.F., Nienow, A.W. (eds.) Mixing in the process industries, 2nd edn., pp. 225–249. Butterworth-Heinemann, London (1992)

Liu, W., Haller, G.: Strange eigenmodes and decay of variance in the mixing of diffusive tracers. Physica D 188, 1–39 (2004)

Thiffeault, J.L.: Scalar decay in chaotic mixing. In: Transport in Geophysical Flows: Ten Years after. Proceedings of the Grand Combin Summer School, Valle d'Aosta, Italy, June 2004. Springer, Berlin (2004) (in press)

Phelps, J.H., Tucker, C.L.: Lagrangian particle calculations of distributive mixing: limitations and applications. Chem. Eng. Sci. (2006) (to appear), doi:10.1016/j.ces.2006.07.008

Dahm, W.J.A., Southerland, K.B., Buch, K.A.: Direct, high resolution, four-dimensional measurements of the fine scale structure of Sc> >1 molecular mixing in turbulent flows. Phys. Fluids A 3(5), 1115–1127 (1991)

Szalai, E.S., Kuhura, J., Arratia, P.E., Muzzio, F.J.: Effect of hydrodynamics on reactive mixing in laminar flow. AIChE Journal 49, 168–179 (2003)

Mohr, W.D., Saxton, R.L., Jepson, C.H.: Mixing in laminar-flow systems. Ind. & Eng. Chem. 49(11), 1855–1856 (1957)

Spencer, R.S., Wiley, R.M.: The mixing of very viscous liquids. J. Colloid Sci. 6, 133–145 (1951)

Ottino, J.M., Ranz, W.E., Macosko, C.W.: A lamellar model for analysis of liquid-liquid mixing. Chem. Eng. Sci. 34, 877–890 (1979)

Wetzel, E.D., Tucker, C.L.: Area tensors for modeling microstructure during laminar liquid-liquid mixing. Int. J. Multiphase Flow 25, 35–61 (1999)

Galaktionov, O.S., Anderson, P.D., Peters, G., Tucker, C.: A global multi-scale simulation of laminar fluid mixing: The extended mapping method. Int. J. Multiphase Flow 28, 497–523 (2002)

Clifford, M.J., Cox, S.M., Roberts, E.P.L.: Measuring striation widths. Phys. Letters A 260, 209–217 (1999)

Giusti, E.: Minimal Surfaces and Functions of Bounded Variation. Birkhäuser, Boston (1984)

Evans, L.C., Gariepy, R.F.: Measure Theory and Fine Properties of Functions. CRC Press, Boca Raton (1992)

Osher, S., Solé, A., Vese, L.: Image decomposition and restauration using total variation minimization and the H^{-1} norm. Multiscale Model. Simul. 1(3), 349–370 (2003)

Andreu, F., Mazón, J.M., Moll, J.S., Caselles, V.: The minimizing total variation flow with measure initial values. Comm. Contemp. Math. 6, 1–64 (2004)

Dopazo, C., Calvo, P., Petriz, F.: A geometric/kinematic interpretation of scalar mixing. Phys. Fluids 11(10), 2952–2956 (1999)

Stroock, A.D., Dertinger, S.K.W., Ajdari, A., Mezic, I., Stone, H.A., Whitesides, G.M.: Chaotic mixer for microchannels. Science 295, 647–651 (2002)

Wiggins, S., Ottino, J.M.: Foundations of chaotic mixing. Phil. Trans. R. Soc. Lond. A 362, 937–970 (2004)

Hoffmann, M., Schlüter, M., Räbiger, N.: Experimental investigation of liquid–liquid mixing in T-shaped micro-mixers using μ-LIF and μ-PIV. Chem. Eng. Sci. 61(9), 2968–2976 (2006)

Bothe, D., Stemich, C., Warnecke, H.J.: Computation of scales and quality of mixing in a T-shaped microreactor. Computers & Chemical Engineering 32, 108–114 (2008)

Mathew, G., Mezic, I., Petzold, L.: A multiscale measure for mixing. Physica D 211, 23–46 (2005)

Rys, P.: The mixing-sensitive product distribution of chemical reactions. Chimia 46, 469–476 (1992)

Missiaen, J.M., Thomas, G.: Homogeneity characterization of binary grain mixtures using a variance analysis of two-dimensional numerical fractions. J. Phys. Condens. Matter 7, 2937–2948 (1995)

Tucker, C.L., Peters, G.W.M.: Global measures of distributive mixing and their behavior in chaotic flows. Korea-Australia Rheol. Journal 15(4), 197–208 (2003)

Part 2: Experimental Methods for Visualization and Measurements in Macro- and Micro-Scale Dimensions of Mixing

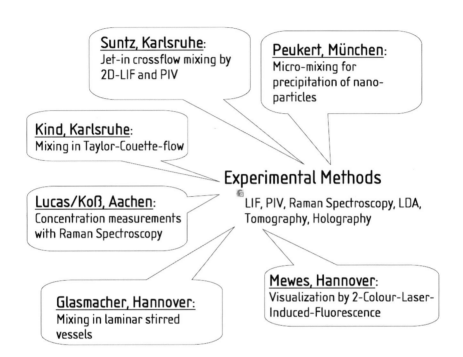

The Two-Color Laser Induced Fluorescence (LIF) Technique has been developed, which enables the measurement of the progress in mixing on macro- and micro scale simultaneously. A Raman scattering test-bench has been set-up to investigate dynamic processes in more than one dimension. The Raman scattering approach allows investigating concentrations of several components simultaneously in reacting and non-reacting fluids with high accuracy. The technique was improved to a pseudo two-dimensional multipoint technique. Particle-Image Velocimetry (PIV) and Laser Induced Fluorescence (LIF) techniques are applied to characterize mixing at different length and time scales. Thereby all relevant flow structures down to the Batchelor length scale are detected by a special high resolution LIF set-up. Nanoparticle precipitation is used to quantify the micro mixing efficiency. The mean size and the width of the Particle Size Distribution (PSD) indicate the effect of mixing efficiency on the chemical reaction and on particle formation.

Time-Resolved Measurement of Concentrations in Mixing Processes Using Raman Spectroscopy

Development of an Optical Measurement Technique to Investigate Reacting and Non-reacting Mixing Processes with Temporal and Spatial Resolution

Birgit Jendrzok, Christoph Pauls, Hans-Jürgen Koß, and Klaus Lucas

Chair of Technical Thermodynamics, RWTH Aachen University
Schinkelstraße 8, 52062 Aachen, Germany
secretary@ltt.rwth-aachen.de

Abstract. Optical measurement techniques are more and more widely used to investigate the mixing processes in small and medium scale mixers. Most of these techniques (including LIF based techniques) suffer from principle problems when investigating and quantifying reacting processes. Within this work a Raman scattering test-bench has been set-up to investigate dynamic processes in more than one dimension. While the Raman scattering approach allows investigating concentrations of several components simultaneously in reacting and non-reacting fluids with high accuracy, it is normally restricted to one-dimensional measurements, which is not sufficient for characterization of mixing processes. So the technique was improved to a pseudo two-dimensional multipoint technique. This article deals in detail with the advantages, disadvantages and limitations of different approaches. Furthermore the technical details of the developed Raman scattering set-up are described.

1   Introduction

The analysis of mixing processes, especially in reacting systems, is of great interest. There are several measurement techniques to determine the composition of different mixtures [8, 11, 12]. While some do not allow measuring in-situ or disturb the investigated system, most of the others are only capable to analyze at most a two component mixture correctly. Especially the established measurement techniques based on sampling by probe heads are not capable to investigate reacting systems, because of a time-shift between assay and analysis, as well as the disturbance of the mixing process. These problems can be overcome by optical measurement techniques.

2   Demands on the Measurement System

The measurement system has to fulfill several requirements to be able to investigate processes like mixing especially in combination with reaction, including high temporal and spatial resolution, quantitative detection of several species at once, and minimal disturbance on the investigated system.

1. The applied measurement technique needs to have a high temporal and spatial resolution to characterize mixing processes, which are highly dynamic and non-symmetric. Temporal resolution in this case means the duration of the measurement process is short in comparison to the characteristic times of the mixing process, while spatial resolution stands for the possibility to measure at several locations determined in all three dimensions simultaneously[1].
2. A lot of different species are involved in reacting and non-reacting mixing processes. Since all of them can influence the following mixing and reactions it is necessary to detect the concentrations of as many species simultaneously as possible.
3. To get experimental data of the origin process unaffected by the measurement technique, the applied technique should be as interaction free as possible. This means that especially no mechanical parts are added to the investigated system. Therefore all techniques based on probe heads are inappropriate. On the other hand optical measurement techniques are the first choice concerning interaction free diagnostics.

3   Comparison of Different Optical Approaches

There are three different widely used approaches for the investigation of concentrations in fluids, i.e. Infra-Red-absorption (IR), Laser Induced Fluorescence (LIF) and Spontaneous Raman Scattering (SRS). All of them have specific advantages and disadvantages discussed shortly in the following sections. Additionally in each of these sections bibliographic references are given, in which these techniques are described in detail.

To anticipate the conclusion of the comparison, from our point of view SRS is the best choice for investigating reacting mixing processes or mixing processes dominated by diffusion, since all of the other approaches suffer from at least one of the requirements mentioned above. For mixing processes and substances with negligible diffusion and quenching effects LIF is the best choice due to the much higher intensities of the measurement signal and therefore higher sensitivity of the set-up.

3.1   IR-Absorption

The IR measurement technique (independent if NIR or MIR) is based on the wavelength dependent attenuation of light due to absorption of photons by molecules in the mixing fluids. The main disadvantage of this approach is the line of sight character. This means the investigated volume is determined only in two dimensions,

---

[1] The locations have to be measured within a time frame, which is short compared to the characteristic times of the investigated process.

while the third is integrated and therefore averaged. This averaging effect makes it impossible to investigate strongly non-symmetric mixing processes. Applying tomographic methods it is possible to overcome the problem of the insufficient spatial resolution at the cost of time resolution and an all around optical access is necessary. Since this measurement technique misses at least one of the demands mentioned above it is not adequate for investigating mixing processes.

The basics of the IR measurement technique are described in detail e.g. by Schrader in [9].

3.2 Laser Induced Fluorescence

Laser induced fluorescence, which is an inelastic scattering process, is a common approach to measure the concentration of a species in two dimensions [5, 10]. After the absorption of a photon with the corresponding wavelength[2], the molecule ends up in an excited state. This excited molecule will after short time, usually in the order of nanoseconds, de-excite and emit light at a wavelength larger than the excitation wavelength. For a lot of different molecules this process yields quite high quantum efficiencies (QE) and relatively large cross-sections. Therefore already small amount of a fluorescent tracer substance added to the fluid is sufficient for a satisfactory signal. Using adequate filters it is possible to suppress the elastically scattered light as well as to separate the fluorescence of a few different species.

In principle the LIF signal is proportional to the amount of tracer molecules and therefore a measure for the concentration of the fluid laced with tracer. Furthermore this approach allows acquiring the concentration in two dimensions. In contrast to the high intensities and the capability to measure two dimensional concentration distributions the LIF technique has some disadvantages.

- The ambient species, pressure and temperature of the molecule can influence the signal intensity due to quenching[3] so only qualitative results can be gained.
- If none of the involved substances is fluorescing a tracer, which possibly influences the reaction process, has to be added.
- Not detecting the fluid directly, but the added tracer, leads to principle problems by investigating mixing processes dominated by diffusion.
- If several fluorescing substances are involved or produced during the reaction process an overlapping of the emission peaks may occur. So a separation of the different species by usage of optical filters is not perfectly possible. To further improve the separation of the LIF signal originating from different species the usage of a spectrometer gaining detailed spectral information is possible. The drawback of this approach is loosing one dimension in spatial resolution, like for normal SRS investigations.

The LIF measurement technique as well as its application is in detail described e.g. by Eckbreth in [6].

---

[2] The energy of the photon equals the energy difference between the actual state and an excited state of the molecule.

[3] Quenching is the reduction of the gained signal due to radiationless relaxation.

3.3 Spontaneous Raman Scattering

Spontaneous Raman Scattering is an inelastic scattering process, too, but in contrast to the LIF applying SRS the molecule is excited into a non-existent, so called virtual, state. This leads to a very short lifetime of the excited state and therefore very fast de-excitation (quasi instantaneous process) into another than the initial state. The change of the frequency of the radiation represents the energy difference between the initial and final state and is called Raman-shift. This shift can be positive (anti-Stokes, the scattered photon has a higher energy, blue shifted) or negative (Stokes, the scattered photon has a lower energy, red-shifted). For low temperatures (below approximately 1000 K) negative Raman shifts are much more probable, therefore in this study the Stokes signal is investigated.

These frequency shifts are species specific and the gained intensities are to the first order independent of the ambient conditions. So the signal intensity is a direct measure for the investigated species. Unfortunately the cross sections are very small and therefore the achieved signals are very low. So high laser pulse energies are required to gain sufficient signal intensities.

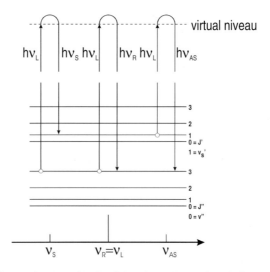

Fig. 3.1. The picture shows a sketch of the absorption and emission process of SRS and Rayleigh scattering. The meaning of the indexes is as follows: L = laser, R = Rayleigh, AS = anti-Stokes, S = Stokes.

Normally Raman measurements are applied only at a single point or on a line[4], which is not sufficient in our case. To be able to acquire the Raman signal in two dimensions a special set-up is required, which is described in detail below.

---

[4] The detection system has at most two dimensions. Since one is needed for the spectral information only one is left for spatial resolution.

The potential of SRS to investigate concentration profiles even under high pressure and high temperature conditions in the gas phase was shown by Hoffmann et al. [7]. In this study the Raman approach allowed to measure the temperature of the mixture simultaneously.

Behmann demonstrated the successful investigation of reacting mixtures using SRS [4], while Bardow et al have applied SRS successfully to measure concentration profiles of ternary mixtures and calculate the corresponding concentration dependent diffusion coefficients [2, 3].

3.4 Comparison of LIF and SRS

Comparing the approaches LIF and SRS the following statements can be given:

- LIF yields high signal intensities even for a small amount of tracer substance, while SRS suffers from small cross sections and therefore low signal intensities. ➜ Usage of high power lasers and sensitive detection systems when applying SRS.
- The fluorescence signal is strongly influenced by the surrounding conditions like pressure, temperature, and mixture composition. SRS has only a low dependence on the surrounding conditions. ➜ SRS can be quantified more accurately, especially for non-isothermal and reacting conditions.
- Using LIF the tracer is measured while SRS allows detecting most relevant fluids directly. ➜ Investigating processes strongly influenced by diffusion, reaction or evaporation LIF may lead to inaccurate results.

4 Experimental Set-Up

A sketch of the **experimental** set-up can be found in Fig. 4.1. The set-up consists of three functional units:

1. Preparation of the laser beam
2. Optical accessible probe volume
3. Data acquisition system

Before the beam from a pulsed Nd:YAG laser tuned to 532 nm (GCR230, Spectra Physics, 8 ns pulse duration, 800 mJ per pulse, 10 Hz rep. rate) is formed to a sheet (approx. width: 10 mm; approx. thickness: 0.3 mm) using spherical and cylindrical lenses the pulse width is increased by one order of magnitude (8 ns ➜ 100 ns) using a pulse stretcher (ESYTEC Energie und Systemtechnik GmbH, Erlangen) reducing the peak laser power to especially avoid damage of the couple in window. Afterwards the beam is guided through the probe volume (the mixer) into a beam dump.

The data acquisition system consisting of a newly developed "filter-and-fiber" box as well as an OMA (Optical Multi-channel Analyzer) is placed orthogonally to the laser sheet. Those "filter-and-fiber" box is the important part to gain the two

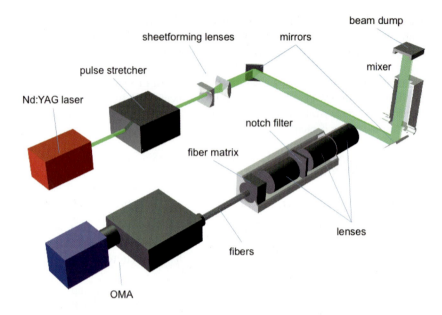

Fig. 4.1. Sketch of the experimental set-up used within this study. Depicted are light source (laser, pulse stretcher, and sheet forming lenses), the measurement object (T-mixer), and the detection system (lenses, "filter and fiber" box, and OMA).

dimensional multipoint resolution and consists of lenses, a notch filter (placed in a region with as parallel beams of light as possible) to suppress elastic scattered light and optical fibers connecting the box with the OMA. The OMA consists of an imaging spectrometer (Acton Research) and an intensified CCD camera (Scientific Instruments, 576 by 384 Pixel). The optic is set-up in a way to measure at several positions in a two-dimensional plane. The detailed mode of operation as well as an alternative set-up is described in the section 4.3.

4.1 Selection of the Laser

In the following sections the advantages and disadvantages of the different laser types available on the market are discussed. After balancing all the points mentioned below, we decided to use a pulsed Nd:YAG laser equipped with a second harmonic generator and a pulse stretcher described in section 4.1.2.

4.1.1 Pulsed vs. CW (Current Wave)

Due to the small scattering cross sections of SRS high laser light energies are required. If a low time resolution is sufficient the usage of a cw-laser is proposed. Investigations of concentration profiles and diffusion coefficients for different systems have been applied within the DFG (Deutsche Forschungsgemeinschaft, German Research Foundation) projects "Model-based Experimental Analysis of

Kinetic Phenomena in Fluid Multi-phase Reactive Systems" (SFB540) and "Determination of concentration and temperature diffusion coefficients in hydrogels using Raman spectroscopy" (LU 250/35-1). Due to the low dynamic of the investigated processes in these cases a cw Argon laser was used.

In the presented work the investigated system is highly dynamic requiring a high temporal resolution. Therefore a pulsed laser is used leading to very high intensities.

4.1.2 Pulse Stretcher

Since the achieved intensities of the formed laser beam may exceed the damage threshold of the couple in glass window or ignite the mixing substances there is a demand to decrease the maximum intensity while keeping the overall energy of one laser pulse needed to achieve a sufficient signal level. Therefore a pulse stretcher is added to the set-up, increasing the pulse duration by one order of magnitude.[5] The set-up of the pulse stretcher is depicted in Fig. 4.2.

To further decrease the maximum laser power density an additional beam homogenizer was included in the optical path of the laser beam. Because the deterioration of the beam shaping quality was larger in comparison to the reduction of the maximum laser power density we decided to remove the homogenizer.

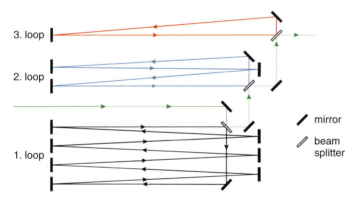

Fig. 4.2. Sketch of the optical path inside the pulse stretcher used within this study. At each beam splitter parts of the laser beam are guided through the different stages of delay loops. The resulting single pulses are overlaid giving a pulse with longer duration and reduced maximal intensity.

4.1.3 Selection of the Wavelength (UV, VIS, IR)

The selection of the right wavelength is in short the question of high measurement signal and low disturbing signal, especially fluorescence from impurities.

The shorter the wavelength the higher is the intensity of the SRS (wavelength to the third power). So an excitation in the UV range (200 nm to 400 nm) would

---

[5] Much higher factors are not practicable due to the long extend of the laser-beam inside the pulse-stretcher and therefore problems with the stability of the adjustment.

be desirable.[6] On the other hand unavoidable impurities in the investigated substances show an increasing tendency to fluorescence with shorter wavelengths leading to unwanted underground in the measurement signals.

However using an excitation wavelength close to or even in the IR range would overcome the fluorescence problem but leads to very low signals. Furthermore up to now only one-dimensional IR detectors systems are available fulfilling the requirements like dynamic and SNR (leading to a zero dimensional resolution), while 2-d detectors suffer from at least one of the required properties.

4.2 Selection of the Recording System

Because short gating times are required, while only very low signal levels are available, an ICCD (Intensified Charge Coupled Device) was chosen for acquisition within this work. In the meantime (since the start of this project) newer CCD systems with very high quantum efficiencies up to 90 % are available leading to a better SNR (Signal to Noise Ratio) compared to the up-to-date ICCD cameras. Furthermore the newly available EMCCD cameras (Electron Multiplying CCD) in which a gain register is placed between the shift register and the output amplifier seam to be a very interesting alternative for the set-up discussed in this work. So for newly designed set-ups one of these new camera types is suggested because of the higher SNR.

Fortunately the authors had the opportunity to fulfill a few measurements with one new back illuminated CCD camera having a QE of over 80 %. The results of these investigations are really promising.

4.3 Multi-line vs. Multi-point

Normally SRS is used with one- or zero-dimensional resolution. This is insufficient to investigate mixing processes. There are several ways to increase the spatial resolution to pseudo two-dimensional. Two of them (the basic ideas are depicted in Fig. 4.3 and 4.4) will be discussed in more detail in the following two sections.

4.3.1 Multi-line
The laser beam is formed to a wide sheet or split up into two or more separated beams. Using appropriate mirror optics it is possible to image several one-dimensional sections staggered onto the slit of the spectrometer. Even though for each of these sections a high spatial resolution is achieved, this approach suffers from two main disadvantages:

---

[6] The usage of even shorter wavelengths leads to significant absorption of the laser beam by the surrounding air. Furthermore the high energy of the photons with very short wavelengths may destroy the excited molecules.

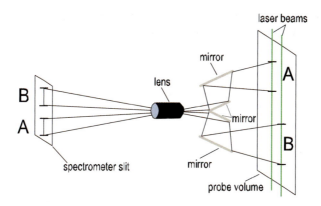

Fig. 4.3. The sketch shows the basic idea of the multi-line approach. The two line sections A and B are imaged to the spectrometer slit using a mirror optics.

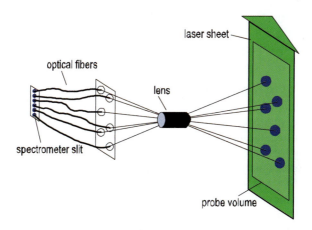

Fig. 4.4. The sketch shows the basic idea of the multi-point approach. The light arising from the probe volume is imaged to a matrix of optical fibers. In front of the spectrometer slit these fibers are lined up, so the information acquired from a two-dimensional plane are reordered to fit on a single row.

- Only a small amount of lines can be imaged without extravagant expenses, while already 4 lines result in a high adjustment effort. So normally one would stay with just two line sections.
- The edges of the imaged sections are not sharp, leading to a certain overlap. Therefore the effective measurement area is reduced.

4.3.2 Multi-point

For the multi-point approach the laser beam is formed to a sheet before being guided through the probe volume. The Raman signal is imaged to a matrix of fiber holders equipped with several (in our case 30) fibers. One end of the fibers is arranged in the matrix, while the other end is arranged in-line with each other in front of the entrance slit of the spectrometer. This set-up allows investigating a number of more or less completely independent measurement points in a plane.

Collective lenses are placed in front of each fiber, in order to improve the coupling efficiency, and therefore the signal intensities.. Up to now the position relative to the fiber entrance is the same for all fibers. This means the placement was optimized on average and not separately for each fiber. To further improve the signal strength such a degree of freedom is desirable even though this would lead to a more complex adjustment of the acquisition system.

5 Experiments

To demonstrate the functionality of the developed set-up, measurements have been conducted for two different mixer geometries. These are a scaled up model of a T-mixer used in the group of Prof. Peukert (Erlangen) and a Taylor-Couette reactor, which is identical to the one used by the group of Prof. Kind (Karlsruhe) except for the optical accesses. In the following sections the course of measurement including the calibration, as well as the data reduction are described. Furthermore a few typical results are shown.

5.1 Course of Measurement

The overall measurement process consists of three steps. Having prepared the mixer and adjusted the optical set-up, the mixing process is started and after the run-in period the acquisition of the SRS signal is conducted. The last step is the data reduction to extract the concentrations out of the acquired profiles.

5.1.1 Acquisition of the Raman Signal

After the run-in time the actual measurement process is started. The camera is connected to the advanced sync signal of the laser and is driven with a repetition rate of 5 Hz (every second laser pulse is recorded, unfortunately the used camera is not fast enough to capture every laser pulse.).

To get a more complete picture of the mixing process the concentration distributions are acquired in several layers. But since these different layers are recorded consecutively the different concentration can only be compared on basis of averaged values.

5.1.2 Data Reduction

Alsmeyer et al. [1] give a detailed description of the data reduction method used in this work, which is called ´Indirect hard Modelling´. Therefore only the basic idea will be resumed here.

The Raman spectrum of a mixture to the first order is a superposition of the pure substance spectra, weighted with the concentrations of the different components. Therefore the concentrations can be determined by minimizing the difference between the measured spectrum and the calculated superposition spectrum by adjusting the weighting factors.

To be able to measure with high precision influences of the surrounding phase leading to variations in the peak shift, height, and shape have to be considered. This is done by using parameterized representations of the pure substance spectra consisting of a certain number of Voigt profiles. This approach allows accounting concentration dependent effects, which can be quantified using measurements of several calibration mixtures. Further more the dark image as well as disturbing broadband underground signals has to be subtracted. The best way to achieve this is to include a so-called "non-component" into the evaluation process.

Based on this technique it is possible to achieve high precision measurements with an accuracy of approximately 1 %.

5.1.3 Calibration

To be able to conduct the data reduction described in the section above an additional calibration is necessary. Therefore for each fiber the calibration factor representing the ratio of signal intensity and mole fraction was determined by usage of various mixtures with known composition. Since the transmission of the fibers differ slightly it is necessary to calibrate all of them separately.

At first the models of the investigated species are created using pure substances, to get the signal intensities of the different species in the calibration mixtures. Applying the fitting algorithm of the data reduction one gains the areas of the peaks and therefore the signal intensities. The ratio of signal percentage and mole fraction of each component is the calibration factor. The error of this factor is further minimized by using the result of a regression calculation applied to the results of different mixture concentrations.

5.2 Typical Measurements of the T-Mixer and Taylor-Couette Reactor

Before the measurement can be started the storing tanks have to be filled and pressurized. Additionally the laser has to warm up for half an hour to get stable and the optics has to be adjusted. Having the test rig prepared in this way the valves of the inlets can be opened to start the mixing process. After the run-in time the acquisition of the SRS spectra is started.

In the Fig. 5.1 a typical result of an experiment conducted in the T-mixer is depicted. This result will not be discussed in detail, since its purpose is to demonstrate the proper operation of the developed set-up. The details of the operation conditions are given in the caption.

Furthermore experiments have been conduced on a Taylor-Couette reactor. A typical measurement result is depicted in Fig. 5.2. To get a better impression of the mixing process a corresponding stray light image is depicted, which was acquired using a simultaneously running video camera.

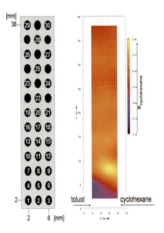

Fig. 5.1. On the right side of the figure a typical single-shot result of a mixing experiment conducted in the T-mixer is depicted. The concentrations between the discrete measurement locations shown on the left side are computed by interpolation. The investigated conditions are 600 ml/min flow rate, $Re_{toluol} = 1230$, $Re_{cyclohexane}$, and t = 61 s after opening of the valves. The dimensions of the mixer are: mixing channel 10 mm by 10 mm by 200 mm, inlet 5 mm diameter and 200 mm length.

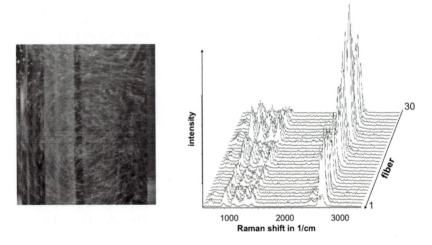

Fig. 5.2. On the left hand side a stray light image of an eddy inside the Taylor-Couette reactor acquired with a video camera is depicted. On the right hand side the SRS spectra of the corresponding region recorded with the new multi-point technique are shown.

Measurements based on the "4. Bourne reaction" have been conducted in addition to the measurements of non-reacting mixing processes. The two reactant solutions consist of DMP[7]-ethanol-water-NaCl-NaOH and Hcl-Water, respectively,

---

[7] DMP = 2,2-dimethoxypropane.

while the products are acetone and methanol. The SRS spectra of the Raman active substances DMP, acetone, methanol, ethanol, and water could be acquired using the developed set-up without problems (pure substances and mixtures). Unfortunately the accurate mole fraction of DMP for the investigated reaction is below 0.002. This led to severe problems detecting DMP and the product species due to the very low concentrations. These are no principle problems of the SRS measurement technique, but dedicated to an insufficient SNR and signal level, respectively. For test purpose the SRS spectrum of the reactant mixture containing DMP was acquired with reduced spatial resolution averaging multiple laser shots. The resulting spectrum is depicted in Fig. 5.3. So the usage of a CCD with increased QE and a train of laser pulses, increasing the coupled in energy, would solve this problems.

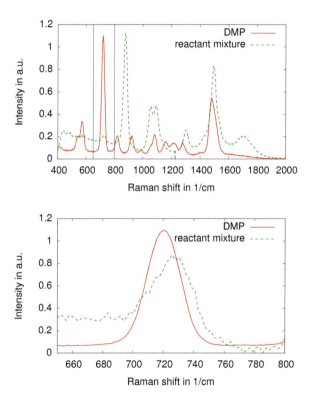

Fig. 5.3. The figure shows the spectra of DMP and the belonging reactant solution. The upper image shows the acquired spectrum, while the lower shows a zoom of the marked area containing a striking peak of DMP. For signal intensity reasons the measurement of the mixture was conducted without spatial resolution and averaged over 20 laser shots.

6 Outlook

As discussed before, there is a problem with the dynamic range and the SNR, respectively, of the investigated system. Newly available EMCCD cameras having an even higher QE and an internal amplifier further improve the SNR keeping the same intensity resolution (16-bit).

A step further to the measurements using fiber optics an optimizing of the collective lens position for each fiber would further increase the gained signal for the set-up.

7 Summary

Within this work a test-bench for the experimental investigation of mixing processes was developed, built, and optimized. In advance different approaches and technical realizations have been considered and their pros and cons have been discussed. As a last point some measurements have been conducted to check for proper operation.

8 Acronyms

CCD	Charge Coupled Device
DMP	2,2-dimethoxypropane
EMCCD	Electron Multiplying CCD
ICCD	Intensified Charge Coupled Device
IR	Infra-Red
LIF	Laser Induced Fluorescence
NIR	Near Infra-Red
MIR	Mid Infra-Red
OMA	Optical Multi-channel Analyzer
QE	Quantum Efficiency
SNR	Signal to Noise Ratio
SRS	Spontaneous Raman Scattering

References

[1] Alsmeyer, F., Koß, H.J., Marquardt, W.: Indirect spectral hard modeling for the analysis of reactive and interacting mixtures. Appl. Spectrosc. 58(8), 975–985 (2004)
[2] Bardow, A., Göke, V., Koß, H.J., Lucas, K., Marquardt, W.: Concentration-dependent diffusion coefficients from a single experiment using model-based Raman spectroscopy. Fluid Phase Equilibria 228, 357–366 (2005a)
[3] Bardow, A., Göke, V., Koß, H.J., Lucas, K., Marquardt, W.: Ternary diffusion coefficients from model-based 1-d-Raman spectroscopy. In: World Congress of Chemical Engineering, Glasgow (July 2005b)
[4] Behmann, M.: Experimentelle und theoretische Untersuchungen am Stoffsystem CHOH-HSO-HO. Dissertation, RWTH Aachen (2002)
[5] Cunge, G., Booth, J., Derouard, J.: Absolute concentration measurements by pulsed laser-induced fluorescence in low-pressure gases: allowing for saturation effects. Chemical Physics Letters 263(5), 645–650 (1996)

[6] Eckbreth, A.C.: Laser Diagnostics for Combustion Temperature and Species. Gordon & Breach Science Pub., Reading (1996)

[7] Hoffmann, T., Hottenbach, P., Koß, H.J., Pauls, C., Grünefeld, G.: Investigation of mixture formation in Diesel sprays under quiescent conditions using Raman, Mie and LIF diagnostics. SAE Technical Paper 2008-01-0945 (2008)

[8] Katz, E : Handbook of HPLC. CRC Press, Boca Raton (1998)

[9] Schrader, B., Bougeard, D.: Infrared and Raman spectroscopy. Wiley-VCH, Weinheim (1998)

[10] Unger, D., Muzzio, F.: Laser-induced fluorescence technique for the quantification of mixing and impinging jets. AIChE Journal 45(12), 2477–2486 (1999)

[11] Wilms, J., Weigand, B.: Composition measurements of binary mixture droplets by rainbow refractometry. Appl. Opt. 46(11), 2109–2118 (2007)

[12] Wilms, J., Roth, N., Weigand, B., Arndt, S.: Determination of the composition of multicomponent droplets by rainbow refractometry. In: 12th International Symposium on Applications of Laser Techniques to Fluid Mechanics (2004)

Measurements of Macro- and Micro-scale Mixing by Two-Colour Laser Induced Fluorescence

Kerstin Kling and Dieter Mewes

Institute of Process Engineering, Leibniz-University of Hannover,
Callinstr. 36, 30167 Hannover, Germany

Abstract. Two-Color Laser Induced Fluorescence (LIF) Technique enables the measurement of the progress in mixing on macro- and microscale simultaneously. This is done by injecting a mixture of an inert and a reacting fluorescent dye into the vessel. The inert dye serves as a tracer for the macro mixing but does not predicate the mixing quality on the nano scale. Since the chemical reaction requires mixing on the molecular scale, the reacting dye visualizes the micro mixing indirectly. Low Reynolds number measurements are performed in a mixing vessel equipped with a Rushton turbine. Repetitive layers form as the impeller blades create new folds each time they pass by the viewing plane. These lamellar structures can be clearly resolved. Areas of micro mixing are detected by calculating the local degree of deviation from the measured concentration fields. They are mainly found in the boundary layer of the lamellas. By choosing a suitable border value for the degree of deviation the lamellas can be classified into a center region which is not micro mixed yet, and the boundary layer with a high degree of micro mixing. The center and border region is separated by means of mathematical filters and their distance is measured. This is the macroscopic length scale of segregation, which is not influenced by micro mixing.

1 Introduction

The homogenization of two multicomponent liquids is often accompanied by a fast chemical reaction. This mixing process is of great technical interest but still is not fully understood. The correlation between the convective transport in the flow field and the diffusive transport as well as the kinetic of the chemical reaction is not predictable yet. For the process of the chemical reaction, complete mixing on the molecular scale is required. Micro mixing models exist for turbulent mixing most of which make the assumption that the fluid is completely mixed from a macroscopic point of view. A review of available models is given by Baldyga and Bourne (1999). For the laminar mixing process the composition of the compounds strongly varies with the position in the vessel. Therefore global, averaged values for the mixing quality or the mixing time, are only of restricted importance. The prediction of the local composition on micro scale still is not possible for complex geometries. Ottino (1994) introduced a lamellar model which describes the interplay between stretching, diffusion and reaction at small scales. It can be applied for laminar and turbulent mixing, but the application is limited to simple geometric flow domains. Recently, Muzzio and coworkers (2003, 2002) extended the one-dimensional model to the complex flow in stirred tanks. The stretching of small fluid elements is calculated in a Lagrangian frame of reference. They found that the stretching field in chaotic flow has a controlling influence on reactive processes, when convection, reaction and diffusion occur on the same time scale.

Despite the progress in predicting reactive mixing there is still a need for experiments, visualizing the local distribution of inert and reactive tracers with high spatial and temporal resolution. Laser Induced Fluorescence has proven to be a suitable measurement technique. The visualization of passive scalars show the convective mixing process (Villermaux et al. (1996), Distelhoff and Marquis (2000), Guillard et al. (2000)). In viscous mixing applications stretching and folding occur simultaneously at different rates in each portion of the flow, creating complex, layered (lamellar) structures. Unger and Muzzio (1999) and Lamberto et al. (1996) have shown that chaotic flows are the only effective way to destroy segregation rapidly in these applications. Reactive tracers, mostly pH-sensitive dyes, are used to measure the micro mixing indirectly (Bellerose and Rogers (1994), Hong et al. (2002)). The drawback of using just one reactive dye for viscous mixing applications is that zones of micro mixing can qualitatively be visualized by detecting the reaction product but the convective transport and the dilution of the dye on the macroscopic scale is unaccounted for: zones without the reaction product can either be not mixed on the molecular scale or not mixed at all. In this article, an experimental technique with high spatial and temporal resolution to determine local measures for the macro- and micro mixing in stirred vessels simultaneously is presented.

2 Micro and Macro Mixing- the Degree of Deviation

To separate the transport phenomena for macro- and micro mixing, namely convection and diffusion, a mixture of one inert and one reacting dyes is injected into the mixing vessel. The convective transport affects both dyes whereas the micro mixing only varies the concentration of the reacting dye since the process of the chemical reaction requires complete mixing on the molecular scale. As a quantitative measure for the progress in micro mixing the local degree of deviation

$$\Delta(\vec{x},t) = 1 - \frac{c_{1,react}(\vec{x},t)}{c_1(\vec{x},t)} \tag{1}$$

is defined. It is obtained by comparing the local concentration of the reaction product $c_{1,reac}$ with the concentration c_1 of the reacting dye which would locally appear without the reaction. The latter is calculated from the local concentration of the inert dye c_2 and the initial concentration ratio:

$$c_1(\vec{x},t) = c_2(\vec{x},t)\frac{c_{1,0}}{c_{2,0}} \tag{2}$$

The local degree of deviation equals the conversion rate, that is the portion of the reacting dye which has not reacted yet. For a completely segregated fluid the local degree of deviation is one. During the micro mixing it decreases to its minimum value of zero for a completely homogeneous fluid.

3 Fluorescent Dyes

The fluorescence intensity emitted by a fluorescent dye I_F is proportional to the intensity of the light absorbed by the dye I, which is calculated by Lambert-Beer's Law. The quantum yield Φ describes the effectiveness of the fluorescent emission:

$$I_F = \Phi I = \Phi I_0 e^{-\varepsilon sc} . \tag{3}$$

I_0 is the intensity of the exciting light, ε is the molar extinction coefficient and s is the length of the measurement volume. For small concentrations eq. (3) can be simplified by a series expansion so that I_F only linearly depends on the concentration c of the dye:

$$I_F = \Phi I_0 K \varepsilon sc \tag{4}$$

K is a parameter depending on the measurement system, considering for example the viewing angle of the detector. For constant parameters $m = \Phi I_0 K \varepsilon s$ only a simple calibration procedure by measuring the fluorescence intensity for known concentrations of the dyes is necessary in order to predict the concentration from measured intensities:

$$I_F = m \cdot c \tag{5}$$

A system of two fluorescent dyes is used for the experiments. The following set of predetermined criteria has to be met: The dyes are required to be excitable at the same wavelength, and their emission characteristics must be distinguishable. Only one of the dyes should undergo a chemical reaction which alters its fluorescent behavior. The kinetics of the chemical reaction must be fast enough for the mass transfer (the mixing) and not the chemical reaction to be the limiting factor. Only in this case the chemical reaction indirectly visualizes the micromixing. Two dyes purchased from Molecular Probes Inc. are used. The fluorescent emission spectra of both dyes are shown in Fig. 1. The reacting dye, fluo-4, is an indicator for Calcium ions which is changing its fluorescent emission characteristics with the formation of a Calcium complex. This is done quasi spontaneously with a time constant of approximately 10^9 1 (mol s)$^{-1}$ (Naraghi (1997)). The second dye, carboxy-SNARF, does not react with Calcium ions and therefore serves as the inert dye. On the other hand its absorption and emission spectrum strongly varies with pH so that a constant pH value has to be adjusted. During the experiments a solution of Tris-buffer (Merck KGaA) of pH = 8.2 is used. For that, carboxy-SNARF is excitable at the same wavelength as fluo-4 but can be detected at much longer wavelengths, which allows a separation of the fluorescent light by means of optical filters.

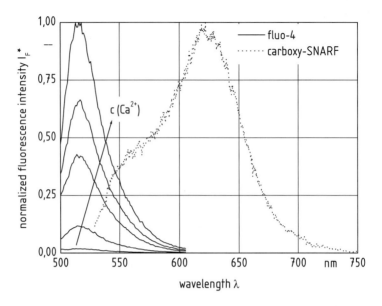

Fig. 1. Normalized fluorescence emission spectrum of fluo-4 and carboxy-SNARF

The reacting dye, fluo-4, is essentially non-fluorescent unless bound to Ca^{2+}. In this case the fluorescence intensity is enhanced depending on the concentration of Calcium ions as shown in Fig. 1. For a concentration of free Ca^{2+} of approx. 40 μmol/l a saturation condition is reached and the fluorescence intensity shows its maximum. Only the saturation condition is used during the experiments since only the concentration of the dye which already reacted with its environment (and therefore is mixed on the molecular scale) is of interest. In order to achieve this a concentration of $c(Ca^{2+}) = 0.6$ μmol/l is adjusted in the mixing vessel. This high surplus of Calcium ions ensures that the saturation condition is reached immediately in the direct surrounding of the injected dye and the reaction can proceed completely. Highly purified fluo-4 can exhibit a fluorescence enhancement from the minimum value I_F' to the saturation condition I_F'' of $\beta = 40 \div 100$ (Haugland, 2002), with

$$\beta = \frac{I_F''}{I_F'} . \qquad (6)$$

Due to impurities of Calcium, which are introduced e.g. by the chemicals for the pH- buffer solution or the glassware, there are always free Calcium ions present in the dye solution. This leads to a background fluorescence of fluo-4 and a decreased value of $\beta < 25$. In order to improve the signal-to-noise ratio the background fluorescence must be reduced. For that free Calcium ions, which are introduced due to impurities, are bound to EGTA (SIGMA-ALDRICH Chemie GmbH). EGTA has a higher affinity to Calcium than fluo-4 so that the Calcium

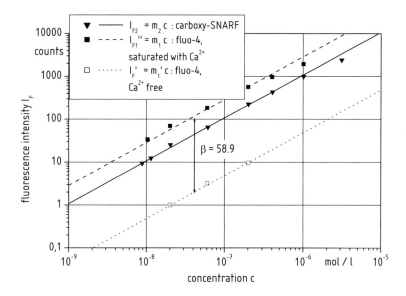

Fig. 2. Fluorescence intensity as a function of dye concentrations

ions are no longer available for a reaction with fluo-4. For a concentration of $c_{EGTA} = 7 \cdot 10^{-6}$ mol/l an enhancement factor of $\beta = 58.9$ can be measured (Fig. 2).

In Fig. 2 the fluorescence intensities for various concentrations of fluo-4 in a Ca^{2+}-free solution and in saturated condition as well as of carboxy-SNARF is presented. The measured values are approximated with a straight line each according to eq. (5). For both dyes the linear assumption holds only up to a maximum concentration of approximately 10^{-6} mol/l. For higher concentrations quenching effects are not negligible any more so that only dye solutions of lower concentrations should be used. With the calibration factors m_1, m_2 and β, which can be extracted from Fig. 2, the concentrations of the inert dye c_2 and the reaction product $c_{1,reac}$ are calculated from measured fluorescence intensities I_{F1} and I_{F2}:

$$c_2\left(I_{F1}, I_{F2}\right) = \frac{I_{F2} - f I_{F1}}{m_2} \tag{7}$$

$$c_{1,reac}\left(I_{F1}, I_{F2}\right) = \frac{\beta}{(\beta-1)\, m_1} \frac{1}{}\left(I_{F1} - \frac{1}{\beta} \frac{m_1}{m_2} \frac{c_{1,0}}{c_{2,0}} \left(I_{F2} - f I_{F1}\right) \right) \tag{8}$$

The parameter $f \approx 0.1$ accounts for the non-ideal behavior of the optical filter for the fluorescence intensity of carboxy-SNARF and has to be determined experimentally. Before the calculation of the concentration fields according to eq. (7) and (8), some corrections are applied. They include the non-uniform illumination of the light sheet due to the intensity profile of the laser beam, the varying intensity of the laser from pulse to pulse and the oblique viewing of the display window from the two apertures of the Double Image Optics (see Chapter "Experimental Set-up"). Details about these corrections can be found in Kling 2003. As a last step the local degree of deviation is calculated according to eqs. (1) and (2).

4 Experimental Set-Up

The Planar Laser Induced Fluorescence technique (PLIF) is used to measure the concentration fields of two fluorescent dyes simultaneously in the mixing vessel. The optical set-up is schematically depicted in Fig. 3. As a light source a pulsed laser of wavelength $\lambda = 495$ nm (NewWave Nd:Yag, Tempest 30 and GWU OPO VisIr) is used. The laser beam is expanded to a thin light sheet using a system of spherical and cylindrical lenses. The divergent light sheet is illuminating an arbitrary plane in the mixing vessel, exciting the fluorescent dye in this area. The emitted light is detected by an intensified CCD-camera (Imager 3 and Image Intensifier, LaVision) which is positioned vertical to the measurement plane. The

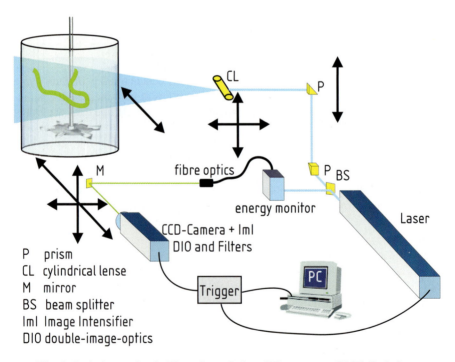

P   prism
CL   cylindrical lense
M   mirror
BS   beam splitter
ImI Image Intensifier
DIO double-image-optics

Fig. 3. Optical setup for the Planar Laser Induced Fluorescence (PLIF) Technique

intensity of the light is measured in 12 bit grey values. The fluorescent light is passing two optical filters (BP523/10 and RG645) which are suitable to separate the fluorescent light of the two dyes. The so-called Double-Image Optics (LaVision) is used to detect the same display window twice at the same time. It consists of two apertures which are equipped with the two filters and a set of adjustable and fixed mirrors. The fluorescent light is reflected by the mirrors onto the light sensitive camera chip such that the same display window is projected side by side on the camera chip with one half each ideally is representing the fluorescent light emitted by one of the fluorescent dyes. The light sheet optics, the mixing vessel and the camera are mounted onto a linear positioning system each in order to allow reproducible adjustments. The exposure of the camera and the pulsation of the laser are controlled by a computer. The maximum measurement frequency is 30Hz and the resolution of the camera is 640 x 480 pixels. With the start of the camera exposure also the injection of the dye is automatically beginning. Each position in the vessel except the position directly underneath the stirrer can be adjusted for the injection. The maximum flow rate amounts 1.25 ml/s but every flow rate below this value can be adjusted.

Measurements are performed in a flat-bottomed, transparent vessel of diameter 100 mm. It is placed inside a rectangular viewbox filled with water in order to minimize reflections and distortions at the cylindrical walls. The vessel is filled to a height of 130 mm and the 6-blade Rushton turbine is placed in the vessel with a bottom clearance of 60mm. In order to achieve low Reynolds' number mixing the viscosity of the liquid is increased. This is done by dissolving different mass fractions (0,5%, 1% or 2%) of carboxy-methyl-cellulose (CRT 10000, Wolff Walsrode) in de-ionized water, which leads to a slightly shear thinning behavior of the liquid.

5 Experimental Results

As an example the course of the macro- and micro mixing for the cellulose solution with a mass fraction of 1% and the injection position close to the stirrer shaft is shown in Fig. 4. A mixture of the inert dye with $c_{1,0} = 2.2 \cdot 10^{-6}$ mol/l and the reacting dye with $c_{2,0} = 1.02 \cdot 10^{-6}$ mol/l is prepared and a volume of 1 ml is injected into the vessel with a flow rate of 0.5 ml/s after the velocity field had achieved a steady state. The stirrer speed is 300 min^{-1} which leads to a Reynolds number in the laminar region. The display window is situated in the symmetry plane of the vessel whereas the injection position is displaced by an angle of approximately 90°. The display window has a size of 35 mm x 60 mm so that one pixel of the camera chip corresponds to approximately 0.13 mm. The macro mixing is represented by the concentration field of the inert dye c_2 and the micro mixing is depicted by the field of the local degree of deviation Δ which is calculated from measured dye concentrations.

Macromixing:
concentration c_2 0   mol/l $4.0 \, 10^{-7}$

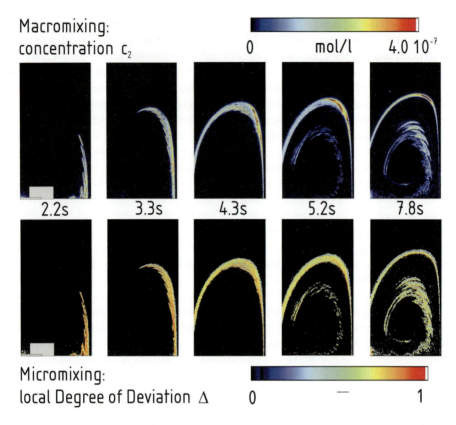

Micromixing:
local Degree of Deviation Δ 0   —   1

Fig. 4. Course of the macro- and micro mixing for the dye injection close to the stirrer shaft

For the injection position close to the stirrer shaft, macro- and micro mixing are performed efficiently as presented in Fig. 4. Since the dye is initially injected in a region of high shear rates and high axial velocities it is rapidly transported through the stirrer plane, generating lamellar structures. The repetitive layers which are visible already after 2s form as the impeller blades create new folds each time they pass by the viewing plane. The local degree of deviation is high in the center of the lamellas and already decreased in the boundary layers. The lamellar folds travel towards the wall, deform further, stretch and fold and after 4s recirculated again through the impeller region forming a second loop. After 8s the multi-layered structure can only be resolved in the second inner loop. In the center regions of some lamellas the local degree of deviation is still high, but in large parts it already decreased to a value of approx. 0.4-0.6.

From Fig. 4 it is visible that during the laminar mixing process lamellar structures are formed which consist of repetitive layers with and without the dye. Depending on the viscosity of the fluid and the velocity at the injection position different lamellar structures are forming. The spacing between the layers, the striation thickness, is the macroscopic length scale of segregation. It depends on the position and the residence time in the vessel. For the evaluation of the global mixing quality depending on a change of e.g. geometric or material properties an average value for the striation thickness is calculated. In the following, the determination of the striation thickness is described and the influence of viscosity, rotational speed and injection position is explained in order to verify the method.

From the concentration field it is usually difficult to differentiate between the layers with and without dye because the boundary is not sharp due to diffusive transport. A simple approach is selecting a suitable filter value for the concentration. Hence, fluid elements with $c < c_{filter}$ are assumed to belong to the layer without dye and only fluid elements with $c > c_{filter}$ are assigned to the layer containing the dye. Since during the mixing process the dye is diluted and therefore the concentration strongly varies it is difficult to define a universal filter value. This problem can be overcome by using the field of the local degree of deviation. One example is shown in Fig. 5 and the marked area is enlarged in Fig. 5a. The lamellar structure is clearly resolved. A central region with a high degree of deviation is surrounded by a region with a small degree of deviation. Small values for the local degree of deviation indicate a high progress of micro mixing. As expected, micro mixing is initiated in the boundary layers.

In Fig. 6 the concentration and the degree of deviation is shown along the line of constant x marked in Fig. 5. For distinct peaks of the concentration profile the local degree of deviation takes a value $\Delta > 0.7$. This value is universal and independent of the dilution. Segregated regions can be separated from regions with a

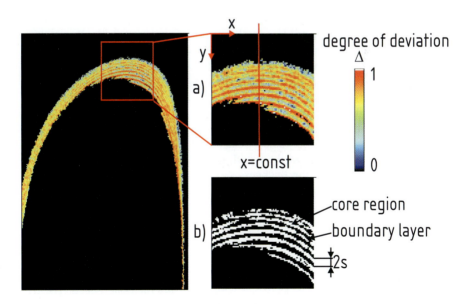

Fig. 5. Field of the local degree of deviation enlarged (a) and binary-coded (b)

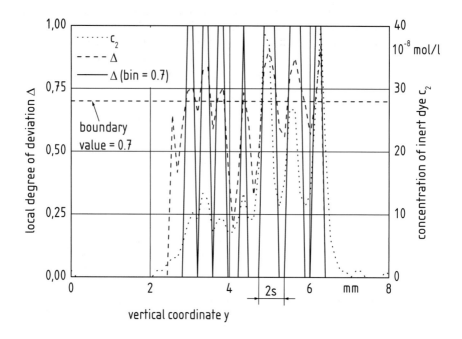

Fig. 6. Profile of the local degree of deviation and the concentration of inert dye

high degree of micro mixing. Hence, the macroscopic length scale, the striation thickness, can be determined unaffected from the micro mixing. For that, the field of the local degree of deviation is binary-coded with the value $\Delta = 0.7$ ($\Delta > 0.7$: $\Delta = 1$; $\Delta < 0.7$: $\Delta = 0$) which is shown in Fig. 5b and results in the compact line in Fig. 6. The striation thickness is half of the distance between two similar layers (with or without dye, respectively).

The striation thickness is determined for various positions and for different times. From that an average value is calculated which is representative for the given settings, eg. viscosity of the liquid, rotational speed and injection position. In Fig. 7 the average striation thickness s_m is presented as a function of the rotational speed with the viscosity and the injection position as parameters. The error bars indicate the standard deviation of the measured results. There are three main conclusions:

1. For the same viscosity of the fluid the striation thickness is decreased with increasing rotational speed.

    For the injection position close to the stirrer shaft (position B) the striation thickness is decreased with increasing rotational speed. The reason is the increased velocity and therefore the faster transport into zones of high shear rates in the impeller region. However, this transport is reduced for the injection position in the centre of the vortex (position A).

The dye is transported mostly in circumferential direction so that the higher rotational speed does not significantly affect the mixing process and the striation thickness remains nearly constant. For fluids of different viscosity there is no consistent trend for the striation thickness as a function of rotational speed. This is due to the fact that for higher viscous fluids the rotational speed has to be increased in order to receive a similar flow field.

2. With increasing viscosity of the cellulose solution the striation thickness is increased.

   The effect of the viscosity is distinct especially for the two higher viscous fluids. Viscous dissipation causes a diminished secondary flow in the upper and lower vortex compared to the flow in circumferential direction. Therefore the mixing effect due to stretching and folding in the secondary flow is decreased for increasing viscosity. This leads to an increasing striation thickness.

3. The striation thickness depends on the injection position.

   As described above the mixing effect of the secondary flow loop has only a small influence for fluid elements injected in the centre of the vortex. Therefore the striation thickness is higher for the injection in the centre of the vortex (position A) compared to the injection in regions of high axial velocity close to the stirrer shaft (position B).

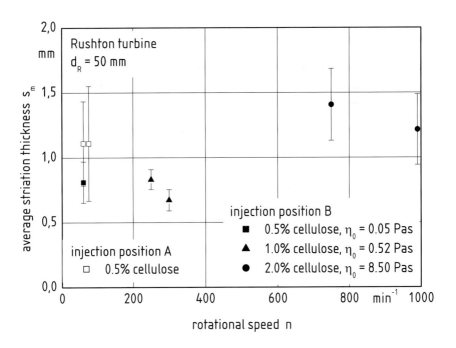

Fig. 7. Average value of the local striation thickness as a function of rotational speed

6 Conclusions

The two-color Laser Induced Fluorescence Technique gives new insight into the mixing process. It is possible to measure the local intensity of segregation at a multitude of points inside the stirred vessel. This is done by injecting a mixture of an inert and a reacting fluorescent dye into the vessel and by measuring their concentration fields. The fluorescence intensity of the inert dye only depends on its concentration and it therefore serves as a tracer for the macro mixing. The fluorescence intensity of the reacting dye instead is enhanced by a chemical reaction which requires mixing on the molecular scale and therefore shows the micro mixing. Low Reynolds number measurements are performed in a mixing vessel equipped with a Rushton turbine. The creation of lamellar structures can clearly be resolved. Areas of micro mixing are detected by calculating the local degree of deviation from the measured concentration fields. They are mainly found in the boundary layer of the lamellas. By choosing a suitable border value for the degree of deviation the lamellas can be classified into a center region which is not micro mixed yet, and the boundary layer with a high degree of micro mixing. From that the macroscopic length scale of segregation, the striation thickness, can be determined unaffected by micro mixing. The striation thickness is decreased for decreasing viscosity, increasing rotational speed and increasing axial or radial velocity at the injection position.

References

Baldyga, J., Bourne, J.R.: Turbulent mixing and chemical reaction. John Wiley & Sons Ltd., Chichester (1999)

Bellerose, J.A., Rogers, C.B.: Measuring mixing and local pH through Laser Induced Fluorescence. Laser Anemometry 191, 217–220 (1994)

Distelhoff, M.F.W., Marquis, A.J.: Scalar mixing in the vicinity of two disk turbines and two pitched blade impellers. Chem. Eng. Sci. 55, 1905–1920 (2000)

Guillard, F., Träghard, C., Fuchs, L.: New image analysis methods for the study of mixing patterns in stirred tanks. Can. J. of Chem. Eng. 78, 273–285 (2000)

Haugland, R.P. (ed.): Molecular Probes- Handbook of Fluorescent Probes and Research Products, 9th edn. (2002)

Hong, S.D., Sugii, Y., Okamoto, K., Madarame, H.: Proceedings of the 10th International Symposium on Flow Visualization, Kyoto, Japan, August 26-29 (2002) paper F0275

Kling, K., Mewes, D.: Quantitative Measurements of Micro- and Macromixing in Stirred Vessels using Planar Laser Induced Fluorescence. Journal of Visualization 6, 165–173 (2003)

Lamberto, D.J., Muzzio, F.J., Swanson, P.D., Tonkovich, A.L.: Using time-dependent RPM to enhance mixing in stirred vessels. Chem. Eng. Sci. 51, 733–741 (1996)

Muzzio, F.J., Alvarez, M.M., Zalc, J.M., Shinbrot, T.: Mechanisms of mixing and creation of structure in laminar stirred tanks. AIChE Journal 48, 2135–2148 (2002)

Muzzio, F.J., Szalai, E.S., Kukura, J., Arratia, P.E.: Effect of hydrodynamics on reactive mixing in laminar flows. AIChE Journal 49, 168–179 (2003)

Naraghi, M.: T-jump study of calcium binding kinetics of calcium chelators. Cell Calcium 22, 255–268 (1997)

Ottino, J.M.: Mixing and chemical reactions - a tutorial. Chem. Eng. Sci. 49, 4005–4027 (1994)

Unger, D.R., Muzzio, F.J.: Laser- Induced Fluorescence technique for the quantification of mixing in impinging jets. AIChE Journal 45, 2477–2486 (1999)

Villermaux, J., Houcine, I., Vivier, H., Plasari, E., David, R.: Planar laser induced fluorescence technique for measurements of concentration fields in continuous stirred tank reactors. Experiments in Fluids 22, 95–102 (1996)

Analysis of Macro- and Micromixing in Laminar Stirred Mixing Vessels Using Laser Optical and Numerical Methods

Martin Faes and Birgit Glasmacher

Institute of Process Engineering, Leibniz-University of Hannover,
Callinstr. 36, 30167 Hannover, Germany

Abstract. During the mixing process the transport of species takes place on different length scales. The convection of several species is identified on large scales, whereas diffusion progresses on molecular scales. For a laminar mixing process these two distinguishable mechanisms of transport are analyzed experimentally and theoretically for two- and three-dimensional mixing systems. The local concentration fields of an injected mixture of two fluorescent dyes are measured by means of the 4D Two-Colour Laser Induced Fluorescence (LIF) in the volume of a mixing vessel. An inert dye is used for the identification of the convective transport. A second dye is reacting with an individual component in the mixing vessel in a fast chemical reaction. The formation of the reaction product indirectly indicates the molecular transport. Both measured concentration fields are used for the calculation of a local mixing quality. Particle Image Velocimetry (PIV) is used for measuring local velocity vectors. The source term of the local energy dissipation is calculated and visualized by means of the velocity gradients. The length scale of the forming lamellas depends on varying local energy dissipation. As a result of increasing energy dissipation local mixing quality enhances. The macro- and micromixing is analyzed depending on Reynolds number and position of injection. The flow and concentration fields are calculated by means of numerical methods. Therefore, a commercial CFD code is used. The geometries are equivalent to those of the experimental investigation. The experimental and numerical results are in good agreement.

1 Introduction

In many industries e.g. chemical and pharmaceutical industries mixing processes of liquid-liquid components are of great technical interest. These industries have a rich faculty of knowledge, but the mixing processes are not yet completely predictable. The operation unit "mixing" is actually characterized by empiricism rather than by basic knowledge. Local mixing quality is a crucial parameter influencing the economics of such processes. Because of this interaction the interest of investigations concentrates on the measure of local mixing quality in regard to the macroscopic energy inserted mostly by stirrers into the mixing process. In case of a superimposed chemical reaction, complete mixing on molecular scale is required. Baldyga and Bourne (1999) give a review for mixing models. The interaction between stretching, diffusion and reaction is described by Ottino (1994). Chaotic flow structures are used by Unger and Muzzio (1999) and Lamberto et al. (1996) reaching a fast homogenization. A seminal paper concerning the concept of chaotic advection is published by Aref (1984). Anderson et al. (2002) presented an extended mapping technique for chaotic flows in a journal bearing flow. Gollub et al. (2003) experimentally determined the mixing for a two-dimensional time-periodic flow exhibiting chaotic mixing. Different experimental set-ups with high spatial and temporal resolutions have still been developed further for visualising the local distributions of tracers. Laser induced fluorescence has proven to be a suitable measurement technique by Kling and Mewes (2003). The macromixing is visualised by using passive scalars. In literature different systems are introduced by Villermaux et al. (1996), Distelhoff and Marquis (2000) and Guillard et al. (2000). The micromixing is indirectly described by reactive pH-sensitive tracers. Bellerose and Rogers (1994) and Hong et al. (2002) gave first impressions of using such tracers. In T-shaped micro-mixers the transport phenomena on different scales are investigated by means of pH-sensitive tracers using scanning methods by Hoffman et al. (2006). Van Vliet et al. (2000) used a laser scanning method for investigating three dimensional concentration fields. The analysis of mixing processes by numerical methods is described by Paschedag et al. (2007).

In this research project, two laser optical measurement systems with high spatial and temporal resolution are presented for investigating the local mixing quality, the local energy dissipation and their correlation in different mixing systems. Therefore, a 4D LIF technique is developed for analyzing the concentration field and a PIV system for the velocity and dissipation field. The experimental results are compared to initial CFD-studies. All investigations are performed in laminar flow regime.

2 Experimental Setup

The laser optical experiments are performed in a cylindrical glass vessel, which has an inner diameter of $d_a = 90$ mm and a height of $h = 130$ mm, resulting in 1.3 l volume. For the investigation different full-glass stirrers are used: (i) cylindrical shaft with a diameter of $d_i = 30$ mm, (ii) Rushton-turbine with a diameter of $d_i = 50$ mm and (iii) multi-stage pitched blade stirrer with a diameter of $d_i = 50$ mm.

The cylindrical full-glass shaft is eccentrically positioned, the Rushton-turbine and the multi-stage pitched blade stirrer are positioned in the centre of the mixing vessel. A rectangular viewbox filled with pure glycerol and positioned round the mixing system allows for an isooptical study and minimizes optical distortions. A motor system, consisting of one motor for the outer glass vessel and one motor for the stirrers, is used for the motion of the fluid. The working fluid is a mixture of pure glycerol and a solution of calcium chloride. At room temperature glycerol has the same index of refraction as the cylindrical glass vessel and the full-glass stirrers. It is thus possible to measure velocity and concentration fields throughout the whole mixing system. The viscosity of the glycerol mixture is 0.93 Pas

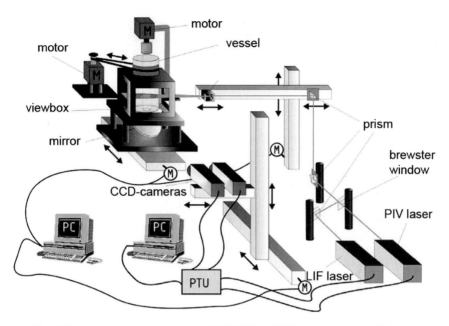

Fig. 1. Laser optical setup consisting of 4D-LIF and PIV measurement techniques

measured with Rheometrics RFS II at room temperature. The experimental setup consisting of the two optical measurement techniques is schematically depicted in Fig. 1. The particle image velocimetry system, a LaVision PIV system, is used to measure the local velocity vectors. The laser is a double pulsed NewWave Nd:YAG laser with a frequency of 10 Hz. It is operating at a wavelength of $\lambda = 532$ nm. The laser beam is expanded to a measurement plane by means of a cylindrical lens. The illuminated laser plane can be adjusted freely by an automated linear positioning system. It is possible to measure vertical and horizontal laser planes. The illuminated horizontal laser plane is projected to the CCD-camera by means of a mirror under 45°, which is placed below the viewbox. The CCD-camera is positioned perpendicular to the laser beam for both laser measurement techniques. The Programmable Timing Units (PTUs) controlled by software consists of computer boards connected to the laser and camera. Control, data acquisition and processing is performed with a software package DaVis® from LaVision. For measuring the velocity field (PIV), the particle paths are visualized by seeding the working fluid with hollow glass spheres of mean diameter of 10 µm. The reflected signals of the glass spheres are captured temporally with a CCD-camera, Imager Intense 1376 x 1040 pixel. For the image processing a cross-correlation algorithm with an interrogation window of 32 x 32 pixel with 50% overlap is selected. In each case 200-500 double images are captured. For measuring phase-modulated a certain number of total images is used for cross-correlation. The second non-intrusive optical laser measurement technique is the 4D two-colour laser induced

fluorescence technique. It is used to measure the concentration fields of two fluorescent dyes simultaneously in the mixing system. The laser, a NewWave
Nd:YAG, Tempest 30 with GWU OPO VisIr, is a pulsed laser configuration with
an adjustable wavelength. For the complex dye system used in the experiments, a
wavelength of $\lambda = 495$ nm is adjusted. The CCD-camera, Imager 3 (640 x 480
pixel) and Image Intensifier delivered by LaVision, is positioned identical to the
PIV camera. The CCD-camera is equipped with a so-called Double-Image Optics
(DIO) which permits to capture the same display window twice at the same time.
The optical DIO consists of two apertures which are equipped with a set of adjustable and fixed mirrors and two optical filters (BP523/10 and RG645) for separating
the fluorescent lights of the dyes (Kling and Mewes, (2004)). The camera exposure
and the injection of the dyes begins simultaneously. The injection system can be
adjusted at each position in the vessel except the position directly below the cylindrical shaft. The flow rate is selected to 0.05 ml/s for the cylindrical shaft with an
absolute volume of 0.5 ml. For the Rushton turbine and the multi-stage pitched
blade stirrer a volume of 2 and 4 ml is injected within 10 s. The illuminated laser
planes have a thickness of 0.5 mm. For receiving three-dimensional images in dependence of time (4D) the cameras, the mixing vessel and the light sheet optics are
mounted onto a linear positioning system. In order to measure the concentration
field, the stirrer is stopped. The three-dimensional images are reconstructed out of
170 two-dimensional plane images captured in a distance of 0.5 mm using Imaris®
from Bitplane. Each two-dimensional plane image consists of an averaged plane
image out of 16 single plane images captured in one second. The scanning procedure from the bottom of the mixing vessel to the free surface of the working fluid
takes 10 minutes including post processing. The laser measurement technique of
PIV and 4D-LIF are used sequentially.

3 Macro- and Micromixing

Transport phenomena of convection and diffusion can be detected simultaneously
in a mixing system using 4D two-colour laser induced fluorescence technique according to Kling and Mewes (2004). Therefore, a system of two fluorescent dyes
is applied. The convection, namely the macromixing, is visualized by an inert
fluorescent dye (carboxy-SNARF). The transport on molecular scale is visualized
by a fluorescent dye (fluo-4) reacting with calcium ions dissolved in the working
fluid. The fluorescent emission characteristics of this dye changes due to a chemical reaction with calcium ions. The progress of the chemical reaction shows indirectly the micromixing because mixing on molecular scale is required for the
progress of chemical reactions. The two dyes are excitable at the same wavelength
but their emission characteristics are distinguishable. After a mixture of both dyes
is injected into the mixing system the concentration fields of the inert and reactive
dyes are measured. The Schmidt number of the tracer is calculated to $3.1 \cdot 10^8$ with
Einstein equation for the diffusion coefficient of calcium ions in high molecular

solutions. By using the Lambert-Beer's law the fluorescence intensity I_F is calculated by

$$I_F = \Phi I = \Phi I_0 \exp(-\varepsilon l c) \tag{1}$$

Φ is the quantum yield, I_0 is the intensity of the exciting light, ε is the molar extinction coefficient and l is the length of the measurement volume. A simplification by series expansion can be done for small concentrations. A linear function for I_F is generated depending of the concentration c of the dye:

$$I_F = \Phi I_0 K \varepsilon l c \tag{2}$$

For constant parameters $m = \Phi I_0 K \varepsilon l$ is only a calibration factor. The system factor K is constant for each measurement; it takes e.g. the viewing angle of the detector resp. the losses of intensities of the mirrors into account. The concentrations in the mixing vessel are measured with calibrated fluorescent intensities, determined beforehand as a calibration

$$I_F = m \cdot c \tag{3}$$

The spectra of the two dyes have a small overlap. The overlap is between the absorption spectra of the inert dye carboxy-SNARF and the emission spectra of the reactive dye fluo-4. The error of the reabsorption is measured to less than 2.5% for the highest concentration and a small illuminated striation thickness of 2.5 mm. Thus, the error in the mixing vessel is always less than 2.5% and under operation conditions it is less than 0.4%. The band-pass filter of the reactive dye has a range of 10 nm permitting to detect only the emission intensity of the reactive dye. The intensity of the reactive dye is also detected through the filter for the inert dye (RG645) and can be determined to 10%. This value will be used for correcting the intensity measured with the filter (RG645) for the inert dye. The calibration of the measurement system and the correction of the shot-to-shot noise of the laser are described in detail in Kling and Mewes (2004).

The ratio β of the minimum value I'_F to the saturation condition I''_F for the reactive dye describes the sensitivity of the chemical reaction; a value between 40 and 100 (Haugland (2002)) gives a good signal-to-noise ratio:

$$\beta = \frac{I''_F}{I'_F}. \tag{4}$$

The diagram in Fig. 2 shows the fluorescence intensities of the dyes as a function of various concentrations. The intensities are measured for fluo-4 in a Ca^{2+}-free solution and in saturated condition as well as for the carboxy-SNARF. For the calibration a linear approximation is applied. The calculated factors can be seen in the caption. Up to a maximum concentration of approximately 10^{-6} mol/l the linear assumption fits correctly the measured values. For higher concentrations quenching effects are no longer negligible so that only dye solutions of lower concentrations should be used. A good signal-to-noise ratio is achieved by the β-factor of $\beta = 58{,}56$. By means of the calibration factors m_1, m_2 and β, which can be extracted from Fig. 2, the concentrations of the inert dye c_2 and the reaction product $c_{1,react}$ are calculated from measured fluorescence intensities I_{F1} and I_{F2}.

The local degree of deviation is a quantitative measure for the quality of micromixing. It is defined as

$$\Delta(\vec{x},t) = 1 - \frac{c_{1,react}(\vec{x},t)}{c_1(\vec{x},t)}. \tag{5}$$

The local degree of deviation can be calculated by comparing the local concentration of the reaction product $c_{1,react}$ with the virtual concentration c_1 of the reacting dye locally appearing without any reaction. Due to the injection as a mixture and transportation in the same manner the local concentration of the reactive dye c_1 without chemical reaction can be calculated by the knowledge of concentration c_2 of the inert dye and the initial concentration ratio to

$$c_1(\vec{x},t) = c_2(\vec{x},t)\frac{c_{1,0}}{c_{2,0}}. \tag{6}$$

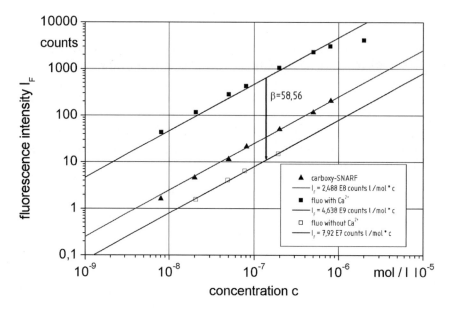

Fig. 2. Fluorescence intensities vs. concentration of the inert and reactive dye, signal-to-noise ratio $\beta = 58.56$

The local degree of deviation is the portion of the reacting dye which has not reacted yet. For a value of the local degree of deviation $\Delta = 1$ the mixture is completely segregated on macroscopic scale. During the mixing process a homogenous fluid on microscopic scale is achieved and the local degree of deviation decreases to $\Delta = 0$ for a complete micromixed fluid.

4 Numerical Calculations

During the last years numerical calculations have reached considerable influence in technical developments. For mixing processes it is desirable to have local information about concentration and velocity fields for optimization mixing geometry or the position of the feeding point. For a theoretical prediction of local mixing quality it is necessary to chose the correct physical models and boundary conditions. In this work the data of the numerical calculations are validated by the experimental investigations. The numerical investigations are performed with a commercial CFD code by ANSYS. To start numerical calculations, a mesh is generated with software from ANSYS-ICEM. These models are equivalent to the glass mixing vessel equipped with the three full-glass stirrers in the experimental set-up. In Fig. 3 the meshes of both stirrer configurations are depicted consisting of 2.4 and 6.0 million finite volumes. For a good convergence during the solving process of the differential equations a high quality mesh is required. This is guaranteed by avoiding cell angles smaller than 20 degrees. The aspect ratio is chosen smaller than 100 and the mesh expansion factor is smaller than 20.

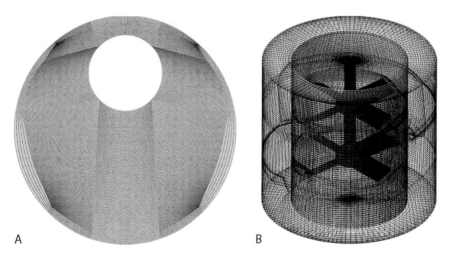

A   B

Fig. 3. Generated meshes for the cylindrical stirrer (A) and the multi-stage pitched blade stirrer (B)

Due to the experiments, which are performed in laminar regime, a laminar model is used. For the multi-stage pitched blade stirrer the multiple frames of reference model is used. Thus, the stationary injection position can be modeled. For the unsteady calculations of the concentration fields the transient rotor-stator model is used. The boundary conditions for the mixing vessel, the bottom and the stirrer are set to a wall with velocity of v = 0 m/s and no-slip condition. The fluid surface is set as free slip. The rotational domain is allocated with several rotational speeds depending on Reynolds number. For the solution of the differential equations for the steady system the advection scheme "high resolution" is used. The velocities and concentrations of the unsteady system are calculated by a second order backward Euler scheme.

5   Results and Discussion

a. Flow Conditions for Different Stirrers

Experimental and numerical investigations are performed in different mixing systems in a laminar region. The different Reynolds numbers and rotational frequencies are listed in Table 1 for each stirrer configuration.

Table 1. Listing of investigated Reynolds numbers and rotational speeds

type of stirrer	diameter [mm]	Reynolds number [-]	speed [min⁻¹]
cylindrical shaft	d_i = 30 mm	Re_i = 1	n_i = 50 min^{-1}
wall of mixing vessel	d_a = 90 mm	Re_a = 1.8	n_a = 10 min^{-1}
wall of mixing vessel	d_a = 90 mm	Re_a = 2.9	n_a = 16 min^{-1}
wall of mixing vessel	d_a = 90 mm	Re_a = 4.8	n_a = 26 min^{-1}
wall of mixing vessel	d_a = 90 mm	Re_a = 6.6	n_a = 36 min^{-1}
	d_i = 50 mm	Re_i = 3	n_i = 53 min^{-1}
	d_i = 50 mm	Re_i = 6	n_i = 107 min^{-1}
Rushton turbine	d_i = 50 mm	Re_i = 10	n_i = 177 min^{-1}
	d_i = 50 mm	Re_i = 3	n_i = 53 min^{-1}
multi-stage pitched blade	d_i = 50 mm	Re_i = 6	n_i = 107 min^{-1}
	d_i = 50 mm	Re_i = 10	n_i = 177 min^{-1}
d_i / d_a - inner/outer diameter; n_i / n_a - inner/outer speed, Re_i / Re_a - stirrer/vessel			

b. Cylindrical Shaft

The degree of eccentricity is defined by the number of eccentricity δ

$$\delta = \frac{e}{r_a}. \tag{7}$$

The degree of eccentricity is the ratio between the distance of the center of the inner full-glass shaft and the outer cylinder and the radius of the outer cylinder $r_a = d_a/2$. The configuration of the system is schematically depicted in Fig. 4. The inner full-glass shaft is positioned with different eccentricity numbers in a range of 0 - 0.5. The inner full-glass shaft rotates constantly with n_i = 50 revolutions per minute (rpm). The speed of the outer wall of the mixing vessel n_a is varied for adjusting different local energy dissipations. The rotational direction of the outer wall of the mixing vessel is also varied. The Reynolds number of the outer cylinder of the mixing system is

$$Re_a = \frac{\rho\, d_a^{\,2}\, n_a}{\eta}. \tag{8}$$

The Reynolds number is given by n_a the outer cylinder speed, d_a the diameter of the cylinder, density ρ and viscosity η of the working fluid. The Reynolds number is adjusted to a maximum of Re = 6.6 for laminar conditions in the mixing vessel filled with a volume of 500 ml of working fluid.

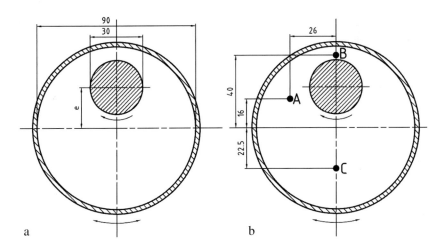

Fig. 4. Schematic of eccentric mixing system (a) and of three injection positions A, B, and C (b)

c. Rushton Turbine and Multi-stage Pitched Blade Stirrer

The speed of each stirrer is adjusted due to the chosen Reynolds number for laminar conditions. The mixing vessel is filled with 500 ml of the working fluid. Due to a complete field of viewing for a uniform secondary flow field the distance from the bottom of the mixing vessel to the middle of the Rushton-turbine is set to 45 mm. For the multi-stage pitched blade stirrer this distance is set to 25 mm depicted in Fig. 5. The height between the centers of the two stages is 30 mm. The chosen positions A, B and C reflect different flow paths of the injected fluid. On their way through the mixing vessel the fluid elements will be stretched and folded. The intensity of these mechanisms depends on the dissipation field, which is measured by PIV technique.

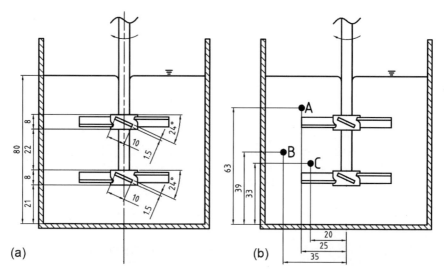

Fig. 5. Schematic of the multi-stage pitched blade stirrer (a), and of the three injection positions A, B and C (b)

d. Velocity and Dissipation Fields

Particle image velocimetry is used for the investigation of dissipation fields. Therefore, local velocity vectors are measured. The local velocities of deformation $\dfrac{\partial u_x}{\partial x}$, $\dfrac{\partial u_y}{\partial x}$, $\dfrac{\partial u_x}{\partial y}$ and $\dfrac{\partial u_y}{\partial y}$ are calculated with the local velocity vectors obtained from the particle paths. For an incompressible Newtonian fluid in two dimensions the function of the dissipation ε is defined according to Bird et al. (2002) as

$$\varepsilon_{xy} = \frac{\eta}{\rho}\left[\left(\frac{\partial v_x}{\partial y} + \frac{\partial v_y}{\partial x}\right)^2 + 2\left(\frac{\partial v_x}{\partial x}\right)^2 + 2\left(\frac{\partial v_y}{\partial y}\right)^2\right]. \tag{9}$$

It can be calculated for every discrete position inside the mixing system. The overall error for the determination of the absolute value of velocity is less than 7%. The measured velocity field and dissipation field are presented in a texture plot in Fig. 6 next to the numerical calculations of both fields. The structure of the particle paths depends strongly on the value of the eccentricity number and speeds. A detailed analysis of these fields can be found in Faes and Glasmacher (2007, 2008). The mixing experiments showed that the fluid volume of the fluorescent dyes is transported along the particle paths of the fluid. The local specific energy dissipation is depicted as a function of the dimensionless radius for several Reynolds numbers in Fig. 7. It is obtained that the highest value of energy dissipation is measured between the small gap of the two rotating cylinders. The value is increasing up to a maximum of 2.5 W/kg for Re = 6.6 in a counter rotating system.

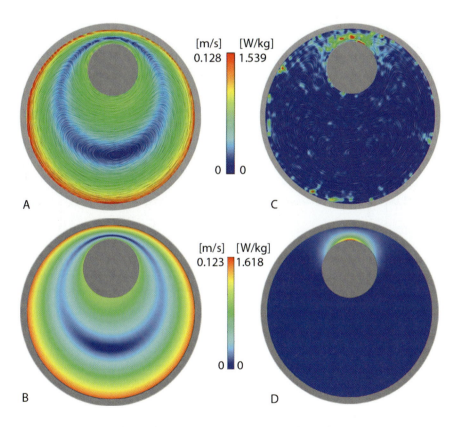

Fig. 6. Comparison of experimental and numerical data - field of local velocities (A - experimental, B - numerical) and local energy dissipations (C - experimental, D - numerical) in the eccentric mixing system for an eccentricity number of δ = 0.5 and a Reynolds number of 4.8 in a counter rotating system

For the applied parameters the experimental and numerical results of the energy dissipation are in agreement according to limitations of the spatial resolution of the camera system. Due to the accordance of the velocity fields of experimental and numerical results further local analysis is done by means of the local energy dissipation from the numerical calculations. Thus, a higher resolution and accuracy is obtained.

The macroscopic energy inserted by the cylindrical shaft and the outer mixing wall is visualized by the field of the function of the dissipated energy. The highest values are measured in the gap between both rotating cylinders. Fluid elements with dye passing the gap are increasing their surface area due to the shear stress. The area extension in dependence of the injection position is analyzed in the next chapter.

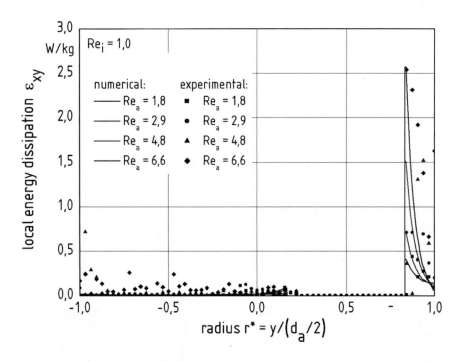

Fig. 7. Local specific energy dissipation as function of dimensionless radius (profile in the center axis) in the eccentric mixing system for an eccentricity number of $\delta = 0.5$ and several Reynolds numbers in a counter rotating system

e. Concentration and Local Degree of Deviation Fields

The concentration fields of the two fluorescent dyes are measured by means of laser induced fluorescence technique. At starting time the concentrations for the inert dye is set to $c_{2,0} = 3 \cdot 10^{-6}$ mol/l and to $c_{1,0} = 1 \cdot 10^{-6}$ mol/l for the reactive dye. Due to a full size image of the eccentric mixing system the observed display window has two different sizes. For the PIV system, the observed display window has a size of 120 mm x 90 mm so that a pixel of the PIV CCD-camera corresponds to approximately 0.087 mm. The spatial resolution of the LIF system corresponds to 0.29 mm/pixel. The observed window is set to 190 mm x 140 mm.

Cylindrical shaft. The progress of the convective mixing is depicted in Fig. 8 for a counter rotating system. Numerical and experimental convective transport show qualitatively a similar lamellar structure. The differences in the structure of the lamellas result from the experimental setup due to non-ideal glass geometry and flow field. Although the diffusion coefficients of the transported species are set to the values from literature, the mixing time is much faster in the numerical calculations. The difference can be explained by numerical diffusion, which cannot be neglected. The analysis of the calculated surface area related to the complete free surface of the mixing system is shown as a function of time in Fig. 9. The injection positions A, B, and C (cf. Fig. 4) belong to regions with different local energy dissipations. For the value of the local energy dissipation of 0.26 W/kg, related to the injection position B, the dimensionless area increases up to a value of 0.11

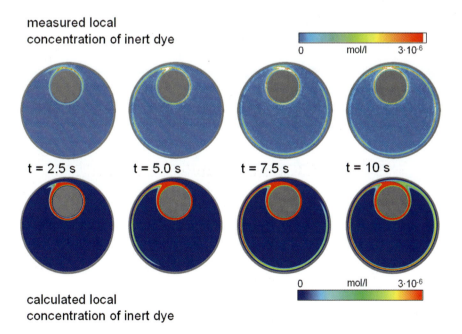

Fig. 8. Experimental and numerical progress of the convective mixing in the eccentric mixing system for an eccentricity of $\delta = 0.5$ in a counter rotating system

during the injection time. In the same period of time the dimensionless surface area reaches only a value of 0.01 for the injection in the death centre C. Here, the local value of energy dissipation is smaller by a factor of 240. The difference of the dimensionless area between these two injection positions after ten seconds can also be obtained in a non-progressed micromixed area for the injection position C.

Multi-stage pitched blade stirrer. The multi-stage pitched blade stirrer induces a complex three-dimensional flow field with several eddies. The structure of the eddies depends of the rotational speed of the stirrer. Here, two different injection positions are analyzed. The injection positions are depicted in Fig. 5. The progress of the macro- and micromixing is shown for injection position B in Fig. 10.

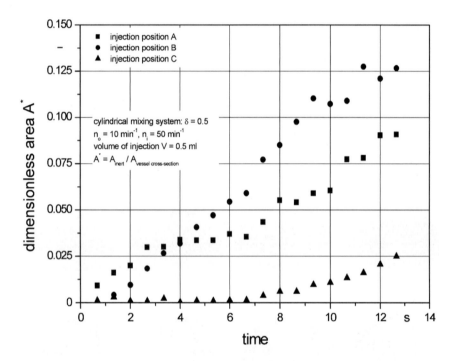

Fig. 9. Temporal progress of the dimensionless area in the eccentric mixing system for an eccentricity number of $\delta = 0.5$ in a counter rotating system

After the injection of the dye mixture, lamellar structures are formed and transported upwards before they reach the two stages of the downward-downward configurated stirrer. The striation thickness of the lamellas decreases due to the convective transport of the volume. The progress of the micromixing proceeds simultaneously. The local degree of deviation decreases from the starting value of 1, which is indicated by red color. The dye volume transported into high regions of dissipation round the blades of the stirrer is micromixed well. The local degree of deviation decreases to a value of 0.25. In Fig. 11 the volume ratio of the local degree of deviation $V^+ = V_{\Delta>k} / V_{\Delta>0}$ is shown as a function of time for different fractions of the local degree of deviation. The volume $V_{\Delta>k}$ is calculated by adding all volume elements of the local degree of deviation with values greater than a factor k. The value of $V_{\Delta>k}$ is related to the complete detected volume of the local degree of deviation $V_{\Delta>0}$. The progress of the micromixing in the vessel is shown by varying the factor k. For k = 0.75 the micromixing is not far progressed and many fluid elements are still completely segregated. A complete micromixed volume is reached for values of k = 0.25. This time depending process is shown for the multi-stage pitched blade stirrer in Fig. 11. For injection positions A and B, the volume ratio of the local degree of deviation is decreasing during the mixing process. After 30 seconds the reached volume ratio of the local degree of deviation with k = 0.75 for injection position B is lower by a factor of approximately 10 compared to injection position A. For a far progressed microscale mixing, according to values of k = 0.5 and 0.25, the difference of the decrease of the volume ratio of the local degree of deviation is greater for injection position B than for A. The phenomena results from the flow path of the injected mixture of the dyes, which is transported through regions of high energy dissipations – namely the regions around the stirrer blades.

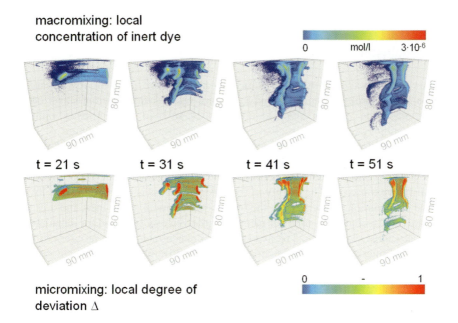

Fig. 10. Progress of macro- and microscale mixing for a multi-stage pitched blade stirrer, Re = 3, injection position B

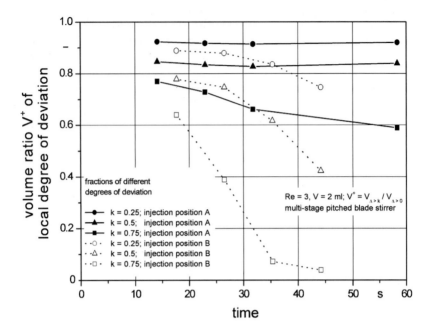

Fig. 11. Temporal progress of the volume ratio of the local degree of deviation for a multi-stage pitched blade stirrer, Re = 3, injection positions A and B

6 Conclusions

In this project experimental and numerical methods are used for investigating macro- and micromixing phenomena in mixing vessels with different stirrer configurations. The experimental set-up consists of two non-intrusive optical measurement systems for visualising local concentrations of a mixture of two fluorescent dyes and local velocities. The experiments are performed in laminar flow regime. The 4D laser induced fluorescence technique, which is enhanced for the use of two different colours, permits to measure simultaneously the convective and diffusive transport. The quality of the mixing process is described by the local degree of deviation, calculated by the local concentrations of an inert and a reactive dye. PIV technique is used for measuring the velocity field and for calculating the local energy dissipation. The investigation concentrates on the progress of the macro- and micromixing as a function of time for different injection positions, so that the injected fluorescent dyes are passing different regions of local energy dissipations. This is performed in a glycerol solution, which provides an isooptic system allowing to measure the whole cross-sectional area.

From the experimental investigations it can be obtained that mixing on molecular scale depends mainly on the flow path through the mixing vessel and the resulting local energy dissipation. The micromixing starts in the boundary layers of the lamellas where the local degree of deviation decreases first. The progress of the dimensionless area of the inert dye depends on the position of injection. For a position of injection in a region of high local energy dissipation the dimensionless area extends by a factor of 10 compared to an injection position in the death centre. The development of a numerical model provides the prediction of macro- and micromixing in different mixing systems. For the cylindrical system the numerical calculations show qualitatively good accordance to the experimental investigations. The comparison of the calculated and measured dimensionless area varies to a maximum ratio of two due to the numerical diffusion, which cannot be neglected without enormous calculating capacity. In a complex flow field induced by a multi-stage pitched blade stirrer the concentration fields are measured and visualised four-dimensionally. The analysis of the volume ratio of the local degree of deviation shows the influence of the injection position to the flow paths of the fluorescent dye and the dependence of the local energy dissipation. As well as for the stationary dissipation field in the cylindrical mixing system the dimensionless volume of the local degree of deviation decreases in the complex flow field in regions of high local energy dissipations. The developed methods and investigations in this project facilitate the understanding of transport phenomena in liquid-liquid mixing systems for laminar conditions.

Acknowledgements. We gratefully acknowledge the scientific support and transfer by the precursor project workers, and express our thanks to Professor D. Mewes and Dr.K. Kling.

References

Anderson, P.D., et al.: Material stretching in laminar mixing flows: extended mapping technique applied to the journal bearing flow. International Journal of Numerical Methods in Fluids 40, 189–196 (2002)

Aref, H.: Stirring by chaotic advection. Journal of Fluid Mechanics 143, 1–21 (1984)

Baldyga, J., Bourne, J.R.: Turbulent mixing and chemical reaction. John Wiley & Sons, Chichester (1999)

Bellerose, J.A., Rogers, C.B.: Measuring mixing and local pH through Laser Induced Fluorescence. Laser Anemometry 191, 217–220 (1994)

Bird, R.B., et al.: Transport Phenomena. John Wiley and Sons Inc., New York (2002)

Distelhoff, M.F.W., Marquis, A.J.: Scalar mixing in the vicinity of two disk turbines and two pitched blade impellers. Chemical Engineering Science 55, 1905–1920 (2000)

Faes, M., Glasmacher, B.: Analyse des Makro- und Mikromischens mit Hilfe von laserdiagnostischen Verfahren. Chemie Ingenieur Technik 79(7), 1059–1065 (2007)

Faes, M., Glasmacher, B.: Measurements of micro- and macromixing in liquid mixtures of reacting components using two-colour laser induced fluorescence. Chemical Engineering Science 63, 4649–4655 (2007)

Gollub, J.P., et al.: Mixing rates and symmetry breaking in two-dimensional chaotic flow. Physics of Fluids 15(9), 2560–2566 (2003)

Guillard, F., Träghard, C., Fuchs, L.: New image analysis methods for the study of mixing patterns in stirred tanks. Canadian Journal of Chemical Engineering 78, 273–285 (2000)

Haugland, R.P. (ed.): Molecular Probes- Handbook of Fluorescent Probes and Research Products, 9th edn. (2002)

Hoffmann, M., et al.: Experimental investigation of liquid-liquid mixing in T-shaped micromixers μ-LIF and μ-PIV. Chemical Engineering Science 61, 2676–2968 (2006)

Hong, S.D., et al.: Proceedings of the 10th International Symposium on Flow Visualization, Kyoto, Japan, August 26-29 (2002) paper F0275

Kling, K., Mewes, D.: Quantitative Measurements of Micro- and Macromixing in Stirred Vessels using Planar Laser Induced Fluorescence. Journal of Visualization 6, 165–173 (2003)

Kling, K., Mewes, D.: Two-colour laser induced fluorescence for the quantification of micro- and macromixing in stirred vessels. Chemical Engineering Science 59, 1523–1528 (2004)

Kling, K.: Visualisieren des Mikro- und Makromischens mit Hilfe zweier fluoreszierender und chemisch reagierender Farbstoffe. Dissertation, Gottfried Wilhelm Leibniz Universität (2004)

Lamberto, D.J., Muzzio, F.J., Swanson, P.D., Tonkovich, A.L.: Using time-dependent RPM to enhance mixing in stirred vessels. Chemical Engineering Science 51, 733–741 (1996)

Naraghi, M.: T-jump study of calcium binding kinetics of calcium chelators. Cell Calcium 22, 255–268 (1997)

Ottino, J.M.: Mixing and chemical reactions - a tutorial. Chemical Engineering Science 49, 4005–4027 (1994)

Paschedag, A.R., et al.: Aktuelle Entwicklungen in der CFD für gerührte Systeme. Chemie Ingenieur Technik 79(7), 983–999 (2007)

Unger, D.R., Muzzio, F.J.: Laser- Induced Fluorescence technique for the quantification of mixing in impinging jets. AIChE Journal 45, 2477–2486 (1999)

Villermaux, J., Houcine, I., Vivier, H., Plasari, E., David, R.: Planar laser induced fluorescence technique for measurements of concentration fields in continuous stirred tank reactors. Experiments in Fluids 22, 95–102 (1996)

Van Vliet, E., et al.: Four-dimensional Laser Induced Fluorescence measurement of micro mixing in a tubular reactor. In: Proceedings 10th European Conference on Mixing, pp. 45–52 (2000)

Experimental Investigation of the Mixing-Process in a Jet-in-Crossflow Arrangement by Simultaneous 2d-LIF and PIV

Camilo Cárdenas, Rainer Suntz, and Henning Bockhorn

Institut für Technische Chemie und Polymerchemie, Universität Karlsruhe
Kaiserstr. 12, 76131 Karlsruhe, Germany

Abstract. A combination of simultaneous planar laser-induced fluorescence and particle image velocimetry measuring-technique is applied to investigate the flow field in a jet-in-crossflow arrangement. Two-dimensional maps of normalized averaged jet-concentrations $<c>$, axial turbulence intensities Tu_x and normalized Reynolds-fluxes $<v_i'c'>/U_{max}$ are measured. In these experiments, the influence of a swirl in the jet as well as varying turbulence levels characterized by the Reynolds number on the mixing efficiency is investigated. A direct relation between Reynolds-fluxes (and -stresses) and the increasing turbulent flows' mass-transfer is observed.

1 Introduction

Mixing is one of the main issues in fluid mechanics and process engineering and relevant for a wide range of applications. A configuration often employed for this purpose is a jet injected perpendicularly to an uniform crossflow, the jet-in-crossflow configuration (JCF). This apparently very simple flow arrangement appears frequently in many industrial applications such as chemical, pharmaceutical, environmental and combustion engineering as well as in the nature.

The clarification of the origin of the complex vortex-structures of this flow configuration and the capability of the JCF to mix two flows intensively with each other make it an object of intense research activities (Margason (1993), Andreopoulos and Rodi (1984), Hasselbrink and Mungal (2001), Kelso and Perry (1996), Camussi et al. (2002)).

The development and validation of appropriate numerical models has fundamental significance in order to understand mechanisms for the mixing processes as well as to search for alternatives for its improvement. This aim can only be achieved, if ample experimental data are available about the actual velocity- and concentration-fields – at best measured simultaneously – of the mixing fluids.

Most of the experimental work found in the literature (Andreopoulos and Rodi (1984), Kelso and Perry (1996), Keffer and Baines (1963), Fric and Roshko (1994)) are based on single-point measurements which have been performed using hot-wire anemometry. This involves the employment of mechanical probes into the flow field which may significantly influence results of these investigations.

Furthermore, the limitation to single-point measurements prevents an estimation of spatial correlations within the three-dimensional flow configuration.

At this juncture, Kelso and Perry (1996) and Fric and Roshko (1994) carried out hot-wire anemometry measurements in combination with flow visualization techniques. From these investigations they obtain insight about the mechanisms responsible for the occurrence of coherent structures and the interaction of both flows.

However, most of the work known from the literature is dedicated to the investigation mechanisms responsible for the formation of the vortex-structures of the JCF rather than the mixing processes. In the literature only a few two-dimensional measurements in a JCF are found, e.g. from Su and Mungal (2004) and Özcan et al. (2001).

In a previous work the authors estimated two-dimensional maps of Reynolds-fluxes and -stresses in a JCF-arrangement by simultaneous planar laser-induced fluorescence (2d-LIF) and particle image velocimetry (PIV) (Cárdenas et al. (2007)). These experimental results were in good agreement with Large Eddy Simulations (LES) of the flow under investigation.

The present work is intended to provide an experimental study by means of simultaneous 2d-LIF and PIV measurements concerned with both: the effect of initial swirl of the jet and the influence of the Reynolds number on the mixing process. Additionally, comparisons with direct numerical simulation (DNS) results from Denev et al (2009) are given. In the last section the relation between Reynolds-fluxes and the mass transfer rate of turbulent flow are discussed.

2 Experimental Setup

The experimental setup is described in detail in Cárdenas et al. (2007). Therefore, only a brief overview is given in this section.

2.1 Flow Arrangement

The experiments were performed in an open-circuit wind tunnel, which allows the setting up of well defined incident flow conditions over a very wide range of Reynolds-numbers for the crossflow (index ∞) $700 < Re_\infty < 360000$ (flow rate: 3.8 $m_n^3/h < Q_\infty < 1930\ m_n^3/h$) as well as the jet (index j) $150 < Re_j < 26500$ (flow rate: $0.05\ m_n^3/h < Q_j < 8.34\ m_n^3/h$). In the literature a "combined" Reynolds-number $Re := U_\infty \cdot D_j / v$, which characterizes the jet and the crossflow simultaneously, is frequently used instead of $Re_\infty := U_\infty \cdot D_\infty / v$ and $Re_j := U_j \cdot D_j / v$, with $D_j = 8$ mm the inner diameter of the jet, $D_\infty = 108$ mm the hydraulic diameter of the crossflow and v the kinematic viscosity.

Fig. 1 shows a schematic illustration of the total arrangement of the JCF. It is comprised of three major parts: a) a feed pipe system for the crossflow, b) a feed pipe system for the jet and c) the wind tunnel.

The feed pipe system for the crossflow is provided with ambient air by means of a fan. The flow rate is controlled by a Rotameter for low throughput cases or by a combination of three parallel Venturi tubes for the cases of high flow rates. The feed pipe system for the jet is provided by both: air from the compressed air network of the building together with NO_2 (1%)-synthetic air mixture from a gas-cylinder. NO_2 serves as molecular tracer for the LIF-technique. The flow-rate of the jet is controlled by means of an arrangement of five calibrated mass flow-controllers, three for the air pipe line and two for the NO_2-cointaining line, respectively. This arrangement allows an appropriate adjustment of the jet and the NO_2-mole fraction in the jet-flow (5000 ppm) for the measuring conditions simultaneously.

   The wind tunnel is comprised of 6 major parts, (1) the measuring section, which is about 500 mm long and has a cross sectional area of 108 x 108 mm. In this section the mixing between jet and crossflow is taking place. Optical access for the measurement techniques is achieved by 4 fused quartz windows inset at each side of the measuring section. (2) The jet-pipe (length of 800 mm) enters into the crossflow in the middle of the measuring section about 300 mm downstream of the nozzle discharge. (3) The diffusor, which connects the air supply pipeline from the fan to the smoothing section (4) without generating additional turbulence. (4) The smoothing section, which has a cross-section 25 times bigger than the one of the measuring section, serves to minimize spatial and temporal fluctuations of the feed flow by a combination of three meshes and a honey-comb structure. (5) The specially shaped contraction nozzle, possessing an optimal contour profile to accelerate the air of the crossflow smoothly into the measuring section forming a unique plug flow. (6) The outflow-section, which is 2500 mm long, represents the connection between the measuring section and the exhaust system. It possesses the same cross-section as the measuring section.

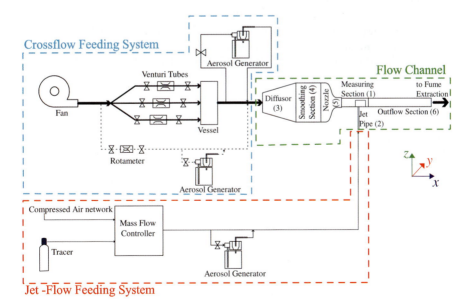

Fig. 1. Schematic illustration of the Jet-in-Crossflow-arrangement

The diffusor, the smoothing section and the contraction nozzle are required to guarantee well defined flow conditions in the measuring section (Prandtl (1990)). The latter is of fundamental importance for the development and validation of appropriate models by comparing experimental results with numerical simulations (Cárdenas et al. (2007)).

To account for density variations in the experiments the temperatures of both flows are measured. The effect of fluctuations of the humidity of the air is neglected as well as the addition of NO_2 in the jet, because this would influence the density of the fluids by less than 0.5 % within reasonable assumptions for these fluctuations.

Aerosol generators, which are connected parallel to the corresponding feed lines, are used to add liquid droplets (Di-Ethyl-Hexyl-Sebacat - DEHS) to both, the crossflow and the jet-flow, respectively. These droplets serve as tracer "particles" for the PIV measurements. By using droplets instead of solid particles a static charging of the particles and a consecutive fouling of the windows for optical access into the measuring section is avoided.

The origin of the coordinate system used in the figures of the following sections is located in the outlet plane of the jet in its center point (see Fig. 1 and Fig. 2). The x- and y-directions correspond to the axial and lateral directions of the crossflow, the z-direction to the axial direction of the jet-flow, respectively.

2.2 2d-LIF-PIV Measuring System

A schematic illustration of the 2d-LIF-PIV system is given in Fig. 2. A frequency doubled, double pulse Nd:YAG-laser (Quantel TWINS Brilliant) is used as a light source for the simultaneous LIF- and PIV-measurements. The laser beam (135 mJ/single pulse energy, 6 mm beam diameter) is expanded into a light-sheet by means of a Galilean-telescope consisting of a cylindrical lens (f_1 = -100 mm) in combination with a spherical lens (f_2 = 800 mm). Thus, the width of the laser beam is 48 mm and the thickness in the detection zone is ~ 200 μm. In order to obtain a more or less homogeneous distribution of the laser intensity across the total detection area, the laser-light-sheet is reduced to 28 mm by an aperture.

The scattered light from the droplet aerosol at 532 nm as well as the broadband Stokes-shifted fluorescence from the molecular tracer NO_2 (540 nm – 700 nm) are imaged perpendicular to the propagation direction of the laser beam onto appropriate CCD-cameras. For PIV an unintensified (Dantec 80C60 HiSense PIV/PLIF Camera, 1280 x 1024 pixels) and for LIF an intensified CCD-camera (Roper Scientific, PI-MAX, 512 x 512 pixels) are used. Both signals are separated from each other by a dichroic plate (LOT CH-Z532-90S, R_{max} @ 532 nm, T_{max} = 545 nm – 750 nm, angle of incidence: 45°), which reflects the wavelength of the frequency doubled Nd:YAG-laser and transmits the wavelengths > 532 nm of the Stokes-shifted NO_2-fluorescence. To suppress spurious scattered light from the LIF-camera, an additional dielectric coated longpass-filter (LOT CH-LP540-70S, T_{max} = 545 nm – 750 nm, angle of incidence: 0°) is used in front of the ICCD-camera.

The PIV-processor (Dantec FlowMap 1500 PIV System) serves as the master for the synchronisation of the laser and both cameras. The PIV-processor triggers the laser, the PIV-camera-controller (DANTEC HiSense-Camera-Controller) and a pulse generator (SRS DG-535). The latter serves for triggering the ICCD-camera controller (Roper Scientific ST-133) and the ICCD-camera.

A personal computer was used for data acquisition. The PIV- and the LIF-signals were analyzed with the software packet Dantec FlowManager Version 3.62 and by an own programme written in MatLab, respectively.

The LIF-signal is calibrated to absolute NO_2-concentrations by measuring the fluorescence intensity in the centre of the jet just upon the exit into the crossflow. At this location any dilution of the NO_2-concentration by the crossflow can be excluded. In pre-experiments carried out in a flow cell up to 10000 ppm NO_2, it was assured that the fluorescence-signal at constant ambient air pressure is proportional to the NO_2-concentration after excitation with a frequency doubled Nd:YAG-laser pulse. A linear relation between the NO_2-concentration and the LIF-signal was also observed by Gulati and Warren (1994) for 488 nm excitation using an Ar^+-ion-laser. These investigations were carried out up to a NO_2-mole fraction of ~ 1000 ppm.

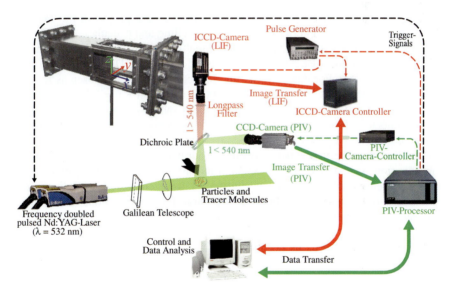

Fig. 2. Experimental set-up of the simultaneous 2d-LIF-PIV measuring technique. In addition to the set-up the measuring section and the location of the coordinate system are given. The x-direction represents the axial (cross-) flow direction, y is the lateral direction and z the vertical direction/flow direction of the jet, respectively

To estimate the NO_2-concentration- and the velocity-field of a complete plane in the JCF and taking into account an overlap of each individual frame with the adjacent frames, 15 single frames were put together. Because of the large dimension of the JCF, the apparatus cannot be displaced for measuring at different locations in the JCF, while keeping the laser and the detection system fixed. Therefore, the laser light-sheet as well as the detection system have to be traversed to carry out measurements at different locations inside the JCF. To avoid time consuming re-focusing of the cameras and the adjustment of the laser-beam for different planes and positions within an individual plane, a 3d-traverse unit made from *item*-aluminium-profiles were assembled (Cárdenas et al. (2007)). Tilted mirrors for the laser-beam, the Galilean-telescope, the dichroic-mirror and the cameras were mounted onto this unit. Within the detection volume the laser-light-sheet as well as the cameras can be traversed with a precision of < 1 mm in x-, y- and z-direction by three spindles.

3 Results and Discussion

The experimental results obtained by simultaneous 2d-LIF-PIV measurements are used in order to investigate the mixing behaviour of both flows and to develop and validate models for turbulent flows. As mentioned above this aim can only be achieved, if the initial flow conditions are well specified in addition to the availability of accurate experimental data.

Therefore, this work was carried out within the research project of the German Research Foundation SPP-1141 ("Analysis, Modelling and Calculation of Mixing Processes with and without Chemical Reaction") in close cooperation with the numerical project from Bockhorn (Cárdenas et al. (2007), Denev et al. (2009), Denev et al. (2006)). For this purpose, the flow conditions were chosen as follow: Reynolds number of the crossflow $Re_\infty = 3000$ and the jet $Re_j = 667$ leading to a bulk velocity ratio of $r = U_j/U_\infty = 3$. This velocity ratio was chosen to ensure that the jet trajectory follows the central axis of the measuring section after being bended horizontally by the crossflow.

As a first step, the experimental results of normalized temporal averaged NO_2-concentrations ($<c>$), normalized temporal mean axial velocities ($<v_x>/U_{max}$) and axial turbulence intensities ($Tu_x = \sigma(v_x)/<v_x>$) were compared with numerical simulations from Denev et al. obtained by Large-Eddy-Simulations (LES). U_{max} represents the maximum axial velocity over the total cross-section area of the crossflow (without jet-flow), $\sigma(v_x)$ the standard deviation of the velocity in x-direction and $c = C/C_{jet}$ the normalized NO_2-concentration. C is the local NO_2-concentration and C_{jet} the pure jet NO_2-concentration in mol/l. Due to the limited space, the results of these experiments carried out at four different horizontal planes ($z/D_j = 1.5$, 4.5, 5.5 and 6.5) and the corresponding numerical simulations are given in Cárdenas et al. (2007) and (2008). From the corresponding publication of 2007 it is noticeable that the agreement between the experimental and numerical results is very good.

3.1 Influence of a Swirled Jet with Respect to the Mixing between Jet and Crossflow

In order to improve the mixing-process inside a JCF-configuration, the influence of a swirled compared to an unswirled jet with respect to the mixing behavior of both flows was investigated.

The swirl nozzle is shown in Fig. 3a. The nozzle consists of 3 different parts: (1) a cover, (2) a swirler and (3) a flow straightener. The cover guides the swirled jet-flow into the crossflow. The vertical flow of the air-NO_2-mixture inside the pipe feeding system of the jet is deflected horizontally by tangential channels of the swirler possessing an eccentricity (e). According to equation (3-1) (Mundus and Kremer (1989)), the swirl number (S) is proportional to the eccentricity

$$S = \frac{\pi \cdot D_j \cdot e}{2 \cdot n \cdot A_c},$$

(3-1)

with n the number of tangential channels in the swirler (six in this case) and $A_c = w \times h$ the cross-sectional area of each of the n channels, (w) width and (h) height. The selected swirl-sense was the counter-clockwise (top view).

Table 3.1. Swirler configurations

Case:	Re_∞	$r = U_j/U_\infty$	$r' = Q_j/Q_\infty$	S	D_j [mm]	e [mm]	n	$A_c = w \times h$ [mm^2]
A	3000	3	1.28×10^{-02}	0.0	8.0	-	-	-
B	3000	3	1.28×10^{-02}	0.2	8.0	0.8	6	48.0
C	3000	3	1.28×10^{-02}	0.4	8.0	1.5	6	48.0
D	3000	3	1.28×10^{-02}	0.6	8.0	2.3	6	48.0
E	3000	4	1.28×10^{-02}	0.4	6.9	1.7	6	48.0
F	3000	6	1.28×10^{-02}	0.4	5.7	1.4	6	31.2
G	3000	8	1.28×10^{-02}	0.4	4.9	0.7	6	13.2

Table 3-1 summarizes the flow conditions of the experiments and corresponding parameters of the different swirlers under investigation. Reference case A corresponds to the standard jet configuration with an unswirled jet.

Mean distributions of the concentration $<c>$ (first row) and normalized Reynolds-fluxes in the axial direction of the crossflow $<v_x'c'>/U_{max}$ (second row) as well as in the lateral direction $<v_y'c'>/U_{max}$ at $z/D_j = 1.5$ are shown in Fig. 4 for cases A (left), C (middle) and D (right). Due to the limited space, corresponding results of configuration B are not shown in the figure.

From Fig. 4 it is obvious that swirl-jet configurations show larger spatial distributions of the NO_2-jet-concentration in the crossflow as well as of the Reynolds-fluxes. Additionally, the absolute values are higher in the swirled compared to the unswirled case. This behaviour is attributed to the increase of the tangential component of the swirl-jet velocity (in x and y direction) and could be interpreted as an improvement of the mixing process at lower heights above the base-plate. It should be noticed, that in the frame indicated by (*) the measured quantities depicted in the figures ($<v_x'c'>/U_{max}$ from case A) are multiplied by a factor 5.

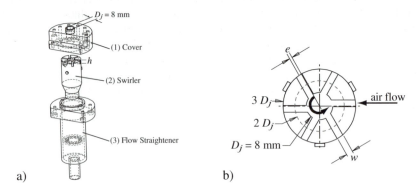

Fig. 3. a) Schematic illustration of the swirl-nozzle of the jet consisting of 3 major parts: (1) Flow straightener, (2) swirler and (3) cover; b) superior perspective of the swirler $S = 0.2$; $e = 0.8$ mm and counter-clockwise swirling sense.

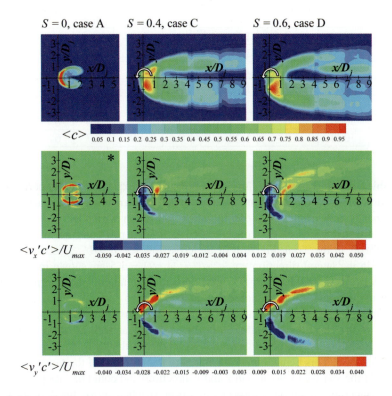

Fig. 4. Mean distributions of the concentration $<c>$ (first row) and normalized Reynolds-fluxes in the axial direction of the crossflow $<v_x'c'>/U_{max}$ (second row) as well in the lateral direction $<v_y'c'>/U_{max}$ (third row) at $z/D_j = 1.5$. (*) For case A $<v_y'c'>/U_{max}$ is multiplied by a factor 5. Flow conditions are given in Table 3-1

In Fig. 5 the jet centre streamline trajectories of the different configurations presented in Table 3-1 are depicted. To obtain dimensionless labelling of the axis, x and z are normalized by a reference diameter of the jet D_j^{ref} = 8 mm. The jet penetrates substantially less into the crossflow, if the swirl is increased (cases A - D) and keeping both, the jet-diameter and the velocity ratio r, constant. This behaviour of the jet is in accordance with the numerical results obtained in Denev et al. (2005). Besides at lowest heights above the base plate discussed in context to Fig. 4 this is counterproductive with respect to an improvement of the mixing process.

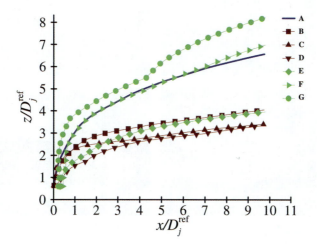

Fig. 5. Centre streamline-trajectories of the different JCF- Configurations

From Fig. 4 and Fig. 5 it can be expected that an improved mixing of a swirled jet can be achieved, if the jet penetrates into the crossflow as deep as in the case without swirl. In this case the total cross-section area of the crossflow is available for the mixing process. Therefore, the velocity ratio r was increased to such an extent, that the trajectory of the swirled jet is comparable to the trajectory without a swirl. To avoid a variation of the throughputs of the flows in this case, the diameter of the jet is decreased for the swirled jet appropriately. Therefore, a flow-rate ratio $r' = Q_j/Q_\infty$ was defined, which was kept constant for all flows under investigation.

Further experiments were carried out (cases E-G in Table 3-1) to find a swirl-jet configuration possessing a jet-penetration comparable to case A without a swirl. The swirl-number for these cases is kept constant at $S = 0.4$. From Fig. 5 it becomes obvious that the penetration of the jet for case F is very similar to the unswirled case.

In a next step, we compared swirl-case F possessing roughly the same jet-trajectory as the unswirled case A with respect to the mixing behaviour. In Fig. 6 two-dimensional maps of the mean jet mole-fraction $<c>$ at two different heights above the baseplate of the channel, z/D_j^{ref} = 1.5 (top) and z/D_j^{ref} =3 (bottom), for cases A (left) and F (right) are shown. From this figure it is obvious for both measuring planes that case F reveals a major dispersion of the jet flow into the crossflow than the reference case A.

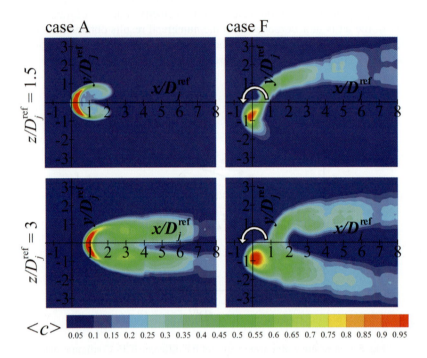

Fig. 6. Two-dimensional maps of mean distributions of the jet-concentration of reference case A (left column) and case F (right column) at $z/D_j^{\text{ref}} = 1.5$ (top) and $z/D_j^{\text{ref}} = 3$ (bottom).

This improvement in the mixing have been also obtained quantitatively by means of the spatial mixing deficiency (*SMD*) (Boss, (1986)), which is one of the most widely applied so-called mixing indices. It is defined as:

$$SMD = \frac{RMS[\langle c \rangle - AVG(\langle c \rangle)]}{AVG(\langle c \rangle)} = \frac{\sqrt{\dfrac{\displaystyle\sum_i^M \sum_j^N [\langle c_{i,j} \rangle - AVG(\langle c \rangle)]^2}{M \cdot N}}}{AVG(\langle c \rangle)}, \qquad (3\text{-}2)$$

in which $AVG(\langle c \rangle) = (M \cdot N)^{-1} \sum_i^M \sum_j^N \langle c_{i,j} \rangle$ denotes spatial averaging between $-1 \leq x \leq 8$ and $-3.5 \leq y \leq 3.5$, and *RMS* root-mean-square in the spatial sense. The indices i and j represent the grid points of the measuring plane in x and y directions and N and M the corresponding number of grid points, respectively. Therefore, the quality of mixing increases with lower values of the index.

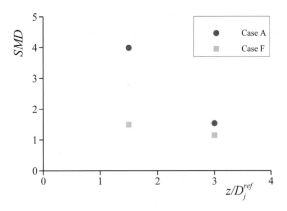

Fig. 7. Spatial mixing index *SMD* in the horizontal planes $z/D_j^{ref} = 1.5$ and 3 for cases A and F

In Fig. 7 corresponding *SMD* values for both cases (A and F) and both measuring planes ($z/D_j^{ref} = 1.5$ and 3) represented in Fig. 6 are given. It becomes obvious that there is a better mixing for case F (green) compared to the reference case (blue) for both heights above the base plate. It can be also appreciated that at $z/D_j^{ref} = 3$ the difference between both cases decreases and becomes relatively small. From these observations it could be concluded that the swirl-jet configurations offer better mixing of the flows than a standard JCF-configuration at lower heights above the base place. However, further investigations have to be done, to determine the effectiveness of the mixing at even higher heights. At lower heights the swirl leads to an increase in the mixing due to an intense interaction between the swirl of the jet and the crossflow. At higher heights the situation could possibly be reversed because the swirl disturbs or even suppresses the formation of the counter-rotating vortex pair (CRVP). The latter is not only the most significant flow feature of the JCF, but also tremendously influences the mixing process (Broadwell and Breidenthal (1984)).

3.2 Influence of the Reynolds Number on the Mixing Process

To offer an experimental data base to develop and validate models further measurements are carried out in the JCF for two different turbulence levels. Therefore, the Reynolds-numbers of the crossflow as well as the jet was varied simultaneously, to keep the velocity ratio *r* (similar configuration) constant.

Table 3-2 shows the operating parameters for both low (Low-Re) and high (High-Re) Reynolds number flow conditions. Even though the Reynolds number of the crossflow in both cases was larger than 2500, both flows can be still considered as laminar, regarding also the effect of the high contraction of the nozzle and the space averaged turbulence intensities of the crossflow without jet-flow which were below 1.2%. The above discussion points that simplified laminar in-flow

boundary conditions can be used for its simulation. In the course of the propagation of the jet into the crossflow turbulent structures are formed as a consequence
of the intense interaction of both flows with each other (Kelvin–Helmholtz instability) (Kelso and Perry (1996)).

Table 3.2. Flow parameters of the investigated similar configurations

Configuration	$Re_\infty = U_\infty \cdot D_\infty / v$	$Re_j = U_j \cdot D_j / v$	$r = U_j / U_\infty$	[1]$Re = U_{max} \times D_j / v$
Low-Re	4120	1070	3.5	325
High-Re	8240	2140	3.5	650

In Fig. 8 instantaneous two-dimensional NO_2-LIF maps in the symmetry plane
($y/D_j = 0$) for both cases, low-Re (left) and high-Re (right), are given. From the figure it is obvious that in the higher Reynolds-number case the transition from laminar to turbulent appears at lower heights above the base-plate of the crossflow.

In order to quantify the development of the mixing process along the trajectory
of the jet in both cases, three different measuring planes were investigated: $z/D_j =$
2.5, 4.0 and 5.5 depicted as dotted lines in Fig. 8.

At the lowest height above the baseplate of the crossflow the jet-flow is laminar
for both Reynolds-numbers. At $z/D_j = 4.0$ the jet-flow for the low-Re number case
is still laminar, whereas for the high-Re number case the jet-flow becomes unstable leading to the formation of coherent structures. At even higher heights ($z/D_j =$
5.5) both flows are unstable. The experimental results for the position of the transition point "laminar-turbulent" is in good agreement with numerical simulation
carried out in Denev et al. (2009).

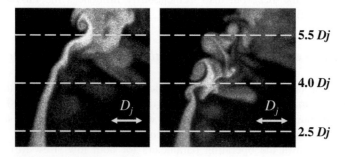

5.5 Dj

4.0 Dj

2.5 Dj

Fig. 8. Instantaneous two-dimensional NO_2-LIF maps in the symmetry plane ($y/D_j = 0$) for
both, low-Re (left) and high-Re (right) cases (see Table 3-2). The dotted horizontal lines
represent the measuring planes

---

[1] Combined Reynolds number used in Denev et al. (2009).

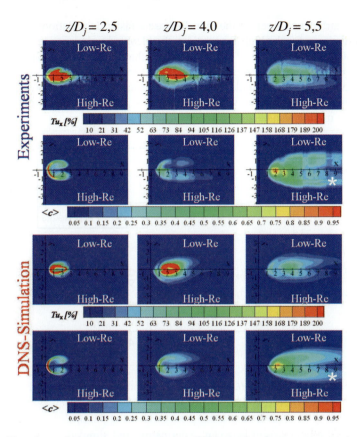

Fig. 9. Experimental (upper two rows) and DNS results (lower two rows) of two-dimensional maps of turbulence intensities in x-direction (Tu_x) and temporal averaged jet mole-fractions ($<c>$) for both cases (upper half of each plot: low-Re, lower half of each plot: high-Re) at the three different measuring planes: $z/D_j = 2.5$ (left); 4.0 (middle) and 5.5 (right). Numerical simulations are carried out by Denev et al (2009). (*) Mean concentration maps $<c>$ for high Re-number cases are multiplied by a factor 2

Fig. 9 shows experimental (upper two rows) and direct numerical simulation (DNS) results from Denev et al (2009) (lower two rows) of two-dimensional maps of turbulence intensities in x-direction (Tu_x) and mean jet mole-fractions ($<c>$) for both Reynolds number cases. The low-Re case is depicted in the upper half of each two-dimensional map, the high-Re case in the lower half, respectively. The results of Fig. 9 are shown for the three different planes depicted in Fig. 8: $z/D_j = 2.5$ (left row); 4.0 (middle row) and 5.5 (right row).

The progress of the mixing process between the jet- and crossflow with increasing z for both Reynolds number cases can be appreciated in this figure.

It is remarkable that there is a somewhat higher turbulence level for the experimental results compared to the simulations. However, besides the results at $z/D_j = 2.5$, which are analyzed below, a good agreement between simulations from Denev et al (2009) and experiments can be seen. In addition, as a consequence of the higher turbulence level for the high-Re case a faster degradation of the jet concentration is seen for the high-Re case, i.e. the elevated turbulence level awards an improvement in the mixing of the flows. Note that the mean concentration maps $<c>$ for high Re-number cases indicated by (*) in the figure are multiplied by a factor of two. This spread in the false color scale accounts for the lower concentrations of this frame compared with the other ones. The afore mentioned improvement of the mixing also becomes obvious from Fig. 10.

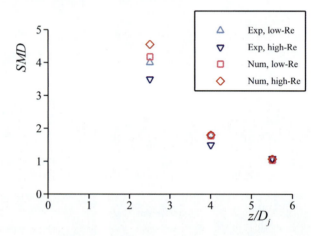

Fig. 10. Spatial mixing index *SMD* for the three horizontal planes (represented in Fig. 8) for both Re-cases (EXP: Experiments, Num: Simulations from Denev et al. (2009)

In Fig. 10 a comparison between experiments and simulations from Denev et al. (2009) of the *SMD*- index (see equation (3-2)) for both cases (low-Re and high-Re) at the three selected planes (see Fig. 8) are given. It is obvious that in the first two planes ($z/D_j = 2.5$ and 4.0), a better mixing (smaller *SMD*-index) is found for the high-Re-number case in the experiment in contrast to the corresponding numerical simulations. This is in accordance with higher turbulence intensities depicted in Fig. 9.

As mentioned above, the experimental results for the lower planes are not in agreement with the corresponding numerical results. In the numerical simulation the flow is still laminar for lower heights above the baseplate ($z/D_j = 2.5$), which leads to a decrease in the mixing process for the higher Reynolds-number. This is a consequence of the inverse proportionality between the mass transfer rate and Reynolds-number for a laminar flow. The elevated turbulence intensity $Tu_x = \sigma(v_x)/<v_x>$ in the experiments and numerical simulations in the experiments and numerical simulation showed in Fig. 9 at $z/D_j = 2.5$ can not be interpreted as a parameter representing the instability of the flows. At low heights above the baseplate the jet propagates more or less solely in z-direction. Therefore, on the one hand $<v_x>$ is very small and on the other hand the intermittent fluctuations $\sigma(v_x)$ are comparably large leading to a large value for Tu_x. The intermittence of the flow is originated from coherent structures, e.g. the counter rotating vortex pair, which are not considered as turbulence. As it becomes obvious from Fig. 8 the flow is still stable and instabilities in the flow are not observed at $z/D_j = 2.5$.

At $z/D_j = 5.5$ the jet flow is almost oriented parallel to the crossflow and therefore the turbulence intensity as well as the distribution of the jet NO_2-mole-fraction have decreased, which is obvious from the experiments and the simulation. For this reason, the corresponding *SMD*-values are similar for both cases (low and high-Re).

3.3 Turbulent Mass Transfer

In Fig. 11 a comparison of the experimental results for the axial (left) and the lateral (middle) Reynolds-fluxes at $z/D_j = 2.5$ for low-Re (first row) and high-Re (second row) case are shown. Additionally, exemplary instantaneous LIF-maps are depicted (right).

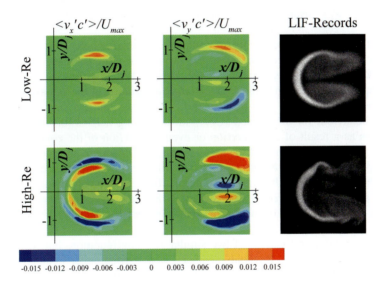

Fig. 11. Comparison of experimental determined axial (left) and lateral (middle) Reynolds-Fluxes and instantaneous LIF-maps (right) at $z/D_j = 2.5$ are given for the cases low-Re (first row) and high-Re (second row)

From the figure, it becomes obvious that the structures formed as result of the interaction between both flows have few more or less well defined contours for the low-Re case, whereas the structures contours for the high-Re case are more frequent and diffuse. Furthermore, it is also obvious that those structures are larger for the high-Re case than for the low-Re case. This primarily indicates that flows possessing higher Reynolds number reveal more instabilities than those with smaller Reynolds-number. Moreover, from a mathematical point of view Reynolds-fluxes (and -stresses) represent non-normalized correlation coefficients.

Larger values for the Reynolds-fluxes for larger Reynolds numbers provide higher correlation between the temporal fluctuation of the velocity in the corresponding direction and the fluctuation of the concentration. Therefore, there is a direct relation between Reynolds-fluxes (and stresses) and the increasing turbulent flows' mass-transfer.

These conclusions are further promoted from the LIF-maps depicted in Fig. 11 (right column). It is obvious that the jet-flow posses a more intensive interaction with the crossflow for the high-Re case. It is noticed that the jet flow at this case is being ripped by the crossflow notably stronger than for the low-Re case.

4 Conclusions

Simultaneous two-dimensional maps of mean velocities and concentrations in addition to quantities reflecting the temporal and spatial fluctuations of the flow, e.g. turbulence-intensities and especially Reynolds-fluxes and -stresses, have been gained in a JCF arrangement by means of simultaneous 2d-LIF-PIV measurements.

In a first series of experiments we compared the mixing behaviour of a swirled with an unswirled JCF configuration. By varying the flow parameters of the jet, we took care that the penetration of the jet into the crossflow as well as the throughput was similar in both cases tolerating a variation of the velocity ratio. At lower heights above the base-plate of the channel we found a significant better mixing behaviour for the swirled JCF. With increasing heights this improvement decreases as a result of the distortion or even destruction of the counter-rotating vortex-pair by the presence of the swirl of the jet. Obviously, the former significantly promotes the mixing process of both flows with each other especially at higher heights above the base plate. To make a final decision of the mixing behaviour of a swirled compared to an unswirled JCF further investigations have to be done at a larger number of measuring planes, especially at higher heights above the base-plate.

Experimental investigations of two similar JCF-configurations but with different turbulence levels, i.e. Reynolds-numbers, were carried out and compared with DNS-simulations form Denev et al. (2009). The experimental results were in good agreement with the numerical simulation. As expected, the comparison of the results shows that the mass transfer exchange increases with the turbulence level. This relation became manifest by an increase in the Reynolds-fluxes and -stresses in both, the magnitude of the spatial distribution as well as the absolute values, respectively.

References

Andreopoulos, J., Rodi, W.: Experimental investigation of jets in a crossflow. J. Fluid Mech. 138, 93–127 (1984)

Boss, J.: Evaluation of a homogeneity degree of a mixture. Bulk Solids Handling 6(6), 1207 (1986)

Broadwell, J., Breidenthal, R.: Structure and mixing of a transverse jet in incompressible flow. J. Fluid Mech. 148, 405–412 (1984)

Camussi, R., Guj, G., Stella, A.: Experimental study of a jet in a crossflow at very low Reynolds number. J. Fluid Mech. 454, 113–144 (2002)

Cárdenas, C., Suntz, R., Denev, J.A., Bockhorn, H.: Two-dimensional estimation of Reynolds-fluxes and -stresses in a Jet-in-Crossflow arrangement by simultaneous 2D-LIF and PIV. Appl. Phys. B: Lasers and Optics 4, 581–591 (2007)

Cárdenas, C., Suntz, R., Bockhorn, H.: Experimentelle Untersuchung von Vermischungsvorgängen in einer Jet-in-Crossflow-Anordnung. Gas-Wärme-International 5(57), 337–342 (2008)

Denev, J.A., Fröhlich, J., Bockhorn, H.: Structure and mixing of a swirling transverse jet into a crossflow. In: Proc. 4th Int. Symp. Turbul. and Shear Flow Phenom, Virginia, USA (2005)

Denev, J.A., Fröhlich, J., Cárdenas, C., Suntz, R., Bockhorn, H.: In: FA- Sitzung "Mehrphasenströmungen" und "Computational Fluid Dynamics" der VDI- Gesellschaft Verfahrenstechnik und Chemieingenieurwesen. Friedrichshafen/Bodensee (2006)

Denev, J.A., Fröhlich, J., Falconi, C., Bockhorn, H.: Direct numerical simulation, analysis and modelling of mixing processes in a round jet in crossflow (2009) (present issue)

Fric, T.F., Roshko, A.: Vortical structure in the wake of a transverse jet. J. Fluid Mech. 279, 1–47 (1994)

Gulati, A., Warren, R.E.: NO2-based laser-induced fluorescence technique to measure cold-flow mixing. J. Propuls. and Power 10, 1–54 (1994)

Hasselbrink, E.F., Mungal, M.G.: Transverse jets and jet flames. Part 1. Scaling laws for strong transverse jets. J. Fluid Mech. 443, 1–25 (2001)

Keffer, J.F., Baines, W.D.: The round turbulent jet in cross-wind. J. Fluid Mech. 15(4), 481–496 (1963)

Kelso, R.M., Lim, T.T., Perry, A.E.: An experimental study of round jets in cross- flow. J. Fluid Mech. 306, 111–144 (1996)

Margason, R.J.: Fifty years of jet in crossflow research. In: AGARD-CP- 534 conf. proc. on a Jet in Cross Flow, Winchester, UK (1993)

Mundus, B., Kremer, H.: Untersuchung der strömungstechnischen Eigenschaften unterschiedlicher Bauformen von Drallerzeugern. Gas-Wärme-International 38(4), 205–211 (1989)

Özcan, O., Meyer, K.E., Larsen, P.S., Westergaard, C.H.: Simultaneous measurements of velocity and concentration in a jet in channel-crossflow. In: Proc. FEDSM, ASME Fluid. Eng. Div. Summer Meeting, New Orleans, USA (2001)

Prandtl, L.: Führer durch die Strömungslehre. Vieweg-Verlag, Berlin (1990)

Su, L.K., Mungal, M.G.: Simultaneous measurements of scalar and velocity field evolution in turbulent crossflowing jets. J. Fluid Mech. 513, 1–45 (2004)

Characterization of Micro Mixing for Precipitation of Nanoparticles in a T-Mixer

Johannes Gradl and Wolfgang Peukert

Institute of Particle Technology, Friedrich-Alexander University of Erlangen-Nuremberg, Cauerstr. 4, 91058 Erlangen

Abstract. Mixing, especially on molecular scale, is a key parameter for controlling rapid chemical synthesis in liquid media. Macro, meso and micro mixing influence significantly reaction kinetics and thus the particle formation as well as the resulting product properties. In this study, mixing in a static T-shaped mixer is investigated experimentally and numerically by different approaches. Firstly, particle-image-velocimetry (PIV) and laser-induced-fluorescence (LIF) techniques are applied to characterize mixing at different length and time scales. Thereby all relevant flow structures down to the Batchelor length scale are detected by a special high resolution LIF set-up (HR-LIF). Secondly, nanoparticle precipitation is used to quantify the micro mixing efficiency. These experimental approaches deliver two global parameters, namely the mean size and the width of the particle size distribution (PSD), showing explicitly the effect of mixing efficiency on the chemical reaction and on particle formation. Based on the experimental results a theoretical model is developed, which couples the fluid dynamics with the solid formation kinetics. This model predicts accurately the complete shape of the resulting PSD for the precipitation of barium sulfate in a T-mixer and visualizes the relevant kinetic data in three dimensions in the mixer.

1 Introduction

Static micro mixers have emerged as a promising technology for continuous production in biological (Kamholz et al. 1999; Sugiura et al. 2005), pharmaceutical (Douroumis and Fahr 2006) and chemical industry (Johnson and Prud'homme 2003; Hessel et al. 2004; Loeb et al. 2004). Small dimensions of the reactors offer many advantages regarding heat transfer and yield due to fast mixing times. The application of e.g. T- or Y-shaped micro mixers improves the product quality in technological processes with fast kinetics and were used under laminar (Kockmann et al. 2006; Li et al. 2008) as well as turbulent flow conditions (Marchisio et al. 2002; Schwarzer and Peukert 2002; Judat and Kind 2004; Schwarzer and Peukert 2004). The precipitation of nanoparticles requires high mixing intensity, which is documented in several publications showing the strong influence of mixing intensity on the particle size distribution of the nanoparticles produced (Gradl et al. 2006; Marchisio et al. 2006; Schwarzer et al. 2006; Baldyga et al. 2007; Lince et al. 2008). The correct and detailed description of mixing at different length and time scales is essential to tailor the product properties.

In liquids, mixing is generally distinguished by macro, meso and micro mixing (Bakker 1996; Baldyga and Bourne 1999; Pope 2000). Macro mixing takes place on large length scales similar to the reactor geometry, brings both liquid phases into contact with each other and thus affects an inertial-convective dispersion of several phases. Viscous convective meso mixing describes the stretching of small vortices in the range of Kolmogoroff and the engulfment of surrounding fluid. Finally, the diffusive transfer of species between the laminae is called micro mixing and occurs on length scales down to molecular scale. These processes, taking place on different length and time scales, are visualized in Fig. 1.

Fig. 1. Schematic visualization of macro, meso and micro mixing (Ottino 1989)

The major objective of this research work is a detailed investigation of the mixing efficiency of a T-mixer. For this purpose, nanoparticle precipitation is a promising method to characterize micro mixing. In contrast to competitive parallel reactions, two characteristic parameters (mean size and width of PSD) quantify the mixing efficiency on the molecular scale. A model is developed to describe the influence of mixing at different length and time scales on the precipitation process. Coupling the fluid dynamics with the solid formation kinetics is the key to predict the resulting particle size distribution (PSD). Additionally, the numerical investigations of the cooperation partner Prof. Manhart are validated by optical measuring methods (Schwertfirm et al. 2007). Thereby the flow and concentration field as well as mixing on molecular scale is investigated experimentally by PIV and two LIF approaches. The strategy of Reynolds similarity is chosen to increase the temporal and spatial resolution and thus an enlarged T-mixer with good optical accessibility is constructed.

2   Optical Methods to Characterize Mixing

The T-mixer was constructed of two circular feeding tubes with a diameter of 40 mm (0.5H) and a length of 960 mm (10H), which are attached at the opposite sides of a main mixing channel with a quadratic cross section width H of 80 mm and a length of 1000 mm (12.5H). The ends of the feeding tubes are flush mounted with the bottom of the main channel. For orientation in the mixer, a co-ordinate system is defined by the mixer axes: the x-axis follows the main channel in the direction of the flow, the y-axis corresponds to the centreline of the feeding tubes and the z-axis completes the right hand coordinate system. As origin, a corner at the bottom of the mixer is chosen. Fig. 2 shows a schematic draft of the T-mixer. Lamellae are implemented to prevent undesired secondary motions inside the feeding tubes, in order to ensure that a fully developed flow profile enters the mixing channel. The main channel is made of coated glass for a maximal optical accessibility and to reduce reflections. The edges of the mixer are strengthened by aluminium to stabilize the apparatus. At the top of the main channel an overflow basin is installed to prevent backstreaming at the outlet. The fluid is provided by two stainless steal 300 l pressure tanks, which are temperature-controlled by heating coils. Pressurized air of 3 bar leads to pulsation-free flow rates and additional flow control units provide an exact flux up to a maximal Re-number of 3000, which is based on the width of the main channel H. The fluctuations in volume flux are less than 1% of the actual flow rate. The overall flow field of the constructed mixer is investigated by particle-image-velocimetry (PIV), the concentration field is measured by two dimensional laser-induced-fluorescence (2D-LIF) and finally small scale mixing is detected by a high resolution laser-induced-fluorescence technique (HR-LIF).

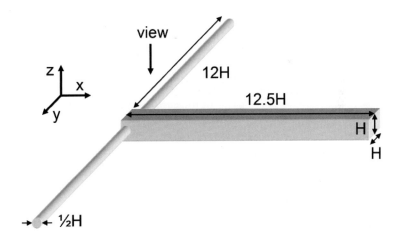

Fig. 2. Schematic draft of the T-mixer

2.1 Particle-Image Velocimetry (PIV)

For measuring the flow field, a 2D-PIV technique was employed. The periphery of the PIV system contains a Nd–Yag double pulsed laser with a frequency of 10 Hz, two lenses for stretching and focusing the laser beam to a 1 mm thin sheet and a CCD camera for recording images in a double frame mode. The evaluation system generated the final vector maps by across-correlation technique. The maximum energy of the laser was 200 mJ and consequently, one pulse with a length of 10 ns has the theoretical power of 20 MW. The scattered laser light is detected under an angle of 90° by a CCD chip. Each singular measurement provides a 49 x 27 vector map, resulting in a spatial resolution of 1.63 mm. All vectors with an absolute value 10 times larger than the mean bulk velocity u_{bulk} have been considered as invalid and replaced by a moving average. These replaced vectors appeared mainly at the wall regions, where reflections of the laser sheet were observed, the number of invalid vectors was less than 3% of the total samples. Polyamide seeding particles of a mean particle size of 20 µm and a density of 1020 kg/m^3 were used. Our tests demonstrated that a concentration of 15 mg/l provided the best results and was consequently used throughout the measurements. PIV measurements were conducted in four planes. Two lie in the symmetry plane spanned by the axis of the main and the feeding tubes at different x positions. The two other planes are displaced by 0.25H and 0.75H from the symmetry plane in the z-direction in order to check the symmetry of the measured flow. The statistical steady state was recorded by taking series of measurements after the 15 min of the complete transient period have elapsed. These series consist of 400 single measurements taken in bursts of three single measurements within one second followed by a break of 9 s. This procedure has been repeated three times thus leading to 1200 single snapshots over a time period of 4000 s.

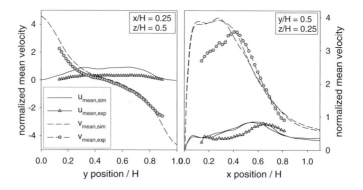

Fig. 3. Validation of DNS by PIV

A quantitative comparison of experimental and numerical results is given in Fig.3 for the time-averaged, normalized (by u_{bulk}) velocity components u_{mean} and v_{mean} in x- and y-direction. The main features of the mean concentration profiles match nicely, although there are some quantitative differences. The main differences occur close to the bottom wall of the main duct. Here, the main duct's axial velocity component u_{mean} at z = 0.5 is overpredicted by the simulation as well as the main duct's azimuthal velocity component v_{mean} at z = 0.25. This means that the momentum of the feeding streams is directed somewhat more to the bottom wall in the DNS than in the experiments. These differences could result from (i) differences in inflow conditions between DNS and experiment or, in principle, from (ii) insufficient statistics in the DNS data. All in all, the experimental data shows that the DNS overall flow field is qualitatively correct. Further results of PIV experiments are shown in (Schwertfirm et al. 2007).

2.2 2D-Laser Induced Fluorescence (2D-LIF)

Disodium fluorescein (uranine AP – C.I.45350 – Merck KGaA) is applied, with a dye concentration of $1.2 \cdot 10^{-8}$ kmol/m^3 as fluorescing tracer molecule in one of the feed streams. Uranine AP shows a high quantum efficiency of 93% and a Sc-number in aqueous media is reported as 1930 (Walker 1987). The low concentration promises the validity of the law of Lambert Beer and that the laser intensity is not weakened significantly by the dye solution. The excitation wave length of uranine is reported as 491 nm (Käss 2004) for a pH-value of 9, which matches the wave length of the laser applied. As the fluorescence intensity of uranine is pH-independent for pH-values larger than 9, a sufficiently high amount of sodium hydroxide is added to the tanks. The maximum of the emission spectra of uranine is detected at a wave length of 512 nm and the fluorescence life-span is determined as 4.2 ns (Käss 2004).

In order to characterize mixing down to the Kolmogorov length scale, a 2D-LIF set-up is built-up. An air-cooled argon-ion laser with a power of 150 mW and a wave length of 488 nm in TEM$_{00}$-mode is applied as light source. The emitted laser beam is extended to a sheet and focused in width by two planar concave lenses. Mirrors lead the laser light from the bottom through the mixer. A digital camera with 2304 x 3456 pixels collects the emitted fluorescence light rectangularly in RGB mode. For the evaluation of the measured concentration patterns, only the green channel of the camera is activated to filter reflected laser light. For a statistical sufficient number, 300 single patterns are taken with a frequency of 0.3 Hz, an exposure time of 10 ms and an optical aperture of 1:1.8 for each measuring run. The spatial resolution is set to 400 µm x 400 µm x 400 µm, which is larger than the Batchelor length scale and in the range of the estimated Kolmogorov length scales. The mixer was investigated by this measuring technique in five different x-y planes (z = 0.1H, 0.25H, 0.3H, 0.4H and 0.5H) from the bottom of the mixer till a length of 3H and for a Re-number of 500, 1100 and 2500. Areas near to the wall did not give satisfying results due to unavoidable reflections and are ignored.

  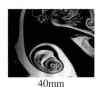

22mm   35mm   40mm   8mm   20mm

Fig. 4. Vortex structures in the flow field visualized by LIF

By mixing both liquids in a T-Mixer, vortex structures are generated, which are stretched during the meso mixing process. The vortices decompose to smaller flow structures, producing more contact area and consequently leading to an intensified micro mixing. The diffusive mass transfer takes place especially between the laminae of the vortices at the interface of both phases. The LIF-technique clearly visualizes the vortices generated in the fluid flow, which are shown in Fig. 4. The original width of the image sections is denoted at the bottom. By increasing the flow rates, the three-dimensional flow structures (vortices) become smaller, indicating that higher power inputs generate more contact area between both liquids and thus micro mixing is improved.

Mean concentration fields deliver a clear impression of the overall flow conditions and the ongoing mixing process in the mixer. Six mean concentration fields of different cross sections along the T-mixer, each determined from 300 single samples are depicted in Fig. 5 exemplary for a Re-number of 1100. Black indicates a minimal concentration of 0, whereas high concentrations are marked by bride colors. In order to reconstruct mean concentration fields of different cross sections by five horizontal x-y mean concentration fields, which are measured in

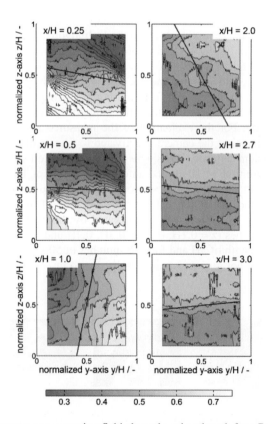

Fig. 5. Different mean concentration field along the mixer length for a Re-number of 1100

the front half side of the mixer, the detected data are interpolated in the yz-plane. Based on the point symmetrical behavior of the mean flow field, complete mean concentration fields of the cross section are reconstructed. The symmetrical behavior of the flow field was identified by previous PIV measurements and DNS calculations as reported in (Schwertfirm et al. 2007).

An increase of the Re-number does not significantly influence the main features of the flow field in the mixer and thus concentration fields for Re = 1100 are discussed. As can be seen in Fig. 5, both inflowing jets do not impinge but pass each other. Thereby a large helical vortex is generated, which fills the whole mixer. The inlet jet angle is about 30° and the rotational direction is absolutely random. The rotational symmetry axis of the vortex is defined as a straight line ("rotational axis") approximated by the isoline of perfect mixing ($c/c_0 = 0.5$), as drawn in each cross section of Fig. 5. In the centre of the vortex, a "mixing zone" is built up, where most of both fluids come into contact with each other. With regard to technical applications, chemical reactions and homogeneous nucleation takes place mainly in this region.

At the inlet (x/H = 0.5) the qualitative flow pattern does not change significantly but the amount of fully unmixed fluid decreases slightly compared to x = 0.25H. Here, the increased mean concentration, which is detected in two opposite positioned corners, indicates dead zones with fluid elements having a longer residence time.

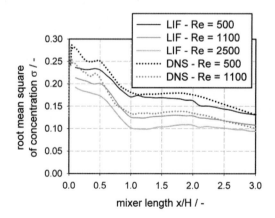

Fig. 6. Influence of the flow rate on the cross-section averaged rms-value σ

The overall picture of the mean concentration field changes extremely above the inlets at x/H = 1.0. The entered fluid elements have reached the opposite side of the mixer. The local concentration gradients over the cross section decay continuously along the mixer length x, indicating advanced mixing. After a distance of 2.7H, the "rotational axis" shows again the same slope as at x/H = 0.25. The fluid elements have passed the complete vortex and have reached again the inlet side, where they have started. By comparing the mean concentration fields for different flow rates, it is remarkable that the size of the vortex cannot be influenced by the Re-number in the observed range.

The rms-value σ of the detected concentration, averaged over the complete cross section quantifies the mixing efficiency along a mixer. Fig. 6 compares rms-values for three different Re-numbers determined experimentally by 2D-LIF and for Re = 500, 1100 and 2500 with calculated data by a semi-DNS (Schwertfirm et al. 2006; Schwertfirm et al. 2007). The simulated data matches the experiments well with the exception of the inlet region. The deviations are caused by an insufficient spatial resolution over the cross section (only 5 measuring planes). Especially in regions with turbulent flow field structures small concentration gradients could not be fully resolved.

As expected from the previously evaluated results, the qualitative progressions of all three flow rates are similar. In the inlet region, unmixed fluid elements are prevailing and σ decreases slowly. Above the feeding tubes, between x = 0.5H and 1.0H, σ is reduced strongly due to intensive convective mixing. After a distance of x = 1.0, no additional dissipation energy is produced externally. Thus the mixing process is dominated by diffusion and the cross-section averaged rms-value decreases slowly.

2.3   HR-Laser Induced Fluorescence (HR-LIF)

Flow field structures below the Kolmogorov length scales determine significantly chemical processes. However the spatial and temporal resolution of the 2D-LIF method is not sufficiently high. Thus a high resolution LIF set-up (HR-LIF) is applied to detect all relevant structures in the flow field. The challenge is to meet the simultaneous requirements of rapid data acquisition and strong amplification. For the optical set-up the argon-ion-laser, which is applied for 2D-LIF technique, is used as well. The laser beam is expanded by the factor of 20 by a beam expander, is led through a custom-made dielectric mirror and focused through the wall of the mixer by an achromatic lens on the measuring volume. The laser focus has an estimated theoretical volume of 2.8 x 2.8 x 7.9 μm^3, according to the Gaussian beam optics (Siegman 1971). Based on a volume equivalent sphere the diameter λ_M of the optical measuring point is estimated as 4 μm. The emitted light is collected confocally and reflected rectangular in direction of the detection unit. Firstly, the collected light intensity is reduced by ten times in diameter and secondly is led through a filter system of a long-pass filter, a short-pass filter and a holographic filter, which prevent that undesired scattered laser light ($\lambda = 488$ nm) reaches the detector. Two pinholes, arranged at a distance of 200 mm to each other filter fluorescence light, which has its origin outside of the focussed measuring point. Finally, a photomultiplier collects the fluorescence light and the signal is recorded via an amplified current-to-voltage converter (in 24bit, measuring frequency 10 kHz). The schematic set-up is shown in Fig.7. The combination of the high spatial and temporal resolution warrants the detection of all relevant flow field structures in the mixer for the operating conditions chosen. 5-6 carboxyl rhodamine (MoBiTec) is applied as dye molecule to inhibit undesired photobleaching or thermo blooming effects (Walker 1987). After a sufficient starting time, fluorescence signals are detected for 10 minutes with 10 kHz and normalized by the maximal signal of pure dye solution, which is measured for each local measuring position. The time-dependant concentration measurement is evaluated by probability density functions (PDF) and by a discrete Fourier transformation (DFT).

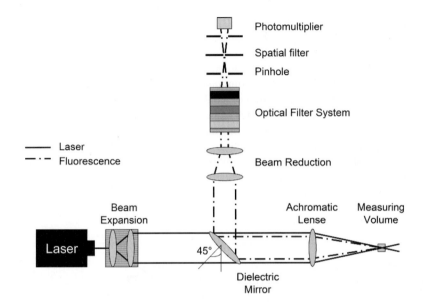

Fig. 7. Schematic draft of the optical bench for the HR-LIF set-up

Table 1. Length and time scales of estimated flow structures

Re	u_{bulk}	ε_{mean}	λ_K	τ_K	λ_B	f
-	mm/s	W/kg	µm	ms	µm	Hz
500	5.56	$1.19 \cdot 10^{-6}$	877	863	20	560
1124	12.5	$1.36 \cdot 10^{-5}$	477	256	11	2300
2500	27.8	$1.49 \cdot 10^{-4}$	262	77	6	9300

The challenge of this measuring technique is to reach a sufficiently small optical measuring volume for the individual flow condition with a sufficient temporal resolution. Otherwise the local concentration data are filtered and micro mixing is overestimated. Table 1 gives an overview of the estimated scales in the flow field. The mean specific power input is obtained indirectly from measured pressure drops of a geometrical similar T-Micro-Mixer, used for the precipitation of nanoparticles and introduced in the next section and subsequently scaled-up. λ_K is the Kolmogorov length, τ_K the life-time of a Kolmogorov vortex, λ_B the Batchelor length for a Sc-number of 1930 and f is the required mean measuring frequency, based on the mean velocity u_{bulk}. For a Re-number of 2500 the optical set-up ($\lambda_M \approx 4\mu m$ and $f_M = 10kHz$) reaches the upper limit of the requirements defined by the fluid dynamical system, but should be sufficient for this application. However, the calculations are applied for the mean velocity and the mean specific power input i.e. that for a high Re-number of 2500, temporarily appearing small scale structures may be filtered.

Fig. 8 shows the influence of the Re-number on the concentration spectra for two different local positions at the centreline of the mixer. On the left side, three spectra at the height of the feeding tubes centre $x/H = 0.25$ and on the right side at the top of the mixer ($x/H = 8.0$) are shown. By increasing the flow rate the amount of small frequencies increases and simultaneously the amount of large frequencies decreases, indicating the generation of smaller structures with higher energy dissipation. The shift to larger frequencies cannot only be explained by a higher velocity u_{bulk}. By multiplying u_{bulk} with the maximal detected frequency, a minimal mean structure length can be determined, which is depicted in Fig. 8 as well. Both diagrams show the tendency that the size of the smallest structures detected decrease with increasing Re-numbers, which is expected as a higher specific power input produces smaller structures.

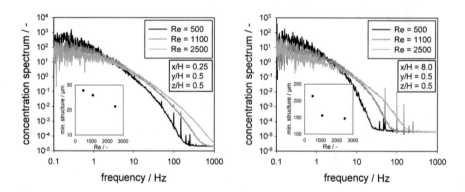

Fig. 8. Influence of the Re-number on the measured concentration spectrum at two different local positions

The PDFs of the normalized concentration at the same positions are shown in Fig. 9, confirming the results of Fig. 8. Closed to the inlet region, the PDFs are widely distributed. Even absolutely unmixed fluid is detected, which is shown as two symmetric peaks with a concentration of 0 and 1, respectively. At the outlet of the mixer ($x/H=8.0$) the PDF shows a sharp peak with a concentration of 0.5, indicating advanced mixing. The generation of small structures, closed to the inlet leads to larger contact areas, which accelerate the mass transfer between both liquids, resulting in more intensive micro mixing. The influence of an increased Re-number can still be observed at a later position in the mixer, where only diffusion dominates the mixing process.

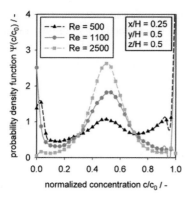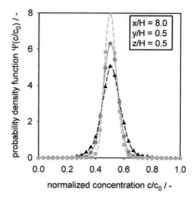

Fig. 9. Influence of the Re-number on the measured concentration PDFs at two different local positions

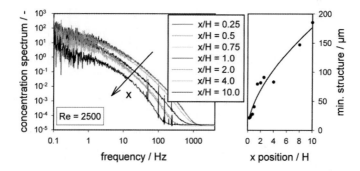

Fig. 10. Measured concentration spectra for a Re-number of 2500 along the center line

Fig. 10 shows several measured concentration spectra for a Re-number of 2500 along the centreline of the mixer. With increasing distance to the inlet region, spectra are shifted to smaller frequencies, indicating an increasing homogenization due to mixing. Additionally, the smallest detected structure length is calculated analogue to Fig. 8 and depicted in Fig. 10 as well. The minimal structure length increases steadily along the mixer length from about 20 μm to almost 200 μm.

Besides characterizing small scale mixing, HR-LIF measurements are also used to validate the micro mixing model, which is used for modelling the solid formation process, which is introduced and discussed later. In order to compare simulated with experimental results, the concentrations measured are converted to a mixing fraction X, which is calculated as follows:

$$X = 1 - \left(2 \cdot \left|0.5 - c/c_0\right|\right) \tag{1}$$

where $X = 0$ stands for an absolutely unmixed and $X = 1$ for a perfectly mixed state. Fig. 11 shows the comparison between theoretically and experimentally determined PDFs of the mixing fraction X at four different local measuring positions for a Re-number of 500. The PDFs are determined by counting mixing fractions calculated by the Lagrangian approach, whose particle paths pass the "measuring volume". X_{Macro} represents the mixing fraction from the DNS, filtered at the Kolmogorov length scale and X_{Micro} stands for the mixing efficiency on molecular scale, which is calculated by the applied micro mixing model (see section Modelling). For each PDF in Fig. 11 between 20 and 80 different Lagrangian particle paths out of the complete set of 700 paths have passed the "measuring volume", which is not sufficient to compare the simulation quantitatively with the LIF-results. However, trends and fluid dynamical phenomena are reflected.

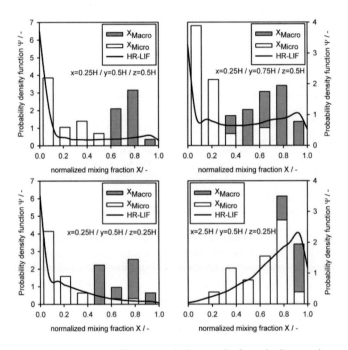

Fig. 11. Comparison between LIF and simulation results from the Lagrangian approach

Overall the simulation matches the experiments, indicating that both the HR-LIF technique resolves all relevant concentration fluctuations in the flow field and the chosen micro mixing approach is applicable to model the subgrid structure for high Sc-numbers (Sc > 1000).

3   Nanoparticle Precipitation in a T-Mixer

Nanoparticle precipitation is a novel significant method to characterize micro mixing in small mixing devices. Precipitation experiments were carried out using barium sulfate as model substance since many relevant material properties are known. Besides the experimental determination of the micro mixing efficiency of the applied T-mixer, a numerical model is developed, which describes all relevant subprocesses and their interactions and can predict the resulting particle size distribution (PSD).

3.1 *Experimental Investigations*

To generate barium sulfate nanoparticles, aqueous solutions of barium chloride and sulfuric acid are mixed continuously at 25°C in a T-mixer. The applied T-mixer is a scale-down version of the T-Mixer investigated by PIV and LIF. It is made from borosilicate glass and consists of two feed tubes of 0.5 mm diameter positioned centrically opposite each other and a main duct of 10 mm length and a quadratic cross-section of 1 mm width. The relatively small geometric sizes of the mixer is chosen to provide intense mixing, necessary for nanoparticle precipitation. Constant pulsation-free flow rates for the continuous experiments are provided by two temperature-controlled syringe pumps. The pressure drop – its value contains information on the total amount of friction and shear in the flow – is also determined. Since the pressure drop does not change with increasing length of the main duct for constant volume flow rates, it can be concluded that the pressure drop arises almost completely from the impinging of the two feeds. From the measured pressure drop, the mean mass-specific power input ε_{mean} in the main duct is obtained and used as an estimate for the mean dissipation rate of the turbulent kinetic energy in the flow.

The generated suspension is collected in a beaker and samples are taken for analysis at steady-state operating conditions. The obtained PSD is measured immediately after precipitation based on quasi-elastic light scattering. Further details on the material and the applied measurement procedures can be found in (Schwarzer and Peukert 2002, 2004; Schwarzer 2005).

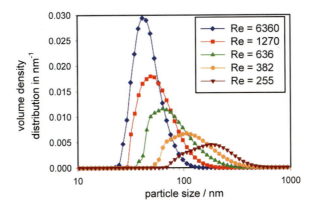

Fig. 12. Influence of increasing flow rate on the PSD

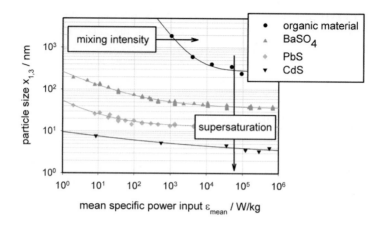

Fig. 13. Influence of the mean specific power input on the mean particle size of several material systems

Measured PSD of five precipitation experiments at different Reynolds numbers, i.e. flow rates through the mixer are presented in Fig. 12 as volume density distributions. The initial educt concentrations were kept constant in these experiments (0.5M $BaCl_2$ and 0.33M H_2SO_4). The measured mean particle sizes as well as the width of the PSD decrease as the Re number and thereby the pressure drop as well as the mixing intensity increase. The reason for the observed decrease with increasing Re number is the different dependencies of nucleation and growth rate on supersaturation and the "generation" of supersaturation through mixing. With increasing supersaturation, the nucleation rate increases stronger than the growth rate and thus more and smaller particles are formed, which grow until supersaturation is completely consumed. For fast micro mixing the local supersaturation built up is steeper and leads to a higher level, which is reduced quickly due to accelerated precipitation kinetics. Low mixing intensities generate a slow increase in supersaturation leading to a slower solid formation. Consequently the resulting mean particle size characterizes the mixing intensity. As long as each particle experiences the same mixing history, the shape of the PSD will be a sharp peak. To turn the argument on its head, precipitation experiments represent a second characteristic mixing parameter, which is the width of the PSD, reflecting fluctuations in the flow field.

Fig. 13 shows the transferability to other inorganic or organic material systems. As the solid formation kinetics are in the same range as micro mixing times scales, mixing determines significantly the product properties. The resulting minimal particle size is different for different materials due to its varying material properties but the dependencies are similar. Thus, the resulting shape of the PSD identifies accurately the mixing process on molecular scale.

3.2 Modelling

In the scope of this work, a coupled DNS-PBE (direct numerical simulation – population balance equation) approach is developed to predict the resulting PSD, using barium sulphate as model system. The cooperation with the group of Prof. Manhart, which carried out the DNS, was decisive for a successful development of a complete model for the rapid precipitation process in a T-Mixer. Details of the concept for the CFD can be found in (Schwertfirm et al. 2007). The method of coupling CFD with the micro mixing model and with the model of the solid formation is introduced in the following.

In addition to solve the Navier-Stokes- and the scalar transport equations by DNS, Lagrangian Particle Tracking is conducted to provide information along the paths of finite volumes (considered as massless particles) through the mixer. This information comprises, besides time and position, the local and instantaneous concentration of the passive scalar and the instantaneous local dissipation rate of the turbulent kinetic energy. In the conducted simulations, particles start from 333 different discrete positions at the inlet tube. Each position represents a fraction of the total flow through the feed tube, calculated from the velocity normal to the plane times the area fraction the starting position stands for. From their starting points the particles are propagated through the flow field by integrating the Lagrangian transport equation for each particle. Once they leave the domain at the outlet, they are reinserted at their starting position and thus at least 700 paths are calculated at the 333 different starting points. The instantaneous dissipation rate is calculated with local velocity fluctuations by evaluating the dissipation rate term as it appears in the turbulent kinetic energy transport equation.

To model the mixing process, the Engulfment model of micromixing is applied which describes the kinetics of micromixing in fully developed turbulent flows based on the assumption that micromixing is the rate-determining step. Details of the model can be found in (Baldyga and Bourne 1999). Since this model was originally derived for the situation of mixing a small volume to a much larger volume, a modified version of the Engulfment model is derived to fit the needs of mixing phenomena in a T-mixer. This modified version is based on the temporal evolution of volume fractions of four ideally mixed compartments. Two of them contain only feed solution A or B (feed zones) and the third fraction (contact zone) is split in two compartments, one containing only feed A and the mixing zone, containing both solutions (A and B). The temporal evolution of the volume fractions is described by a system of coupled ordinary differential equations (ODE). The driving force for the mixing process is the specific power input, which is integrated in the ODE system. Further details on the used modified micromixing model are given in (Schwarzer and Peukert 2004; Schwarzer 2005; Gradl et al. 2006; Schwarzer et al. 2006).

Current state-of-the-art modeling of polydisperse particulate processes is based on the population balance equation (PBE). To simulate the formation of particles, the following 1D population balance equation is applied, characterizing the particles by the diameter x of the volume-equivalent sphere:

$$\frac{\partial n(x)}{\partial t} = B_{hom} \cdot f(x) - \frac{\partial (G(x) \cdot n(x))}{\partial x} + B_{agg}(n,x) - D_{agg}(n,x) \qquad (2)$$

In this equation n(x) is the number density of particles of size x. The first term on the right hand side accounts for nucleation at a rate B_{hom} and the second for particle growth with the size-dependent linear growth rate G(x). B_{agg} as well as D_{agg} represent source and sink terms due to agglomeration/aggregation. As precipitation under stable conditions is investigated in this work, the agglomeration/aggregation terms can be neglected. The function f(x) is a density function describing the size distribution of the formed nuclei.

Under the investigated conditions, primary homogeneous nucleation and transport controlled growth are considered as the dominant mechanisms. Thermodynamic driving force for nucleation and growth is supersaturation, which is a function of the current local concentrations of cations and anions. Details on the calculation of supersaturation and on the modeling of these mechanisms have been published previously (Schwarzer and Peukert 2004; Schwarzer et al. 2006; Gradl and Peukert 2008).

In order to solve the PBE, the commercial solver PARSIVAL by CiT–Computing in Technology GmbH is applied. The numerical algorithm implemented in PARSIVAL uses a finite element type Galerkin h-p-method. The main advantage of the Galerkin h-p method is that the PSD is given as a polynomial expansion with a variable step size h and degree of polynominial p which are both adjusted automatically to keep errors below a given tolerance. No further assumptions are made, for example concerning the PSD shape as in moment methods.

Examples of the temporal evolution of the specific power input are depicted in Fig. 14 (left). The individual histories differ strongly so that different micro mixing kinetics are obtained for each path. Fig. 14 (center and right) shows simulation results of the supersaturation build-up due to mixing and its reduction by solid formation as well as the obtained individual PSDs for the same paths and the precipitation from 0.5m BaCl$_2$ and 0.33m H$_2$SO$_4$. Although the evolutions of supersaturation look relatively similar, clear and significant differences in the obtained

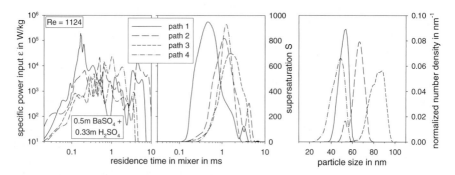

Fig. 14. Evolution of the specific power input and the supersaturation for randomly chosen paths through the mixer and their resulting PSD

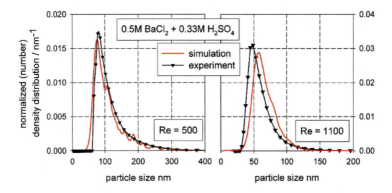

Fig. 15. Comparison between simulated and experimentally determined PSD

PSDs can be seen, i.e. that small changes in the mixing history lead to a strong impact on the precipitation kinetics. Hence, nanoparticle precipitation is an accurate tool to quantify micro mixing.

Fig. 15 shows a comparison between simulated and experimentally determined PSDs for both Re-numbers, 500 and 1100, considering nucleation and growth as solid formation processes for the precipitation of barium sulphate from 0.5M $BaCl_2$ and 0.33M H_2SO_4. Thereby 700 single PSDs are calculated for each run based on the DNS-flow field information and independent mixing histories through the mixer. Furthermore, all 700 single PSDs are weighted according to their respective flow rates to obtain the resulting final PSD by averaging. For both flow rates, the complete shape of the simulated PSD matches the experimental PSD precisely.

The calculated time dependent local information of the specific power input, mixing level, supersaturation, nucleation and growth rate as well as mean particle size have been determined for each path until a mixer length of 3 mm is reached. By interpolating the calculated data of all 700 trajectories through the mixer on a grid of 100 x 100 x 300 nodes, a spatially resolved three dimensional time-averaged parameter-field in the inlet region of the mixer is reconstructed. The interpolation is accomplished by a Delaunay triangulation that uses the Quickhull-method. For further details on the algorithm of the interpolation see (Barber et al. 1996). Fig. 16 shows slices in the symmetry plane for the precipitation from 0.5M $BaCl_2$ and 0.33 H_2SO_4 and a Re-number of 500 (left) and 1100 (right). The major part of the kinetic energy dissipates where both inlet jets impinge on each other. Supersaturation increases where a high specific power input leads to intensive micro mixing and is reduced rapidly due to the ongoing phase transition. As nucleation is a highly non-linear function of supersaturation, increasing nucleation rates appear where both liquids are well-mixed on molecular scale. Most of the nuclei are generated in highly supersaturated regions from above the inlets to a distance of 2 mm. After a distance of 3 mm the mean particle size reaches a value around 60 nm for a Re = 500 and 45 nm for a Re = 1100, which is about half of the final mean particle size. By increasing the flow rate, the local position of the strongest

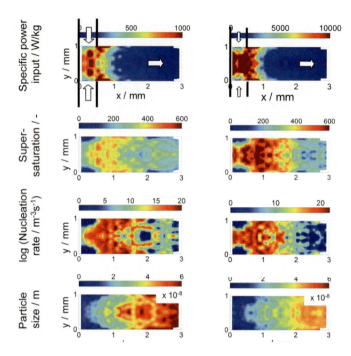

Fig. 16. Visualized characteristic kinetic parameter for a Re-number of 500 (left) and 1100 (right) in the symmetric plane of the mixer

dissipation of the kinetic energy does not change significantly but is shifted to a higher level. The higher the energy input, the faster the mixing takes place due to the formation of smaller flow structures. Consequently more contact surface between both liquid phases is generated, enhancing micro mixing on the molecular scale, leading to higher nucleation rates. With increasing flow rates, the residence time of fluid elements is lower but the solid formation process is accelerated due to increasing nucleation and growth rates, leading to smaller particles, which can be observed in Fig. 16 as well. Further results of the spatial visualization of kinetic parameter in the mixer are shown in (Gradl and Peukert 2008).

4   Conclusion

The detailed characterization of mixing at different length and time scales, especially of micro mixing is represented in this work. Thereby two different experimental approaches are chosen to measure the mixing efficiency in a T-mixer. Firstly, optical measurement methods are applied to characterize the flow and concentration fields in the mixer. A special feature of the LIF set-up is that flow structures smaller than the Kolmogorov length scale are resolved and thus this method can be used for the validation of a subgrid model for high Sc-numbers.

Besides the validation of the flow field simulations of the cooperation partner Prof. Manhart *(A numerical approach for simulation of turbulent mixing and chemical reaction at high Schmidt numbers)* the experimental procedure determines the macro, meso and micro mixing performance of a T-Mixer.

Secondly, precipitation of nanoparticles is applied to characterize the micro mixing efficiency in a T-mixer as an integral method. The more intensive the specific power input and consequently the micro mixing is, the smaller and narrower the produced PSD is. In addition to the experiments, a theoretically model without any free parameter is developed to predict the resulting PSD and to visualize relevant characteristic kinetic parameters inside of the mixer. Considering mixing effects on the molecular level via DNS simulation of Prof. Manhart and coupling them with the solid formation kinetics is the key for an adaptive model for particle synthesis in liquid media.

Acknowledgments

We acknowledge Deutsche Forschungsgemeinschaft (DFG) for the financial support of this research project (SPP1141).

References

Bakker, R.A.: Micromixing in Chemical Reactors: Models, Experiments and Simulations. PhD, TU Delft, NL (1996)

Baldyga, J., Bourne, J.R.: Turbulent Mixing and Chemical Reactions. John Wiley, Chichester (1999)

Baldyga, J., Makowski, L., Orciuch, W.: Chemical Engineering Research and Design 85(5A), 745–752 (2007)

Barber, C.B., Dobkin, D.P., Huhdanpaa, H.T.: ACM Trans. on Mathematical Software 22(4), 469–483 (1996)

Douroumis, D., Fahr, A.: European Journal of Pharmaceutics and Biopharmaceutics 63(2), 173–175 (2006)

Gradl, J., Peukert, W.: Chemical Engineering Science 64(4), 709–720 (2009)

Gradl, J., Schwarzer, H.-C., Schwertfirm, F., Manhart, M., Peukert, W.: Particulate Processes A Special Issue of Chemical Engineering and Processing 45(10), 908 (2006)

Hessel, V., Hofmann, C., Loewe, H., Meudt, A., Scherer, S., Schoenfeld, F., Werner, B.: Organic Process Research and Development 8(3), 511–523 (2004)

Johnson, B.K., Prud'homme, R.K.: Australian Journal of Chemistry 56(10), 1021–1024 (2003)

Judat, B., Kind, M.: Journal of Colloid and Interface Science 269(2), 341–353 (2004)

Kamholz, A.E., Weigl, B.H., Finlayson, B.A., Yager, P.: Analytical Chemistry 71(23), 5340–5347 (1999)

Käss, W.: Geohydrologische Markierungstechnik. Gebrüder Bornträger, Berlin (2004)

Kockmann, N., Kiefer, T., Engler, M., Woias, P.: Sensors and Actuators, B: Chemical 117(2), 495–508 (2006)

Li, S., Xu, J., Wang, Y., Luo, G.: Langmuir 24(8), 4194–4199 (2008)

Lince, F., Marchisio, D.L., Barresi, A.A.: Journal of Colloid and Interface Science 322(2), 505–515 (2008)

Loeb, P., Loewe, H., Hessel, V.: Journal of Fluorine Chemistry 125(SPEC. ISS. 11), 1677–1694 (2004)

Marchisio, D.L., Barresi, A.A., Garbero, M.: AIChE Journal 48(9), 2039–2050 (2002)

Marchisio, D.L., Rivautella, L., Barresi, A.A.: AIChE Journal 52(5), 1877–1887 (2006)

Ottino, J.M.: The kinematics of mixing: streching, chaos and transport. Cambridge Univ. Press, Cambridge (1989)

Pope, S.B.: Turbulent Flows. Cambridge University Press, Cambridge (2000)

Schwarzer, H.-C.: Nanoparticle precipitation - An experimental and numerical investigation including mixing. PhD, Friedrich-Alexander Universität Erlangen-Nürnberg (2005)

Schwarzer, H.-C., Peukert, W.: Chem. Eng. Technol. 25(6), 657–661 (2002)

Schwarzer, H.-C., Peukert, W.: AIChE J. 50(12), 3234–3247 (2004)

Schwarzer, H.-C., Schwertfirm, F., Manhart, M., Schmid, H.-J., Peukert, W.: Chemical Engineering Science 61(1), 167 (2006)

Schwertfirm, F., Manhart, M., Gradl, J., Peukert, W., Schwarzer, H.-C.: Proceedings of ASME Fluids Engineering Division Summer Meeting, FEDSM 2006 (2006)

Schwertfirm, F., Gradl, J., Schwarzer, H.-C., Manhart, M., Peukert, W.: International Journal of Heat and Fluid Flow 28, 1429–1442 (2007)

Siegman, A.E.: Introduction to Lasers and Masers. McGraw-Hill Education, New York (1971)

Sugiura, S., Oda, T., Izumida, Y., Aoyagi, Y., Satake, M., Ochiai, A., Ohkohchi, N., Nakajima, M.: Biomaterials 26(16), 3327–3331 (2005)

Walker, D.A.: Journal of Physics E: Scientific Instruments 20(2), 217–224 (1987)

Mixing in Taylor-Couette Flow

Anna Racina, Zhen Liu, and Matthias Kind

Thermal Process Engineering / Thermische Verfahrenstechnik,
Karlsruhe Institute of Technology (KIT),
Kaiserstr. 12, 76131 Karlsruhe, Germany

Abstract. The mixing in Taylor-Couette flow was studied with 2D laser-optical methods (Particle Image Velocimetry PIV and Laser Induced Fluorescence LIF). Mixing processes are grouped into three stages (macro-, meso- and micromixing) according to the characteristic length scale of the inhomogeneities under consideration. Results of this work are correlations which describe the macro-, meso- and micromixing times as a function of the geometry and process parameters with the mixing iso- and non-isoviscous fluids. The calculation of the micromixing times is shown as a function of the angular velocity of inner cylinder, the fluid viscosity and the geometry. With the fluid viscosity of $1 \cdot 10^{-6}$ m²/s the mixing times are investigated as a function of the radial Reynolds number for three length scales. The macromixing times are calculated with help of two correlations for the axial dispersion coefficient in the turbulent and wave regime. The boundary between the regimes is applied to $Re_\phi = 20\ Re_{\phi,crit}$ at $Re_{ax} = 1$. The macromixing times are the longest from the mixing times. Mesomixing times are shorter and the micromixing runs fastest here. The Mesomixing limits the velocity therefore the micromixing and the macromixinge with which occurs the mesomixing.

1 Introduction

A Taylor-Couette reactor (TCR) consists of two concentric cylinders. The inner cylinder is rotating. The flow induced by the rotation of inner cylinder, with or without a superimposed axial flow, is characterized by the existence of axisymmetric counter rotating vortex cells spaced regularly along the cylindrical axis, Kataoka (1986), see Fig.1. The Taylor-Couette reactor provides certain advantages for a number of practical applications. Among them is the narrow residence time distribution, which corresponds to an almost plug-flow performance under certain conditions. Another characteristic of the Taylor-Couette flow is the rather homogeneous distribution of the mixing intensity in the reactor volume, compared with e.g. conventional stirred tanks, combined with mild local shear rates. These properties are desirable for most of the chemical reactions, including fast reactions resulting in solid products or polymers. Furthermore, this type of reactor exhibits a high surface-to-volume ratio, which makes it preferable for highly exothermal or endothermal reactions.

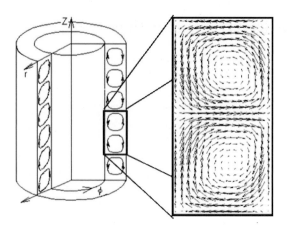

Fig. 1. Taylor-Couette Flow

2 Experimental Set-Up

The following results of an investigation of the mixing in Taylor-Couette flow is investigated by using a reactor with two concentric cylinders of 390 mm length. Three different inner cylinders, with radii R_i of 31.5 mm; 37.9 mm; 44.3 mm have been used. The outer cylinder had a radius R_o of 50 mm.

The flow conditions are described by means of the following four dimensionless numbers: the radial Reynolds number Re_ϕ Eq.(1), the critical radial Reynolds number $Re_{\phi,crit}$ Eq.(2), the axial Reynolds number Re_{ax} Eq.(3) and the Taylor number Ta Eq.(4)

$$Re_\phi = \frac{\omega \cdot R_i \cdot d}{\nu} \tag{1}$$

The primary dimensionless group describing the hydrodynamic conditions in the gap is the Re_ϕ, combining information on geometry (the gap width d and radius of the inner cylinder), angular velocity ω and fluid kinematic viscosity ν.

$$Re_{\phi,crit} = \frac{\omega_{crit} \cdot R_i \cdot d}{\nu} = \frac{1}{0.1556^2} \cdot \frac{(1+\eta)^2}{2\eta \cdot \sqrt{(1-\eta) \cdot (3+\eta)}} \tag{2}$$

$$Re_{ax} = \frac{u_{ax} \cdot 2d}{\nu} \tag{3}$$

$$Ta = \frac{\omega \cdot R_i \cdot d}{\nu} \cdot \sqrt{\frac{d}{R_i}} = Re_\phi \cdot \sqrt{\frac{d}{R_i}} \tag{4}$$

In the radius ratio η in Eq.(2) is used to characterize the geometry of the gap:

$$\eta = \frac{R_i}{R_i + d} \tag{5}$$

Mixing processes are grouped into macro-, meso- and micromixing according to the characteristic length scale of the inhomogeneities under consideration. The macroscopic mass exchange happens between the vortex cells along the gap, and the content of the vortex cell is assumed to be perfectly mixed. In reality, the inside of the vortex cell is not well mixed. Mesomixing is defined as the mixing processes which lead to homogenisation inside the vortex cells. With help of PIV (Particle Image Velocimetry) and LIF (Laser Induced Fluorescence) methods it is possible to investigate the local mixing conditions from the macroscopic scale down to mixing on the molecular level (micromixing).

In Fig. 2 a non-stationary mixing process is visualised. The flow structure intuitively approves the above concept of macro- and mesomixing.

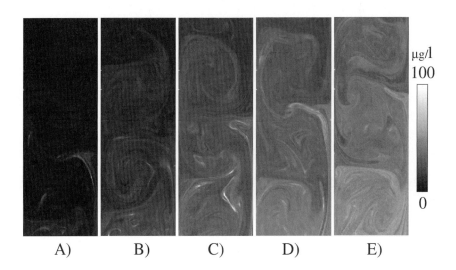

Fig. 2. Visualization: Concentration field from non-stationary LIF-experiments Time after the beginning of the tracer addition: A) 20s B) 60s C) 120s D) 180s E) 240s

In the literature the flow in a Taylor-Couette reactor is frequently described with help of a model of axial back mixing (Croockewit et al. (1955), Tam and Swinney (1987), Enokida et al. (1989), Pudjiono et al. (1992), Moore and Cooney (1995)). The flow is assumed well mixed vertical to the symmetry axis of the reactor. The mixing kinetics in axial direction is considered with help of the dispersion coefficient D_{ax}. The model equation is as follows:

$$\frac{\partial C}{\partial t} = D_{ax} \cdot \frac{\partial^2 C}{\partial z^2} - u_{ax} \cdot \frac{\partial C}{\partial z} \tag{6}$$

Here D_{ax} is the axial dispersion coefficient, u_{ax} is the velocity in axial direction z, and C is fluid concentration. This description does not fully correspond to a Taylor-Couette flow and its actual flow structure, which shows no continuity along the flow direction. So the model of the axial dispersion can not be applied directly. The vortex cells require a discretization of the model, while the flow is split in single ideally mixed areas with the size of a vortex cell, between which a mass exchange takes place. This is therefore a variation of the cell model with back mixing in which the number of cells N is fixed by the flow structure (Kafarov and Glebov (1991)). In the model the two parameters f and N appear. But only one is unknown – the ratio of back mixing f. In Fig. 3 the flow structure according to the cell model is shown.

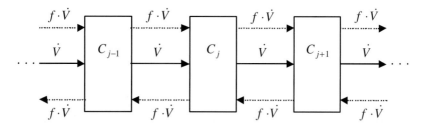

Fig. 3. Cell model with back mixing

According to this cell model the concentration equation in an arbitrarily cell j reads as follows:

$$\frac{\partial C_j}{\partial t} = \frac{1}{t_V} \cdot [(1+f) \cdot C_{j-1} + f \cdot C_{j+1} - (1+2f) \cdot C_j] \tag{7}$$

Here t_V stand for the volume-related residence time of the flow in the vortex cell, which can be calculated from the axial volume stream and the cell volume. By comparison of the Eq.(7) with the discretization from Eq.(8) the relation between the dispersion coefficient and the ration of back mixing is found to be:

$$D_{ax} = \frac{f \cdot d^2}{t_V} \tag{8}$$

The dispersion coefficient D_{ax} characterizes the macromixing between the vortex cells.

3 Experimental Results

3.1 Macromixing

3.1.1 Correlation between Dispersion Coefficient and Process Parameters
The distribution of the residence time in the reactor becomes wider and wider with increasing rotational speed. Accordingly, the axial dispersion coefficients increase with increasing rotational Reynolds number Re_ϕ.

Furthermore, macromixing depends on the flow regime: whereas, wavy flow was observed at all axial Reynolds numbers for the big inner cylinder ($\eta = 0.886$). It was only found at low axial Reynolds numbers for the small ($\eta = 0.63$) and the middle ($\eta = 0.758$) inner cylinders. In the wavy flow regime, the dispersion coefficient depends on the axial Reynolds number only:

$$\frac{D_{ax}}{\nu} = 2 \cdot 10^{-3} \cdot \eta^{-1.75} \cdot Re_{\phi}^{1.25} \tag{9}$$

In the turbulent regime the axial dispersion coefficient also depends on the axial Reynolds number:

$$\frac{D_{ax}}{\nu} = 2.4 \cdot 10^{-1} \cdot \eta^{-1} \cdot Re_{\phi}^{0.57} \cdot Re_{ax}^{0.25} \tag{10}$$

In the turbulent regime the dependence of the dispersion coefficient on the rotational Reynolds number is weaker than in the wavy regime, because there is convective mass transport between the vortex cells via wavy flow. In Fig. 4 the measured data and the above correlations are represented.

The boundary between the wavy and turbulent regimes can approximated with the following equation:

$$Re_{\phi}/Re_{\phi,crit}(Re_{ax}) = 20 + 0.24 \cdot Re_{ax} \tag{11}$$

In this figure the critical Reynolds number is shown as a function of the axial Reynolds number. In general, most working points lie above this boundary, and Eq.(10) is applicable. Below this boundary Eq.(11) should be applied. For spiral flow at $Re_{ax} > 20$ and $Re_{\phi}/Re_{\phi,crit} < 10$ the above equations are not valid. Here, even higher axial dispersion coefficients appear.

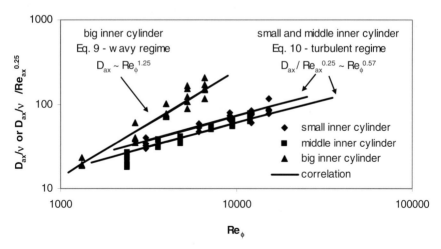

Fig. 4. Correlation between dispersion coefficients and rotational Reynolds number for both regimes and all the cylinder sizes ($0 < Re_{ax} < 120$)

3.1.2 Macromixing Time

In Taylor-Couette flow the macromixing time is the time, during which a flow element on average covers a distance of two gap widths in the axial direction. This movement occurs only on account of the axial dispersion, but does not depend on the axial flow. It can be calculated from the cross sectional area of vortex cells and the axial volume flow. It is assumed that a vortex has a square cross sectional area. Therefore, in the mean a fluid element will be moving the distance equivalent to a pair of vortices along the gap during the macromixing time. Therefore the axial macromixing time t_m can be defined as follows:

$$t_m = \frac{2 \cdot d^2}{D_{ax}} \qquad (12)$$

Experimentally determined macromixing times lie between 10^{-1} and 10^2s.

3.1.3 Influence of Viscosity Differences on Macromixing

In the following the case of mixing of a low viscosity liquid into a liquid of higher viscosity is considered. This case is relevant for continuous polymerisation reactions. If the rotational Reynolds number $Re_\phi/Re_{\phi,crit}$ is set between 1 und 13, Eq.(9) is appropriate to describe the macromixing, if in this equation the higher of the two viscosities is used (validated up to $\mu = 1 \cdot 10^{-2}$ Pa·s). Under these circumstances the dimensionless values of the axial dispersion coefficients are independent of the viscosity of the flow. As a consequence, at constant Reynolds numbers the axial dispersion coefficient increases with viscosity, and viscosity differences of the mixed fluids do not influence the macromixing. The above results hold as long as the added amount in the low viscosity liquid is little compared to the main flow. So the correlations Eq. (9) and (10) for isoviscous mixing can be used for the prediction of the non-isoviscous mixing.

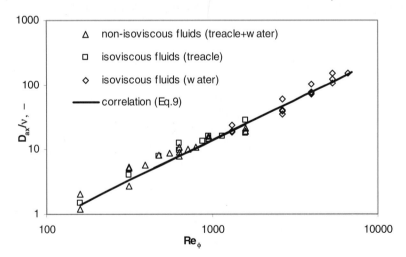

Fig. 5. Axial dispersion coefficients as a function of the Reynolds number for the isoviscous or non-isoviscous mixing in the wavy regime

3.2 Mixing Inside the Vortex Cell

3.2.1 Specific Contact Interface and Segregation Index

The specific contact interface between the segregated liquids and the segregation index are used to quantify the mixing quality within the vortex cells. The mass specific contact interface between the liquids to be mixed is a magnitude to appraise the mixing quality (Bothe et al. (2004), Schlüter et al. (2004)). The contact interface can be calculated according to the following equation:

$$\Phi = \frac{\Delta x \cdot \Delta y}{A} \cdot \sum_i \sum_j \left[\frac{\left((f(x_{i+1}y_j) - f(x_{i-1}y_j)\right)^2}{4 \cdot \Delta x^2} + \frac{\left((f(x_i y_{j+1}) - f(x_i y_{j-1})\right)^2}{4 \cdot \Delta y^2} \right]^{\frac{1}{2}} \tag{13}$$

Here Φ is potential for diffusive mixing, A is the contact interface, and f is dimensionless concentration which is always related to the maximum concentration. The course of this function along the cylinder axis characterizes the progress of mixing from one vortex cell to the next. Experimentally, the contact interface can be estimated from concentration fields measured with LIF.

At the first instance, when the liquids come into contact, their contact interface is small. During the first phase of the mixing process the contact interface quickly increases due to interference and distribution of the liquids into each other. First, the stream in the added solution has a small diameter, and it distributes quickly in the surrounding liquid. This happens in the vortex cell, into which the addition takes place, and the production of contact interface takes place in this first vortex cell. In all other vortex cells the contact interface decomposes until a limiting value, which corresponds to the completion of mixing. In theory this limiting value for the contact interface should be 0. Because of the imprecision of the measurement this limiting value was not found to be 0, and circa 2 m²/m³.

The general trend which follows from Fig. 6 adverts to an increasing decomposition of the contact interface with increasing Reynolds number.

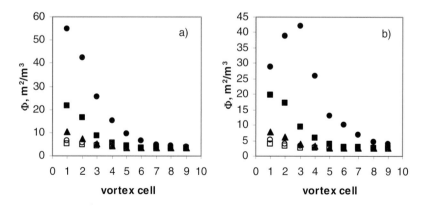

Fig. 6. Specific contact interface with $\eta = 0.886$ a) $Re_{ax} = 14.5$ b) $Re_{ax} = 37.7$ (caption see the Table 1)

Table 1. Caption of the Fig. 6

Re$_\phi$ with	$\eta = 0.886$
●	1322.1
■	2644.3
▲	3966.4
○	5288.6
□	6610.7

From the Fig. 6 b it can be seen that with the lower Reynolds number of 1322.1 die specific contact interface in the No. 9 vortex cell is the same as the in the No. 1 vortex cell with the higher Reynolds number of 6610.7. Furthermore, in Fig. 6 b it can be seen, for the two lowest Reynolds numbers how the contact interface first appears and then disappears. In the first two vortex cells it increases, before in the other vortex cells it decreases.

The segregation index SG is defined to be the variance of the local concentrations:

$$\sigma^2 = \overline{(f')^2} = \overline{(f - \bar{f})^2} \tag{14}$$

The one of the liquids to be mixed has the dimensionless concentration f = 0, and the other has the concentration f = 1. The variance is not an absolute magnitude for the perfectness of mixing, because it depends on the average concentration. To define a general magnitude which allows to compare with different mixing processes the variance is related to the variance at zero mixing:

$$\sigma_0^2 = \bar{f}(1 - \bar{f}) \tag{15}$$

This relation is valid only for the case that all concentrations are related to the maximum concentration. With this, the segregation index can be defined. It is an absolute criterion for the perfectness of mixing.

$$SG = \sqrt{\frac{\sigma^2}{\sigma_0^2}} = \sqrt{\frac{\overline{(f - \bar{f})^2}}{\bar{f}(1 - \bar{f})}} \tag{16}$$

From LIF-measurements in each vortex cell the variance of the local concentrations is calculated with Eq.(14). From this the segregation index can be calculated with help of the Eq.(15) and (16).

The results are shown in Fig. 7. The segregation index has its highest value in the first vortex cell and decreases downstream. There always exists a lower limiting value, which corresponds to complete mixing.

As can be seen, specific contact interface and segregation index follow the same trend. Although their physical meaning is different.

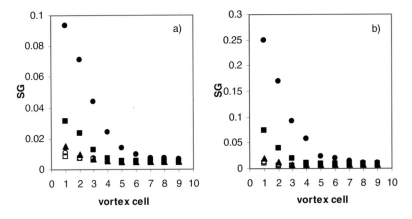

Fig. 7. Segregation index with $\eta = 0.886$ a) $Re_{ax} = 14.5$ b) $Re_{ax} = 37.7$ (caption see the Table 1)

At identical rotational Reynolds numbers the segregation index is dependent on the axial Reynolds number and the specific contact interface is not. This means that besides the rotational Reynolds number the residence time plays a major role. The higher the axial Reynolds number, the shorter is the residence time, which a fluid element spends in a vortex cell, and the time, which is available for the decomposition of the differences of concentrations, is smaller. Therefore, the resultant segregation index in the same vortex cell is higher. From that it follows, that the dissipation rate ε is dependent on the rotational speed only, and that the axial Reynolds number affects the progress of this dissipation in every vortex cell within the respective residence time.

3.2.2 Dissipation Rate and Mesomixing Time

The dissipation of the variance of concentrations in turbulent flows is described with the follows equation:

$$\frac{d\sigma^2}{dt} = -\frac{\sigma^2}{t_d} \tag{17}$$

The differences of the concentration exponentially decrease with the time at a characteristic time constant. This time constant corresponds to the quantity of the dissipated energy in the flow. This time constant is the mesomixing time t_d:

$$\sigma^2(t) = \sigma_0^2 \cdot e^{-(t/t_d)} \tag{18}$$

σ_0^2 is the initial variance of concentrations at the time $t = 0$. Eq.(18) describes the decomposition of the variance of concentrations in a concluded liquid volume. It is assumed that the variance of concentrations is a conservative magnitude which can be balanced (Rosensweig (1964), Baldyga et al. (2001)). Therefore, variance of concentration is convected into and out of volume elements by the flow. The balance for the variance of concentrations in the liquid volume can be written as follows:

$$V \cdot \frac{d\sigma^2}{dt} = \dot{V}_{in} \cdot \sigma_{in}^2 - \dot{V}_{out} \cdot \sigma^2 \tag{19}$$

Under the condition that $\dot{V}_{in} = \dot{V}_{out} = \dot{V}$ Eq.(17) can be solved to

$$\dot{V} \cdot \sigma_{in}^2 - \dot{V} \cdot \sigma^2 - V \cdot \frac{\sigma^2}{t_d} = 0 , \tag{20}$$

or

$$\sigma^2 = \frac{\sigma_{in}^2}{\left(1 + \dfrac{t_V}{t_d}\right)} . \tag{21}$$

Assuming that the maximum possible variance of concentrations is constant at any point of the reactor, the segregation index can be formulated as follows:

$$\sqrt{\frac{\sigma^2}{\sigma_0^2}} = \sqrt{\frac{\dfrac{\sigma_{in}^2}{\sigma_0^2}}{1 + \dfrac{t_V}{t_d}}} \quad \text{or } SG = \frac{SG_{in}}{\sqrt{1 + \dfrac{t_V}{t_d}}} \tag{22}$$

Eq. (22) can be used for the determination of the mesomixing time t_d from experimental data. If a vortex cell is set equal to the volume V,

$$V_{cell} = \pi \cdot (R_o^2 - R_i^2) \cdot (R_o - R_i) \tag{23}$$

The segregation index SG_{in} is the one from the neighbouring vortex cell in upstream direction. The vortex cell, into which the feed liquid is added, is set to be the first vortex cell. The segregation index SG_{in} of this cell is 1. In this vortex cell the liquid is partly mixed and the segregation index at the outlet is less than 1. With help of Eq.(22) and the assumption $SG_{in} = 1$ the mesomixing time t_d can be calculated. In Table 2 the values of the residence times in the vortex cell are summarised.

Table 2. Residence times in the vortex cell

Geometry:	$\eta = 0.63$	$\eta = 0.758$	$\eta = 0.886$
Volume of the vortex cell V_{cell} [L]	0.08763	0.04043	0.009625
Total volume flow $\dot{V}$ [L·h^{-1}]	Residence time in the vortex cell t_V [s]		
8.23	38.3	17.7	4.2
20.12	15.7	7.2	1.7
31.82	9.9		

From these data the following two correlations for the dimensionless mesomixing times θ_d are deduced:

$$\theta_d = \frac{t_d}{t_i} = 2.5 \cdot 10^5 \cdot \frac{1-\eta}{\eta^{1.6}} \cdot Re_\phi^{-1.48} \text{ if } Re_\phi > 20 \cdot Re_{\phi,crit} \tag{24}$$

$$\theta_d = 1.4 \cdot 10^4 \cdot \frac{1-\eta}{\eta^{1.6}} \cdot Re_\phi^{-0.8} \text{ if } Re_\phi < 20 \cdot Re_{\phi,crit} \tag{25}$$

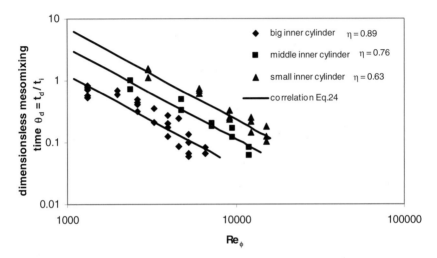

Fig. 8. Experimental data of the mesomixing time and the correlation according to Eq.(24)

The correlations tell that the mesomixing time is dependent on the size of the inner cylinder, and it applies for all the three studied geometries. Time interval $t_i = 60/n$ is the time required for one revolution of the inner cylinder. The mesomixing time decreases with increasing rotational speed, and is not dependent on the axial Reynolds number.

Eq.(25) represents experimental data with an average deviation of 19.2%. It only applies for a radius ratio of $\eta = 0.63$. No further variation of the geometry was carried out.

3.2.3 Mesomixing Times for the Non-isoviscous Mixing

From LIF-measurements, see Fig. 10, it can be seen, that at identical distances from the feed inlet and at low rotational Reynolds numbers the characteristic size of inhomogeneities is smaller for isoviscous mixing than for non-isoviscous mixing because the evolution of the contact interface at the beginning of the mixing process is slower in the isoviscous case than in the non-isoviscous case. In the non-isoviscous case the incorporation of the feed fluid elements into the main flow and their deformation is faster than in the isoviscous case.

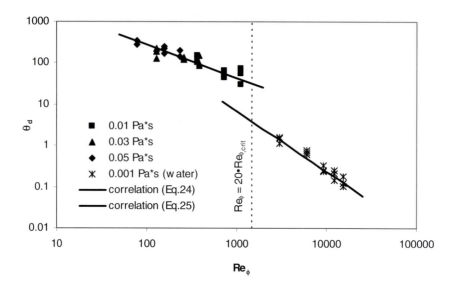

Fig. 9. Mesomixing times for the isoviscous mixing at η = 0.63

At high rotational Reynolds numbers the specific contact interface is smaller for isoviscous mixing than for non-isoviscous mixing, because the decomposition of the contact interface by molecular diffusion is faster in the isoviscous case than in the non-isoviscous case.

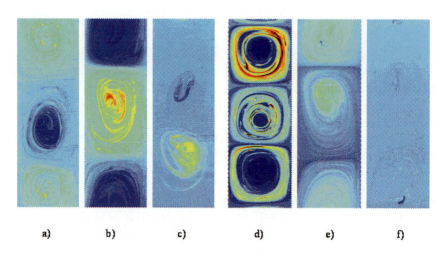

Fig. 10. Concentration fields η = 0.63, $\mu_1 = 30 \cdot 10^{-3}$ Pa·s $\mu_2 = 1 \cdot 10^{-3}$ Pa·s (non-isoviscous) $Re_{ax} = 2.2$: a) $Re_\phi = 141.6$ b) $Re_\phi = 283.1$ c) $Re_\phi = 424.7$; $\mu_2 = 30 \cdot 10^{-3}$ Pa·s (isoviscous) $Re_{ax} = 2.0$: d) $Re_\phi = 128.5$ d) $Re_\phi = 257.1$ f) $Re_\phi = 385.6$

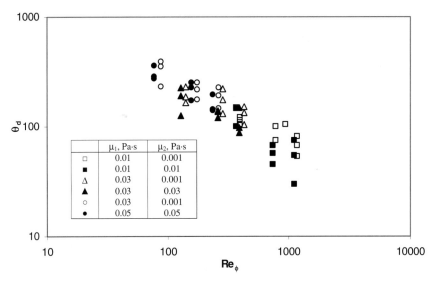

Fig. 11. Mesomixing times for the iso- (black symbols) and non-isoviscous (white symbols) mixing with $\eta = 0.63$ and the different viscosities of the flow

In Fig.11 a comparison between the values of dimensionless mesomixing times for the isoviscous mixing (black symbols) and the non-isoviscous mixing case (white symbols) is given for $\eta = 0.63$. As already shown in Fig. 9 the mesomixing times decrease with increasing rotational Reynolds number. Furthermore, it can be seen that mesomixing times in the non-isoviscous mixing case are higher than in the isoviscous mixing case.

3.3 Micromixing

The characteristic time scale for micromixing on the molecular level t_μ is the Kolmogorov time scale (Baldyga and Bourne (1999)):

$$t_\mu = \sqrt{\frac{\nu}{\varepsilon}} \qquad (26)$$

Micromixing times are local characteristics of the flow. Depending on the location, they can take values, which are orders of magnitude different from each other and from the overall average value. According to Eq. (26) local values of the energy dissipation rate ε_{loc} need to be measured for the calculation of local micromixing time t_μ. These measurements were done by PIV.

The mean dissipation rate of the kinetic energy $\bar{\varepsilon}$ can be calculated as follows:

$$\bar{\varepsilon} = \frac{G \cdot v^2 \cdot \omega}{\pi \cdot d \cdot (R_o + R_i)} \qquad (27)$$

Here G is dimensionless rotary moment of cylinders. The following two correlations for the prediction of the rotary moment were found from torque measurements with liquids of the kinematic viscosity from $1 \cdot 10^{-6}$ m$^2 \cdot$s^{-1} (e.g. water) to 5.3810^{-6} m$^2 \cdot$s^{-1} (e.g. 50 ma.% glycerol mixing with water) at 20°C:

$$G = 2.13 \cdot \frac{\eta^{3/2}}{(1-\eta)^{7/4}} \cdot Re_\phi^{1.445} \text{ if } 800 < Re_\phi < 10^4 \tag{28}$$

$$G = 0.113 \cdot \frac{\eta^{3/2}}{(1-\eta)^{7/4}} \cdot Re_\phi^{1.764} \text{ if } 10^4 < Re_\phi < 3.4 \cdot 10^4 \tag{29}$$

In Fig. 12 the distribution of non-dimensionalized micromixing times is shown in a comprehensive way. The data indicate the volume fraction of the flow, in which the micromixing times take certain values (referred to the volume average). The theoretical line for the homogeneous distribution of the micromixing times throughout the fluid is shown for comparison. The distribution function approaches this homogeneous situation with increasing rotational Reynolds numbers.

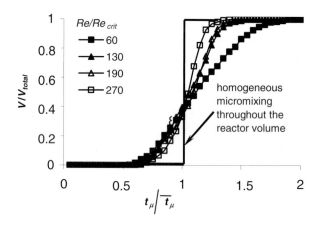

Fig. 12. Cumulative distribution of micromixing time (referred to the average) $\eta = 0.63$

4 Conclusions

The characteristic mixing times in Taylor-Couette flow have been determined for three length scales corresponding to the macro-, meso- and micromixing. The macromixing times take values of 2 – 60s. The mesomixing times lie between 0.02 and 2s, and the micromixing times take values from 50 down to 5 ms. When comparing all mixing times, macromixing proved to be the limiting stage for the whole mixing process in Taylor-Couette reactor. The local rate of the micromixing is however lower towards the center of the vortex cell. In order to predict accurately the mixing behavior of TCR for a wide range of Reynolds number, all three mixing stages should be taken under consideration.

References

Kataoka, K.: Taylor vortices and instabilities in circular Couette flow. In: Encyclopedia of Fluid Mechanics, vol. 1, pp. 236–274. Gulf Publishing Company, Houston (1986)

Croockewit, P., Honig, C.C., Kramers, H.: Longitudinal diffusion in liquid flow through an annulus between a stationary outer cylinder and a rotating inner cylinder. Chem. Eng. Sci. 4(3), 111–118 (1955)

Tam, W.Y., Swinney, H.L.: Mass transport in turbulent Couette-Taylor flow. Phys. Rev. A36, 1374–1381 (1987)

Enokida, Y., Nakata, K., Suzuki, A.: Axial turbulent diffusion in fluid between rotating co-axial cylinders. AIChE Journ. 35(7), 1211–1214 (1989)

Pudjiono, P.I., Tavare, N.S., Garside, J., et al.: Residence time distribution from a continuous Couette flow device. Chem. Eng. Journ. 48, 101–110 (1992)

Moore, C., Cooney, C.L.: Axial Dispersion in Taylor-Couette Flow (R&D Note). AIChE Journ. 41(3), 723–727 (1995)

Kafarov, V.V., Glebov, M.B.: Mathematische Modellierung der Hauptprozesse in chemischer Verfahrenstechnik. Vysschaya Schola, Moscow (1991)

Bothe, D., Stemich, C., Warnecke, H.J.: Theoretische und Experimentelle Untersuchungen der Mischvorgänge in T-förmigen Mikroreaktoren – Teil1: Numerische Simulation und Beurteilung des Strömungsmischens. Chem. Ing. Techn. 79(10), 1480–1484 (2004)

Schlüter, M., Hoffmann, M., Räbiger, N.: Theoretische und experimentelle Untersuchungen der Mischvorgänge in T-förmigen Mikroreaktoren – Teil2: Experimentelle Untersuchung des Strömungsmischens. Chem. Ing. Techn. 76(11), 1682–1688 (2004)

Rosensweig, R.E.: Idealized theory for turbulent mixing in vessels. AIChE Journ. 10(1), 91–97 (1964)

Baldyga, J., Henczka, M., Makowski, L.: Effects of mixing on parallel chemical reactions in a continuous-flow stirred-tank reactor. IChemE A 79, 895–900 (2001)

Baldyga, J., Bourne, J.R.: Turbulent Mixing and Chemical Reactions. John Wiley & Sons, Chichester (1999)

Part 3: Theoretical Methods for Modelling and Numerical Calculations of Mixing Processes

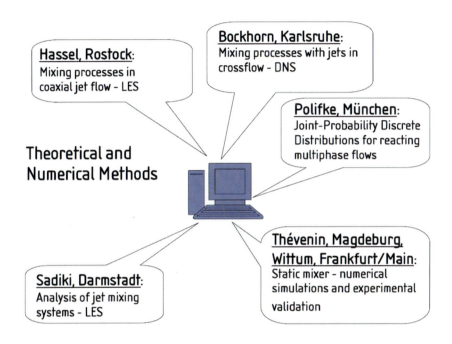

The application of Direct Numerical Simulation (DNS) methods allowed the identification of regions with good mixing by tracing streamlines of the averaged flow field. It was found that regions of good mixing correlate with helical movement of fluid particles. Regions where the streamlines diverge from each other also exhibit good local mixing. The results from DNS allowed also quantifying and comparing mixing hypotheses in the Reynolds-averaged context. Two new subgrid scale models denoted WALES (Wavelet-Adapted-LES) are presented. Wavelets are able to represent multi-scale physical phenomena like turbulence and turbulent mixing in a natural way.

Direct Numerical Simulation, Analysis and Modelling of Mixing Processes in a Round Jet in Crossflow

Jordan A. Denev[1], Jochen Fröhlich[2], Carlos J. Falconi[1], and Henning Bockhorn[1]

[1] Institute for Technical Chemistry and Polymer Chemistry, University of Karlsruhe (TH), Kaiserstrasse 12, D-76128 Karlsruhe, Germany
[2] Institute of Fluid Mechanics, Technical University of Dresden, George-Bähr-Strasse 3c, D-01062 Dresden, Germany

Abstract. Direct Numerical Simulations (DNS) for the flow with transport of passive scalars at different Schmidt numbers and chemical reactions at different Damköhler numbers has been carried out. As a result, a comprehensive database for studying mixing phenomena and chemical reactions in a jet in crossflow has been generated. Results obtained concerning instantaneous mixing structures and laminar to turbulent flow transition has been compared to companion experimental data showing a good agreement. The database obtained was used to perform various analyses of the mixing including a priori testing of mixing models (hypotheses) used in the Reynolds-averaged approach. By means of Lagrangian methods flow regions of intensive mixing have been identified. For the present configuration they have been found to be attributed to helical movement of fluid particles. Also regions, where the streamlines of the averaged flow diverge from each other downstream exhibit good local mixing. In the last part of the work bi-orthogonal wavelets are employed to construct novel, multiscale models for Large Eddy Simulations (LES). These are validated and tested with DNS of turbulent channel flow as well as with the present DNS database for the jet in crossflow. Unlike the Smagorinsky model, the new wavelet models have been found to produce no eddy-viscosity in the laminar inflow region of the round pipe from which the jet originates.

1 Introduction

Direct Numerical Simulation (DNS) accounting for scalar transport and chemical reactions is a method allowing to gain comprehensive and detailed multiscale information about mixing phenomena. It is applied in the present paper to the configuration of a round jet in a crossflow which is a typical situation encountered in industrial applications as it is known to possess favourable mixing capabilities. The geometrical complexity of the setup and the flow requires a fairly complex grid. This rules out the application of a spectral method and makes Finite Volume methods appropriate (Muldoon (2004), Muppidi and Mahesh (2005)). While the focus in these publications is mainly on issues of the flow, the present project focused on mixing phenomena - with and without chemical reactions.

The strategy of the project was the following. First, a comprehensive set of DNS and LES data was generated (Fröhlich et al.(2004), Denev et al.(2007b), Denev et al.

(2008b), Denev et al.(2009a)) conducting parameter variations in order to obtain detailed information about the flow, the mixing phenomena and the impact of these on chemical reactions. The second step was the detailed analysis of these large data sets (Denev et al.(2008a), Denev et al.(2009b)). This activity is still being pursued. The third step was to apply the mathematical apparatus provided by the wavelet framework for analysis and most of all for subgrid-scale modeling.

Right from the outset the present project was conceived to be run in parallel with an experimental project, directed by R. Suntz at the same institute Cárdenas et al.(2009), Cárdenas et al.(2007)) . This turned out to be very fruitful as the continuous discussion between experimentalists and modelers helped to solve many problems and to generate new ideas. It allowed coordination of the investigated parameters together with a combined numerical-experimental look into the phenomena of turbulent mixing, published in Cárdenas et al. (2007) . As a result of these efforts the present database for studying turbulent mixing phenomena (with and without chemical reactions) was designed as described in chapter 2. Very recent results from the various analyses of the DNS data are presented in chapter 3. Earlier results concerned with the impact of swirl on mixing and its quantification by mixing indices as well as results on chemical reactions are reported elsewhere, (Denev et al.(2009a)), Denev et al.(2005a), Denev et al.(2005b), Denev et al.(2007a)). Finally, results and analysis of data from the present DNS served as a basis for the development of two new SGS models employing orthonormal wavelet decomposition. The reason behind developing these models is to connect the present research with more practical methods like LES. Presentation of the new ideas and models for application of wavelets to the complex processes of turbulent mixing, together with first results is done in the last part, chapter 4. Application of the two wavelet-based SGS models developed in the present project to other flows, but without mixing, is presented in Denev et al.(2009b).

2   A Database for Mixing and Chemical Reactions in a Jet in Crossflow Obtained by DNS

For the purpose of analyzing turbulent mixing phenomena a large database for the jet in crossflow with transport of non-reacting and reacting scalars was created within the present collaborative program of the German research association. It consists of the present set of DNS results supported by experiments from a companion experimental project described in Cárdenas et al.(2009). In the following a detailed description of the features of this database is presented together with issues concerning comparisons with the experiments.

2.1   Characteristics of the Present DNS Database for Mixing and Chemical Reactions

The present DNS database for mixing and chemical reactions was configured based on the following desiderata:

- *Variation of a wide range of parameters.* The parameters varied comprise features of the fluid flow, of the mixing of passive scalars and of the chemical reactions between the species.
- *Compatibility with experiments conducted within the present priority program.* This required taking into account the existent experimental equipment as well as its limitations, see Cárdenas et al.(2009).
- *Simple, unambiguous inflow boundary conditions.* There exists a variety of numerical methods for generating turbulent inflow boundary conditions. However, quite often their performance depends on particular implementation details. Consequently such methods are difficult for reproduction by other groups.

After analyzing the above points, the present mixing database was finally designed to offer several advantages:

- *A large set of parameters varied with focus on combustion applications.* The present setup includes the solution of *9* passive scalar equations (*3* non-reacting and *6* reacting) with variation of Reynolds, Schmidt and Damköhler numbers, well suitable for testing of combustion models. The values of the dimensionless numbers are presented in Table 1. Results from the one-step chemical reactions were presented in Denev et al.(2005a), Denev et al.(2006), Denev et al.(2007a)) and will not be reported here. To optimize CPU-time usage, all *9* passive scalars, listed in Table 1 were computed simultaneously for each Reynolds number. Results from computations with a lower Reynolds number than the one reported in Table 1 were presented earlier in Denev et al.(2006).
- *Support by experimental data.* Coordination of boundary conditions and parameters between the present numerical solution and experiments was carried out to the largest possible extent. A more detailed discussion is given below. While the DNS results cover simultaneously the whole computational domain, the experimentally measured horizontal planes (*z=const*) and vertical planes (*y=const*) are presented in Cárdenas et al.(2009).
- *Presence of a transition phenomenon.* Laminar to turbulent flow transition occurs inside the computational domain due to the Kelvin-Helmholtz instability. Therefore the present database in its fluid flow part becomes suitable for testing of low-Reynolds number turbulence models.
- *Repeatability of the results: easy-to-use set of laminar inflow boundary conditions.* For the boundary layer of the crossflow a simple algebraic equation was formulated (see equation (2) below). For the sake of reduction of additional parameters, and due to the lack of detailed experimental measurements required for this purpose, the same boundary layer shape was used independent of the Reynolds number. For the jet inflow a fully developed laminar parabolic pipe profile was prescribed (equation (1), see also Denev et al.(2006)).
- *Fine grid resolution.* The same fine numerical grid used for *Re=650* was used for the lower Reynolds number as well. A grid independence study with coarser grids showed results which are very close to the finest grid used in the present computations thus proving grid independency of the solution.

Table 1. Parameters varied in the DNS and boundary conditions for the scalars. One-step chemical reaction with species A and B is defined as: $A_j + B_j \rightarrow P_j$ with 3 cases: $j=1,2,3$. The reaction rate for each case is defined as: $R_j = Da_j \cdot c_{Ai} \cdot c_{Bi}$ with c_i being the mass fraction of the $i\text{-}th$ scalar.

Scalar equation i	Reynolds number Re	Schmidt number Sc	Damköhler number Da	Mass fraction equal to $c_i=1$ in:	Reaction case j
1	325 & 650	1.0	-	jet	-
2	325 & 650	0.5	-	jet	-
3	325 & 650	2.0	-	jet	-
4	325 & 650	1.0	1.0	jet	1
5	325 & 650	1.0	1.0	crossflow	1
6	325 & 650	1.0	0.5	jet	2
7	325 & 650	1.0	0.5	crossflow	2
8	325 & 650	2.0	1.0	jet	3
9	325 & 650	1.0	1.0	crossflow	3

2.2 Geometry, Parameters and Boundary Conditions

In Fig. 1 the flow configuration and the Cartesian coordinate system are presented. The origin of the coordinate system coincides with the center of the jet and the plane of the wall. The reference length is the pipe diameter, so in non-dimensional units $D=1$; for real dimensions used in the experiments see Cárdenas et al.(2009). Lengths, given in the Figure are dimensionless and their values are $L_x=20$, $L_y=L_z=13.5$, $l_x=3$ and $l_z=2$. The length of the pipe, l_z, is large enough to impose fully developed pipe flow at $z = -l_z$ (for details see Denev et al.(2006)). The crossflow velocity U_∞ in the middle of the channel (see the Figure) serves as reference velocity and the Reynolds number from Table 1 is defined as $Re= U_\infty D/\nu$. When using the hydraulic diameter and the bulk velocity of the crossflow channel in the experiments Cárdenas et al.(2009), the Reynolds numbers in Table 1

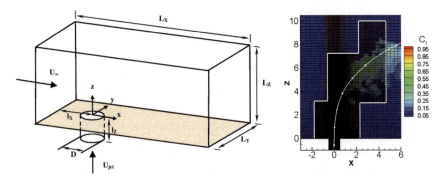

Fig. 1. Geometry of the computational domain and the origin of the coordinate system (left). Locally refined grid near the jet exit (right).

translate to *4122* and *8243*, respectively. The velocity ratio in all simulations is $R=U_{jet,\ bulk}/U_\infty=3.3$ which ensures that the jet trajectory remains remote from the walls.

Boundary conditions for the scalar variables at the inflow (either jet or cross-flow) are presented in Table 1. Since the density is constant and normalized to *1*, the concentration variables c_i represent simultaneously mass fraction and volume concentration. Dalton's law is fulfilled for each scalar/reaction separately so that finally two computations (one for each Reynolds number) were carried out. Neumann boundary conditions were applied for all scalars at the walls. In the present work, only results with the first scalar will be presented and discussed.

The inflow condition for the pipe flow at $z = -l_z$ is:

$$w(r)=2U_{jet,\ bulk}\left(1-\left(\frac{r}{D/2}\right)^2\right)$$ (1)

where r is the coordinate along the radius. The boundary layer at the channel inflow (left hand side of Fig. 1) is given as a function of the distance from the closest channel wall d_n according to:

$$u_{in}(y,z)=1.0-exp(-4.5\ d_n)$$ (2)

This equation which results in a boundary layer thickness of $\delta_{99}=1.03D$ was applied for both Reynolds numbers computed. No-slip boundary conditions were used at all solid walls and convective outflow conditions for all quantities at the exit of the domain.

Despite all efforts to synchronize the present DNS with the experiments, it was still not possible to achieve a perfect match for all parameters within the constraints of the present collaborative project. One of the challenges is constituted by the shape of the boundary layer in the crossflow channel presented, here modeled according to equation (2). The DNS solution was initiated prior to the experimental measurements, so that the shape of the boundary layer at the middle of the walls was only deduced from preliminary data at a lower Reynolds number available at this time Denev et al.(2006). These measurements had been carried out only at one wall of the channel, and data for the flow in the corners of the square channel were not available at all, so that finally the profile according to equation (2) was prescribed in the simulations. Detailed measurements of these profiles close to the walls and in the corners of the channel at a later stage were not possible for diverse reasons. A discussion of the role of the boundary layer on the jet trajectory is provided in Muppidi and Mahesh (2005), where it is argued that deviations up to 0.8 D in the vertical position of the jet can appear for different boundary layer thicknesses.

In the experiments the bulk velocity of the crossflow channel was controlled. In the numerical simulation the velocity in the middle of the channel was prescribed. Thus, in the simulation the flowrate results from the prescribed inflow profile. Table 2 shows the values for the bulk velocity and the maximum velocity in the middle of the channel for both investigations.

Table 2. Conditions in the crossflow channel for the experiments and the simulations

	Re	U.	U_{bulk}	U_∞ / U_{bulk}
Exp	325	0,607 [m/s]	0,562 [m/s]	1.081
Exp	650	1,195 [m/s]	1,120 [m/s]	1.067
DNS	325 & 650	1.0 [-]	0,939 [-]	1.064

2.3 Numerical Method

The Finite Volume Code LESOCC2 (Hinterberger (2004)) was used for the in-compressible flow simulations. A predictor-corrector scheme in time employing a Poisson equation for the pressure-correction and a second-order accurate Runge-Kutta scheme for the predictor were used. The spatial discretization of all terms was accomplished by second-order central schemes except for the convection term of the species equation where the bounded HLPA scheme Zhu (1991) was used to maintain the physically correct interval *[0;1]* for the species concentration. The collocated, block-structured numerical grid is composed of *219* numerical blocks and uses curvilinear coordinates and hexahedral cells. Advantage was taken from local grid refinement: the numerical blocks near the jet-exit and those following the jet trajectory near the exit were refined with a factor of *3* in all three directions. Figure 1 (right), presenting a *2D* cut at *y=0*, gives an idea of this refinement technique. This way a considerable amount of grid nodes could be saved on the present block-structured grid containing for the DNS 22.3 million points. The accuracy of this tangential refinement technique was studied in detail and is available in Fröhlich et al.(2007).

After an initial start-up period with a duration of *60* dimensionless time units $t=D/U_\infty$, the higher Reynolds number simulation was averaged over a period of *t=258.9* and the lower one over *t=227.7*. As a result of the various scalar equations calculated and the related turbulent fluxes, *70* quantities (most of them averages in time) were obtained during each simulation.

2.4 Sensitivity of the Jet Trajectory with Respect to the Velocity Ratio

The DNS solution for the velocity ratio *R=3.3* and the corresponding experiments (denoted EXP1) are compared in Fig. 2, both at *Re=650*. The computed trajectory exhibits a somewhat larger distance from the bottom wall than in the experiment, with a maximum deviation of about *0.8D*. To understand the possible reasons for this difference, the sensitivity of the jet trajectory with respect to the velocity ratio *R* (observed first experimentally, see also Cárdenas et al.(2009)) was studied. For this reason in both experiments and simulations the velocity ratio (by increasing or decreasing the speed of the jet) was varied within the range of *10%*. Hence, two additional ratios (R = 2.97 and R = 3.63) were considered. Taking respect to the large resources, required for DNS, the sensitivity of the numerical results was tested by LES, with the dynamic Smagorinsky model and a numerical grid of 3.6 million control volumes. This was encouraged by the good quality of the LES results: as it can be seen on the Figure, LES for *R=3.3* (denoted LES1 in the Figure) coincides well with the DNS results. The LES results are also presented in Fig. 2.

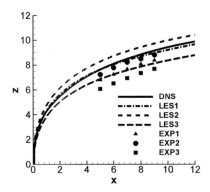

Fig. 2. Sensitivity of streamlines with respect to the velocity ratio R, case Re=650 (Experimental results courtesy Cárdenas et al.(2009)). DNS, LES1, EXP1: R=3.3; LES2, EXP2: R=3.63; LES3, EXP3=2.97).

For the experiments the location of the maximum of the streamwise velocity component *<u>* is shown in the Figure, while for the simulations the streamline originating from point *(0;0;0)* has been presented for convenience. It was checked and confirmed that both definitions yield virtually the same trajectory in the present case and range of *x*-positions of interest (5 < *x/D* < 9). The response to *10%* change in the velocity ratio is quite similar for the experiments and the simulation, being a little bit larger for the simulations. The results show that the trajectory is very sensitive to the velocity ratio, so that small uncertainties in the experimental conditions may cause sizable differences in the position of the trajectory. Further discussion on matching trajectories between numerical simulations and experiments are given in Yuan et al.(1999), who found reasons to compare simulations with ratio *R=3.3* with experiments with ratio *R=4.0*.

3   Analysis of Mixing Based on Present DNS Results

3.1   Laminar-Turbulent Transition, Scalar Fields

Transition characteristics of the current jet in crossflow were studied and are reported in (Denev et al.(2007a), Denev et al.(2008a)). It has to be stressed that, first, this transition is a 3D phenomenon and that secondly the location along the jet-trajectory in the simulations varies in time within a distance of ±0.6D. Animations of the companion experimental data show as well that the first coherent structures vary up to ±0.75D.

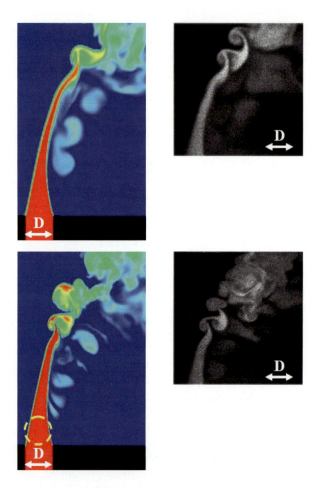

Fig. 3. Instantaneous scalar distributions: comparison of DNS (left) with experiment (right). Top: Re=325, bottom: Re=650. (Experimental photographs courtesy Cárdenas et al.(2009)).

Comparison of instantaneous pictures for the passive scalar ($i=1$ in Table 1) between simulations and experiments is presented in Fig. 3 and shows very good agreement; similarity of flow structures is also obvious. For the experimental results the camera was focused on the region of transition which explains the narrower part of the flow domain in the right-hand side of the Figure. Otherwise experimental and numerical pictures from Fig. 3 are synchronized by their scale, their aspect ratio and their position so that the reader can directly compare the results.

When the averaged scalar value is used for assessing the transition, DNS-values show sudden enlargement of the jet just behind the transition point Denev et al.(2007a). However, as reported in Denev et al.(2008a), this enlargement is not so abrupt in the case of LES which consequently raises uncertainties in its proper location. Therefore in Denev et al.(2008a) instead of the passive scalar, the turbulent kinetic energy is suggested as a suitable transition criterion as presented in subsection 4.4 below.

3.2   Assessment of the Mixing Process by Means of Streamlines

In this section information about the quality of the local mixing intensity is collected based on a Lagrangean technique carried out in the post-processing stage. For the purposes of the present analysis, *3D* streamlines of the time-averaged velocity field were followed for the case with *Re=650*. The target was to locate the origin of the zones of high concentration which appear in the downstream regions of the jet. This allows to gain further understanding of the mixing phenomena and how they can be enhanced.

The Lagrangean methodology used is as follows. Streamlines are traced from various points at the end of the pipe (which has a coordinate *z=0*) downstream and traced in vertical planes with *x=const*. By this process a geometrical mapping between the jet outlet plane and these vertical planes is established. Special attention is given to the mapping of the high-concentration zones in these vertical planes as they indicate potential for improvement of the mixing.

An analysis using this approach is shown in Fig. 4a for eight points at *z=0* distributed equally on the perimeter of a circle with radius *0.45D*; the central point (No *9* in the Figure) is also considered. The distribution of these points on the first downstream plane studied (*x/D=0.725*) can be seen in Fig. 4b. Points *1, 9* and *5* originally located on the line (*y=0, z=0*) move close together, remaining on the symmetry line *y=0*. Points *3* and *4* (also *6* and *7*) approach and almost completely overlap. These points mark two ends of the high-concentration area at *x/D=0.725* which has a horseshoe shape. The streamline starting from point *2* (correspondingly, point *8*) goes also backwards, i.e. against the mainstream of the crossflow. Therefore this streamline crosses three times the plane *x/D=0.725* (in points 2a, 2b and 2c in Fig. 4b); the backward movement (i.e. the streamline direction being against the crossflow stream) appears exactly in point *2b* (respectively in point *8b*). Note that by this triple crossing the streamline transports fluid with higher concentration from the peripheral part of the jet towards its middle as seen in Fig. 4b. This reveals one mechanism for the good mixing observed generally in the jet in crossflow configuration.

A complex helical movement of the above streamlines in planes located downstream is another mechanism responsible the good mixing of the jet in crossflow. This helical movement is well recognized on a *3D* presentation of the streamlines (not shown). Here it can be deduced from the relative height of point *2* compared to points *3* and *4* in Figs. 4c, 4e and 4f. In these Figures point *2* first starts from the highest location with respect to *z* at *x/D=1.45* and goes towards the lowest one at *x/D=2.90* and again moves higher at *x/D=4.35*. Such intense change in the relative position and distance between the points indicates increased local mixing. As a result of the helical movement the high-concentrations in the side arms of the above mentioned horseshoe scalar structure subsequently vanish. By this process the highest concentration at *x/D=4.35* remains close to the upper part of the jet which

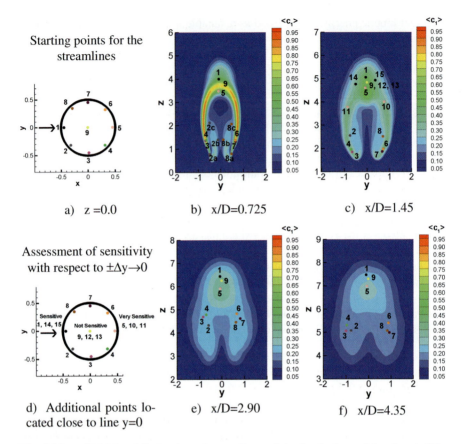

Fig. 4. Map of initial and final points of streamlines originating in the peripheral part of the jet pipe from a radius equal to 0.45 D

is close to the location of point *5*. Further downstream, at $x/D>5$ this point and the central one (point *9*) move above the zone of highest concentration area which is a known phenomenon for the jet in crossflow (Yuan and Street (1998), Denev et al.(2007a)).

Further analysis concerns sensitivity of the streamlines with respect to small changes in the position of the starting location at z=0. Points within a distance of $\pm0.01D$ from points *1, 5* and *9* were added as shown in Fig. 4d and their streamlines investigated. If considering for a moment the flow being stationary and equal to the time-averaged one so that particle trajectories and streamlines are identical. In such a case the analysis shows whether average trajectories diverge (i.e. the distance between them grows downstream), contributing locally to a better mixing, or move close to each other, showing no good mixing on their downstream movement. Results are presented in Fig. 4c. They show that in the middle part of the jet (points *9, 12* and *13* in Fig. 4d) the streamlines are not sensitive towards small changes in the initial location indicating poor mixing. The opposite is valid for points from the windward zone (points *1, 14* and *15* in Fig. 4d) which downstream location is quite sensitive toward Δy-perturbations indicating good mixing in the upper part of Fig. 4c. Best mixing potentials offers the leeward zone presented in Fig. 4d by points *5, 10* and *11*, showing largest sensitivity/divergence in Fig. 4c and therefore largest mixing capabilities in this zone of the jet.

The present analysis was extended also to other points at $z=0$ which are positioned along lines connecting points *2* and *6, 4* and *8*, or lines connecting points *1* and *5, 3* and *7*, but this is omitted for the sake of brevity. Here, we only would like to notice one more interesting feature from this extended analysis: the fact that some points with highest concentration, e.g. in the plane $x/D=1.45$ actually originate from the crossflow and not from the pipe. The reason for this is as follows. Due to the helical motion, discussed above, fluid particles from the crossflow are entrained into the jet, first passing upstream planes ($x/D<1.45$) in zones, where the concentrations are very high, almost reaching that in the pipe. The amount of fluid entrained is relatively small so that no substantial dilution of the jet can occur. When these particles travel downstream, they transport this high-concentration fluid to the plane of consideration.

From this analysis we deduce the general suggestion for improvement of the mixing process: helical shape of streamlines should be aimed at. This is supported by the example of the streamline originating from points *2* and *8* which shows best mixing and high reduction of the concentration when followed downstream.

3.3 Apriori Testing of Mixing Models Based on the Present DNS-Results

DNS datasets comprising complete *3D* information of all flow and scalar variables offer valuable opportunities to obtain reference data required for turbulence modelling. Such an investigation was carried out in Denev et al.(2008a), where the turbulent kinetic energy k and its dissipation rate ε were obtained and presented. Here the analysis is extended further showing how DNS results can be used to assess assumptions regarding the modelling of turbulent mixing.

Reynolds Averaged Navier-Stokes (RANS) models involve different levels of complexity with respect to the treatment of the turbulent scalar fluxes. The most widely used approach is based on the "Simple gradient diffusion hypothesis" (SGDH) equation, which for the incompressible case is:

$$\langle u_i' c' \rangle = -\frac{\langle v_t \rangle}{\sigma_t} \frac{\partial \langle c \rangle}{\partial x_i} = -\frac{0.09 \, \langle k \rangle^2 / \langle \varepsilon \rangle}{0.6} \frac{\partial \langle c \rangle}{\partial x_i} \tag{3}$$

This equation presents the most common and most simple approach for modelling the mixing in turbulent flows. Here the brackets denote Reynolds-averaging, u' and c' - the velocity and scalar fluctuations with respect to that average. As usual, k and ε denote the turbulent kinetic energy and its dissipation rate, respectively, and c – the mass fraction or concentration. The more sophisticated "generalized gradient diffusion hypothesis", (GGDH) has been introduced in Daly and Harlow (1970) as

$$\langle u_i' c' \rangle = -c_t \cdot \frac{\langle k \rangle}{\langle \varepsilon \rangle} \cdot \left(\langle u_i' u_j' \rangle \cdot \frac{\partial \langle c \rangle}{\partial x_j} \right) \qquad (4)$$

where $c_t = (3/2 \, c_\mu / \sigma_t) = 0.22$ and summation over repeated indices is assumed.
The main advantage of equation (4) is that it, unlike equation (3), gives a vertical
scalar flux driven by a horizontal mean scalar gradient in the presence of shear
Hanjalic and Vasic (1993). The GGDH has been shown to produce more accurate
results for non-isothermal cavity flows than the SGDH model Ince and Launder
(1989).

In the following the two models (3) and (4) are compared using the present set
of time-averaged DNS data for all computed quantities at $Re=650$. The results for
the streamwise turbulent flux and the vertical turbulent flux, $<u'c'>$ and $<w'c'>$,
respectively, are presented in Figure 5. As can be seen from these data, both scalar

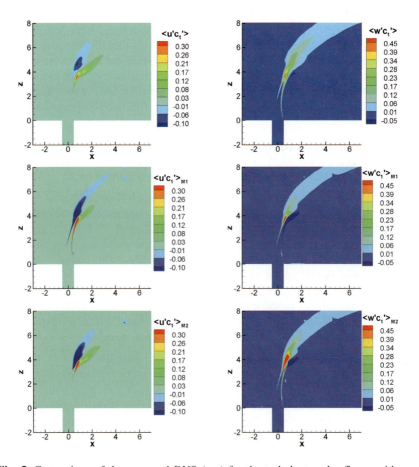

Fig. 5. Comparison of the averaged DNS (top) for the turbulent scalar fluxes with computed results according to two modelling assumptions: the SGDH (middle, denoted M1)
and the GGDH (bottom, denoted M2) for Re=650.

fluxes exhibit largest values in the vicinity of the laminar-to-turbulence transition point. Both modelling hypotheses represent this feature accurately. However, when viewed as a whole, the GGDH delivers more accurate results - the overall size and shape of the isolines for both turbulent fluxes is closer to that of the DNS data.

The above a priori test for RANS models can be extended to the LES context. For instance, the GGDH can be viewed as a special form of the model of Huai and Sadiki (2007) when some of the nonlinear terms are neglected.

4 The Wavelet Approach to LES Modelling (WALES)

LES requires appropriate sub-grid scale (SGS) modelling. Two widely used classes of models are the eddy-viscosity (eddy-diffusivity) models and the scale-similarity models. As wavelets are designed to deal with multi-scale effects, they appear as a natural framework for modelling turbulence and mixing phenomena. In the present study, wavelet-based SGS models belonging to both types of models were designed and tested. This approach was named Wavelet-Adapted LES (WALES).

4.1 The Wavelet Basis Employed

A wavelet basis as defined in Daubechies (1992) and similar references is composed of functions localized both in scale and space. In this framework it is straightforward to decompose a function locally into contributions on different scales. The present work employs the multiresolution approach by Harten (1993) which provides a bi-orthogonal representation of the original signal. In the following we briefly denote the algorithm in one dimension. It employs values at the hierarchical grid points

$$x_i^k = \frac{i-1}{2^k} + \frac{1}{2^{k+1}} \qquad i = 0,...,2^k - 1 , \ k = 0,...,K .$$

These can be viewed as cell centers in a Finite Volume method with 2^k cells on the interval [0;1]. The values given at these points are denoted ϕ_i^k and would then correspond to the averages of the surrounding cell. The values on a certain grid level are decomposed into coarse scales and details. The coarse scales are given by the cell averages on the next finer grid

$$\phi_i^k = \frac{1}{2}\left(\phi_{2i}^{k+1} + \phi_{2i+1}^{k+1}\right) \qquad i = 0,...,2^k - 1 .$$

The next step is to prolongate the coarse solution to the fine-grid values by an interpolation procedure of desired order

$$\tilde{\phi}_{2i,2i+1}^{k+1} = \phi_i^k \pm \sum_{l=1}^{s} \gamma_l \left(\phi_{i+l}^k - \phi_{i-l}^k \right) \qquad i = 0,...,2^k - 1 \qquad (5)$$

which is used here with $s = 1$, $\gamma_1 = 1/8$ yielding third order accuracy, as the maximal order of the scheme is $r = 2s + 1$. The fine-scale details are then given by

$$d_{2i,2i+1}^{k+1} = \phi_{2i,2i+1}^{k+1} - \tilde{\phi}_{2i,2i+1}^{k}$$

The signal ϕ_i^{k+1}, $i = 0,...,2^{k+1} - 1$, on level $k + 1$ is hence decomposed into a coarser signal ϕ_i^k, $i = 0,...,2^k - 1$, on level k and details d_i^{k+1}, $i = 0,...,2^{k+1} - 1$. It can be recombined by a similar algorithm. Here, the decomposition has been applied in a two-grid fashion to start with. Later work will be concerned with extending this. The above decomposition was applied in three dimensions so that, to simplify the notation we use $d\{\phi\}$ to designate the *3D*-details of the quantity ϕ on the finest level. The above algorithm is applied with unmodified coefficients on the *3D* Finite Volume grids used for the LES below. Since grid stretching is usually very low with LES, this issue is disregarded here, as generally done with the filtering in the dynamic procedure, e.g. near boundaries the solution is currently prolongated by zero values sacrificing the order to gain simplicity.

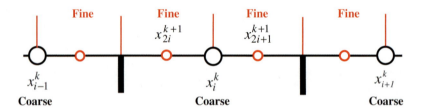

Fig. 6. Arrangement of hierarchical grid points in relation to a Finite Volume scheme in 1D, illustrating the wavelet decomposition algorithm used for the WALES approach

4.2 WALES Subgrid-Scale Models Developed

WALES-E: We now define an eddy-viscosity SGS model by using the step size of the grid as a length scale and the details on the grid level as a velocity scale

$$v_t = C_{WALES-E} \cdot \Delta \cdot \sqrt{(d\{u\})^2 + (d\{v\})^2 + (d\{w\})^2} \qquad (6)$$

where $\Delta = (Vol)^{1/3}$ is obtained from the cell volume as usual. The model constant was calibrated to a value of *0.02*. This model bears some similarity with the structure function model Lesieur et al.(2005). The new model however is insensitive to large-scale features, since by construction $d\{\phi\}=0$ for linear and quadratic functions. This is similar to the filtered models developed in France Lévêque et al.(2007) and arises naturally with the wavelet approach. Scalar transport was modelled with an eddy diffusivity employing a turbulent Schmidt number of *0.6*.

WALES-S: Similarly to the scale-similarity model of Bardina et al.(1980), another WALES model has been developed which is not of eddy-viscosity type. It is defined as

$$\tau_{ij} = -C_{WALES-S}\left(d\left\{u_i'u_j'\right\} - d\left\{u_i'\right\}d\left\{u_j'\right\}\right) \tag{7}$$

where the prime indicates fluctuations with respect to the mean in time which is computed along with the LES solution itself. This removed oscillations in the resulting mean flow field otherwise obtained and is similar to the decomposition of Schumann (1975). After extensive tests for channel flow, the value of $C_{WALES-S}$ was *0.2* for $i = j$ and *0.4* else. This model showed to be less stable than the WALES-E model, so that the SGS term was left unchanged between the Runge-Kutta steps which reduces the order but improves stability. In this case, no SGS model was used for the SGS scalar transport.

4.3 WALES Modelling Applied to Plane Channel Flow

Simulations with $Re_\tau=180, 395,$ and 590 were carried out. Simultaneously, the transport of a passive scalar with $Sc=1.0$ and Dirichlet conditions at the walls was represented. Central schemes for all terms were employed. In Fig. 7, the flow field is compared with the results from Moser et al.(1999), while the scalar field is compared to the DNS results of Schwertfirm and Manhart (2005). Simulations of these cases by the present authors using other SGS models and local grid refinement are reported in Fröhlich et al.(2007). In all cases the same grid was employed

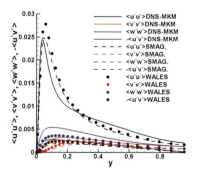
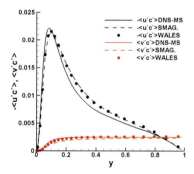

Fig. 7. Turbulent stresses for the WALES-E model compared with DNS and the Smagorinsky model for Re$_\tau$=395 (left); scalar fluxes for the WALES-E model compared with DNS and the Smagorinsky model for Re$_\tau$=180 (right). MKM denotes data from Moser et al.(1999), DNS-MS denotes data from Schwertfirm and Manhart (2005).

with $\Delta x = 0.1, \Delta y = 0.015, \Delta z = 0.067$ near the wall. The lower Reynolds number had to be chosen for the scalar transport since only for this case appropriate DNS data were available.

The result for the turbulent transport obtained with the WALES-E model is practically the same as obtained with the Smagorinsky model (SM), better by a tiny bit. The Reynolds stresses for the higher Re are somewhat better than those obtained with the SM, except for the streamwise stresses $<u'u'>$. Various tests were undertaken to investigate the sensitivity of the result to the influence of the value of the constant, the additional use of van Driest damping, and the treatment near the wall, but these showed little impact on the result. The same holds with $s=0$, in equation (5) reducing the basis to Haar wavelets, see Daubechies (1992).

With the WALES-S model, similar tests were undertaken, further supplemented by variations of the constant between diagonal and off-diagonal stresses as well as addition of a diffusive SM term as used for the mixed model in Bardina et al.(1980). The results with this model, reported in Fig. 8, are of DNS quality for the lower Reynolds number and the scalar fluxes and very good as well for the corresponding Reynolds stresses (not shown here). For the higher Reynolds number, most of the Reynolds stresses improve compared to the SM and the WALES-E model, except $<u'u'>$ which deviates more from the DNS data.

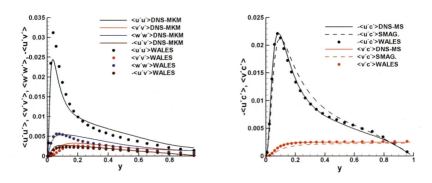

Fig. 8. Turbulent stresses for the WALES-S model compared with DNS for Re_τ=395 (left); scalar fluxes for the WALES-S modelling compared with DNS and the Smagorinsky model for Re_τ=180 (right).

4.4 Results with the WALES Applied to the JICF

Results with the WALES-E model were obtained for the jet in crossflow configuration with Reynolds number equal to *650*. The numerical grid had *4.1* million control volumes. Each numerical block contains an even number of control volumes in each spatial direction thus guaranteeing the continuous distribution of the eddy viscosity at the block boundaries for the two-level wavelet decomposition.

Results from the DNS above presented are used as reference. Furthermore, simulations with the Smagorinsky model, with the dynamic Smagorinsky model as well as without any SGS model were also undertaken. Simulations with the WALES-S model diverged.

Figure 9 shows the turbulent kinetic energy at $y=0$ for the DNS and the three SGS models; for the sake of clarity only part of the computational domain is represented. The shape and the level of the turbulent kinetic energy is reasonably well predicted by the present WALES-E and the classical Smagorinsky model while the values computed with the Dynamic Smagorinsky model are somewhat too low.

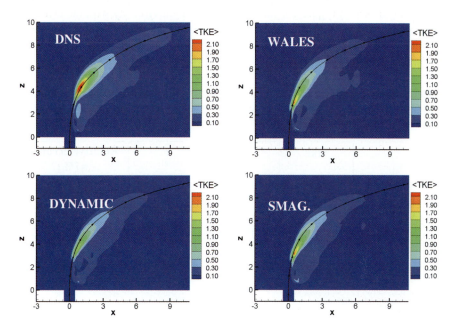

Fig. 9. Turbulent kinetic energy for WALES compared to DNS and other SGS-models

The streamlines, originating from the center of the pipe are plotted in this Figure as well. For all three SGS-models they almost overlap. A direct comparison of these streamlines with DNS would repeat the results from Fig. 2 (see the streamline for the LES1-model there) and is therefore omitted here. The distance along this central streamline - from the origin of the coordinate system (0;0;0) to the point where the turbulent kinetic energy reaches 10% of its maximum value in the plane $y=0$, has been calculated. This distance is then regarded as the transition length. In Table 3 it is denoted as "s_{TKE}". The Table shows that the WALES-E model exhibits the shortest transition length which is less accurate than obtained with the other two models. Ironically, the explanation lies in the lowest turbulent viscosity predicted by this model prior to transition which in this laminar region is obviously the physically most accurate one. However, this correct viscosity appears not so efficient in damping of the numerically arising oscillations which leads to the earlier transition predicted.

Comparison of the turbulent viscosity produced by the WALES-E model and the Smagorinsky model is shown in Fig. 10. Although viscosity values for the Smagorinsky model appear slightly higher behind the transition point, results from the two models are very close to each other. The viscosity of the WALES-E model exhibits lower smoothness compared to the other model. This results from the discontinuity nature of the wavelet details between neighbor control volumes, compared to the other model, but does not affect any other computed quantity of the flow as can be deduced from the line-plots in Fig. 11 below and also from the experience gained with higher Reynolds number flows, Denev et al.(2009b). Although not shown in Figure 10, values of the turbulent viscosity with the Dynamic Smagorinsky model are up to an order of magnitude higher. This may present a reason for stronger damping of the turbulent fluctuations with this model and hence for the lower values for the turbulent kinetic energy it shows in Fig. 9.

Table 3. Distance from the end of the pipe point (0;0;0) to the transition point estimated from the turbulent kinetic energy for different SGS models and the DNS. x and z present the coordinates of the transition point. All data are for Re=650.

	S_{TKE}	x	z	Relative error of s [%]
DNS	3.38	0.52	3.32	reference
WALES-E	2.51	0.23	2.50	-26
SMAG	2.66	0.27	2.64	-21
DYN_SMAG	2.81	0.31	2.79	-17

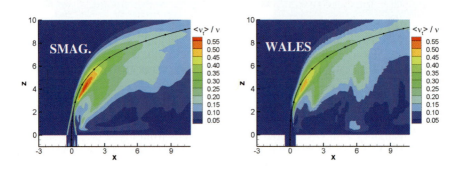

Fig. 10. Time-averaged turbulent viscosity for the Smagorinsky model and the WALES-E model in the plane y/D=0

In the laminar pipe, with exception of small values observed in cells directly affected by the boundary conditions, WALES-E produces zero eddy viscosity. It is due to the numerical feature of Harten´s wavelets to return zero details in regions, where the signal has a constant, linear or parabolic shape. This correct physical behavior of the WALES-E model in the pipe was not reproduced by the Smagorinsky model as shown in the Fig. 10. The same also holds for the jet region prior to transition ($z/D<2$).

Figure 11 shows a quantitative comparison of the two eddy-viscosity models: Smagorinsky and WALES-E. Shown are sample results of different variables on a vertical line in the plane $y=0$. Both models yield very similar values for all mean-flow and turbulence quantities. One exception is the region close to the pipe exit ($z\sim0$). In this region WALES-E, similar to a computation without any SGS model, yields quite high fluctuating quantities (e.g. $<u\acute{}u\acute{}>$, $<w\acute{}c\acute{}>$ etc.) due to the almost vanishing turbulent viscosity prior to transition.

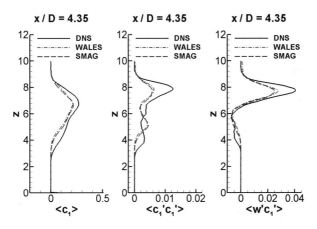

Fig. 11. Comparison of time-averaged passive scalar quantities: mass fraction, scalar fluctuations and Reynolds flux in the vertical direction

Finally, the CPU-time necessary to reach the same dimensionless physical time is compared between the three SGS-models. The Smagorinsky model was taken as a reference i.e. as 100%. As expected, the Dynamic Smagorinsky model required more CPU-time: 117%. The simulation with the WALES-E model was expected to give almost the same CPU-time as the Smagorinsky model. The reason for such an expectation is that the computation of the wavelet details is comparable (in floating point operations) to the computation of the rate of strain tensor required for the Smagorinsky model. Practically, the WALES-E model required slightly higher CPU-time: 107%. Another issue is the variable time step adjusted for stability and the different number of iterations in the solver for the pressure-correction equation. To confirm this, a simulation was carried out without application of any SGS model. The computation required 112% of the CPU-time of the simulation using the Smagorinsky model which is also higher than the CPU-time demand with the WALES-E model and is attributed to the numerical issues mentioned.

5 Conclusions

A comprehensive description of a series of DNS forming a database for mixing and chemical reactions in a jet in crossflow was given. Issues like co-ordination of efforts with companion experiments were discussed. Comparison with experimental data were presented and show good agreement.

The versatile possibilities for analyses of mixing based on the present DNS mixing database are demonstrated. More specifically, the analyses allowed identifying regions of good mixing by tracing streamlines of the averaged flowfield. It was found that regions of good mixing correlate with helical movement of fluid particles. Regions where the streamlines diverge from each other also exhibit good local mixing. The DNS mixing database allowed also quantifying and comparing mixing hypotheses in the Reynolds-averaged context.

Two new sub-grid scale models denoted WALES (Wavelet-Adapted-LES) are presented. The first one is based on an eddy-viscosity/eddy-diffusivity concept and the second one – on the scale-similarity concept. They introduce the new idea of using wavelet details for modelling of turbulent mixing phenomena. Results with the present models are shown for the straight channel flow and for the jet in crossflow. Due to the numerical features of the type of wavelets used, the eddy-viscosity WALES model predicted zero eddy-viscosity in the laminar pipe of the present jet in crossflow, thus being more physically sound than the classical Smagorinsky model. This makes the present WALES models suitable to calculate flows in which both laminar and turbulent flow- and mixing regions are present.

Basically, wavelets are able to represent multi-scale physical phenomena like turbulence and turbulent mixing in a natural way. Possible extensions of the present work concern formulation of subgrid-scale models based on a larger number of grid levels, as well as the application of the dynamic procedure to the developed models.

Acknowledgements

The simulations were performed on the national super computer HP XC4000 at the High Performance Computing Center Stuttgart (HLRS) under the grant with acronym "DNS-jet". The authors would like to thank their colleague Olivier Roussel for detailed explanations and fruitful discussions on the multiresolution wavelet decomposition.

References

Bardina, J., Ferziger, J., Reynolds, W.: Improved subgrid-scale models for large-eddy simulation (1980) AIAA-Paper No 80-1357
Cárdenas, C., Suntz, R., Denev, J.A., Bockhorn, H.: Two-dimensional estimation of Reynolds-fluxes and -stresses in a Jet-in-Crossflow arrangement by simultaneous 2D-LIF and PIV. Applied Physics B - Lasers and Optics 88(4), 581–591 (2007)

Cárdenas, C., Suntz, R., Bockhorn, H.: Experimental investigation of the mixing-process in a Jet-in-Crossflow Arrangement by simultaneous 2d-LIF and PIV (2009) (present volume)

Daly, B.J., Harlow, F.H.: Phys. Fluids, A 13, 2634–2649 (1970)

Daubechies, I.: Ten Lectures on Wavelets. SIAM, Philadelphia (1992)

Denev, J.A., Fröhlich, J., Bockhorn, H.: Evaluation of mixing and chemical reactions within a jet in crossflow by means of LES. In: Proc. of European Combustion Meeting, Louvain-la-Neuve, Belgium, 3-6.04. CD-ROM (2005a)

Denev, J.A., Fröhlich, J., Bockhorn, H.: Structure and mixing of a swirling transverse jet into a crossflow. In: Humphrey, J.A.C., Gatski, T.B., Eaton, J.K., Friedrich, R., Kasagi, N., Leschziner, M.A. (eds.) Procs. of 4th Int. Symp. on Turbulence and Shear Flow Phenomena, Williamsburg, Virginia, June 27-29, vol. 3, pp. 1255–1260 (2005b)

Denev, J.A., Fröhlich, J., Bockhorn, H.: Direct Numerical Simulation of mixing and chemical reactions in a round jet into a crossflow - a Benchmark. In: Nagel, W.E., Jaeger, W., Resch, M. (eds.) Trans. of the High Performance Computing Center Stuttgart (HLRS) 2006, pp. 237–251. Springer, Heidelberg (2006)

Denev, J.A., Fröhlich, J., Bockhorn, H.: Direct Numerical Simulation of a transitional jet in crossflow with mixing and chemical reactions. In: Friedrich, R., Adams, N.A., Eaton, J.K., Humphrey, J.A.C., Kasagi, N., Leschziner, M.A. (eds.) Procs. of 5th Int. Symp. on Turbulence and Shear Flow Phenomena, TU-Munich, Garching, Germany, August 27-29, vol. 3, pp. 1243–1248 (2007a)

Denev, J.A., Fröhlich, J., Bockhorn, H.: Direct Numerical Simulation of a Round Jet into a Crossflow – Analysis and Required Resources. In: Nagel, W.E., Kröner, D., Resch, M. (eds.) High Performance Computing in Science and Engineering 2007, Transactions of the High Performance Computing Center, Stuttgart (HLRS), pp. 339–350. Springer, Heidelberg (2007b)

Denev, J.A., Falconi, C., Fröhlich, J., Bockhorn, H.: DNS and LES of a jet in crossflow – Evaluation of turbulence quantities and modelling issues. In: Proceedings of the 7th International ERCOFTAC Symposium on Engineering Turbulence Modelling and Measurements, ETMM7, Limassol, Cyprus, June 4 – 6, vol. 2, pp. 587–592 (2008a)

Denev, J.A., Fröhlich, J., Bockhorn, H.: Two-Point Correlations of a Round Jet into a Crossflow – Results from a Direct Numerical Simulation. To appear in High Performance Computing in Science and Engineering 2008, Transactions of the High Performance Computing Center, Stuttgart, HLRS (2008b)

Denev, J.A., Fröhlich, J., Bockhorn, H.: Large Eddy Simulation of swirling transverse jets into a crossflow with scalar transport. Phys. Fluids 21(1), 015101 (2009a)

Denev, J.A., Falconi, C., Fröhlich, J., Bockhorn, H.: Wavelet-adapted subgrid scale models for LES. Submitted to the Second International Conference on Turbulence and interaction, Sainte-Luce, Martinique, May 31 - June 5 (2009b)

Fröhlich, J., Denev, J.A., Bockhorn, H.: Large eddy simulation of a jet in crossflow. In: Neittaanmaki, P., Rossi, T., Majava, K., Pironneau, O. (eds.), Rodi, W., Le Quere, P. (assoc. eds.) Proc. of the 4th European Congress on Computational Methods in Applied Sciences and Engineering, ECCOMAS 2004, Jyvaskyla, Finland, July 24-28, vol. 1 (2004) ISBN 951-39-1868-8

Fröhlich, J., Denev, J.A., Hinterberger, C., Bockhorn, H.: On the impact of tangential grid refinement on subgrid-scale modelling in large eddy simulation. In: Boyanov, T., Dimova, S., Georgiev, K., Nikolov, G. (eds.) NMA 2006. LNCS, vol. 4310, pp. 550–557. Springer, Heidelberg (2007)

Hanjalic, K., Vasic, S.: Computation of turbulent natural convection in rectangular enclosures with an algebraic flux model. Int. J. Heat Mass Transfer 36(14), 3603–3624 (1993)

Harten, A.: Discrete multi-resolution analysis and generalized wavelets. J. Appl. Num. Math. 12, 153–193 (1993)

Hinterberger, C.: Three-dimensional and depth-averaged Large–Eddy–Simulation of flat water flows. PhD thesis, Inst. Hydromechanics, Univ. of Karlsruhe, p. 296 (2004)

Huai, Y., Sadiki, A.: Large eddy simulation of mixing processes in turbulent liquid flows with chemical reactions. In: Friedrich, R., Adams, N.A., Eaton, J.K., Humphrey, J.A.C., Kasagi, N., Leschziner, M.A. (eds.) Procs. of 5th Int. Symp. on Turbulence and Shear Flow Phenomena, TU-Munich, Garching, Germany, August 27-29, vol. 3, pp. 1137–1142 (2007)

Ince, N.Z., Launder, B.E.: On the computation of buoyancy-driven turbulent flows in rectangular enclosures. Int. J. Heat and Fluid Flow 10(2), 110–117 (1989)

Lesieur, M., Métais, O., Comte, P.: Large-Eddy Simulations of Turbulence, p. 219. Cambridge University Press, Cambridge (2005)

Lévêque, E., Toschi, F., Shao, L., Bertoglio, J.-P.: Shear-improved Smagorinsky model for large-eddy simulation of wall-bounded turbulent flows. J. Fluid Mech. 570, 491–502 (2007)

Moser, R.D., Kim, J., Mansour, N.N.: Direct numerical simulation of turbulent channel flow up to Re_τ=590. Phys. Fluids 11, 943–945 (1999)

Muldoon, F.: Numerical Methods for the Unsteady Incompressible Navier-Stokes Equations and Their Application to the Direct Numerical Simulation of Turbulent Flows. PhD Thesis, Louisiana State University (2004)

Muppidi, S., Mahesh, K.: Study of trajectories of jets in crossflow using direct numerical simulations. J. Fluid Mech. 520, 81–100 (2005)

Schumann, U.: Linear stability of finite difference equations for three-dimensional flow problems. J. Comput. Phys. 18, 465–470 (1975)

Schwertfirm, F., Manhart, M.: ADM Modelling for Semi-Direct Numerical Simulation of Turbulent Mixing and Mass Transport. In: Humphrey, J.A.C., Gatski, T.B., Eaton, J.K., Friedrich, R., Kasagi, N., Leschziner, M.A. (eds.) Procs. of 4th Int. Symp. on Turbulence and Shear Flow Phenomena, Williamsburg, Virginia, June 27-29, vol. 2, pp. 823–828 (2005)

Yuan, L.L., Street, R.L.: Trajectory and entrainment of a round jet in crossflow. Phys. Fluids 10(9), 2223–2335 (1998)

Yuan, L.L., Steer, R.L., Ferziger, J.H.: Large-eddy simulations of a round jet in crossflow. J. Fluid Mech. 370, 71–104 (1999)

Zhu, J.: A low-diffusive and oscillation-free convection scheme. Communications in applied numerical methods 7, 225–232 (1991)

Analysis of Mixing Processes in Jet Mixers Using LES under Consideration of Heat Transfer and Chemical Reaction

Egon Hassel[1], Nikolai Kornev[1], Valery Zhdanov[1], Andrei Chorny[1,2], and Matthias Walter[1]

[1] Institute of Technical Thermodynamics, University of Rostock, Albert. Einstein Str.2, 18059 Rostock, Germany
[2] Heat and Mass Transfer Institute, National Academy of Sciences of Belarus, 15 P. Brovka Str., 220072 Minsk

Abstract. The paper presents an overview of the results on turbulent mixing of inert and reacting flows in a co-axial jet mixer at high Reynolds and Schmidt numbers. Flow modes without a recirculation zone (j-mode) and with intense separation and development of a recirculation zone (r-mode) are considered. The study has been performed using numerical (RANS, LES) and experimental methods (LDV, PLIF). Performances of various numerical models to predict passive scalar mixing and reacting flows are discussed. The influence of various parameters on mixing, the structure of a scalar field and its dissipation field, multifractal properties of the dissipation field and isotropy of the scalar field are studied for the case of passive scalar mixing. Concentrations of the neutralisation reaction product and the mixture fraction are measured simultaneously using the PLIF. A few RANS and LES models are validated by comparing with the PLIF measurements. The influence of various parameters on the selectivity of competitive-consecutive reaction is analyzed using the RANS.

1 Introduction

Mixing in co-axial confined jets has been investigated for a long time because of needs of engineering facilities, e.g., combustion chambers, injection systems, chemical mixing devices, and many others. In this paper, a co-axial jet mixer comprising a nozzle of diameter d positioned along the centreline of a pipe of diameter D has been considered (see Fig. 1). A fast internal jet (water) with a bulk velocity U_d is confined by a slower external co-flow (water) with a velocity $U_D \ll U_d$. The most important parameters for the mixing process are the flow rate ratio $\dot{V}_D / \dot{V}_d$, the diameter ratio D/d, and the Reynolds number $Re_d = d \cdot U_d / \nu$. Most of the publications on confined jets are devoted to the study of large-scale influences in configurations where the internal jet moves much slower than the co-flow $U_d / U_D \ll 1$ (see, for instance, [16], [19] and [21]). These investigations have been motivated by two important applications:

stabilization of flame fronts by swirl burners and saturation of the air co-flow with molecules of substances transferred by the internal jet. Surprisingly, the case $U_d / U_D \gg 1$ has attracted less attention although this flow mode is of much significance for homogenization devices and free jet reactors.

Two different flow modes can be observed in jet mixers depending on the flow rate ratio $\dot{V}_D / \dot{V}_d$ (see [2]). If $D / d < \beta(1 + \dot{V}_D / \dot{V}_d)$ where $\beta \approx 1$ is an empirical constant found from a simple flow entrainment model, the flow is similar to a free jet (henceforth, referred to as the jet-mode or the j-mode for short). Once $D / d > \beta(1 + \dot{V}_D / \dot{V}_d)$, separating a strong flow from pipe walls causes a recirculation zone to develop behind the nozzle (see Fig. 10 in [23]) (henceforth, referred to as the recirculation mode or the r-mode for short). A qualitative description of the r-mode is given in [2]. Such flow mode finds use in various devices designed for liquid homogenization. The recirculation zone enhances the mixing efficiency drastically so that the homogeneous stage is reached after four or five pipe diameters downstream of the nozzle. Its practical application is seen in many engineering devices such as injection systems, etc. The j-mode is often involved for chemical reactors to control competitive-consecutive chemical reactions (see Sect. 4.3). Engineering aspects of jet mixer applications are discussed in [8] and [22].

This paper is an overview of the results obtained by the research group of Rostock University in 2002-2008 within the framework of the research program SPP 1141 supported by the German Research Foundation (DFG). Both numerical (Reynolds Averaged Navier Stokes (RANS) and Large Eddy Simulations (LES)) techniques and experimental methods (Laser Doppler Anemometer (LDA), Planar Laser Induced Fluorescence (PLIF)) were used to quantify the mixing process in inert and chemically reacting flows. Topically, the work is divided into three stages of research. The focus of the first one (see Sect. 3.1-3.3) is the study of large-scale flow phenomena in passive scalar mixing using both numerical methods and experiments. This stage of research deals with investigating performances of various numerical models to predict passive scalar mixing in jet mixers (see Sect. 3.1) and the influence of various parameters on mixing (see Sect. 3.2). One of the most important results of this stage is outlined in Sect. 3.3. The study has shown that the flow in the r-mode is highly unsteady and the time averaged results obtained here do not describe properly the true nature of flow phenomena. The presence of long-period temporal oscillations with a sort of opposition-of-phase of the flow near the mixer walls is revealed, quantified, and explained using the PLIF and LES. A detailed investigation of passive scalar statistics using the high resolved PLIF has been the subject of the second stage (see Sect. 3.4 and 3.5). The research of formation and dynamics of fine scalar structures in liquids at high Schmidt, $Sc \sim 1000$, and Reynolds numbers belongs to the most complicated fields of fluid mechanics. The structure and isotropy of the scalar field (see Sect. 3.4) and also the scalar dissipation rate field and its multifractal properties (see Sect. 3.5) are studied. Theoretical and experimental studies of reacting flows in the jet mixer are performed during the third stage of research (see Sect. 4.1-4.3). The main results of this stage are connected with experiments on the neutralisation reaction in the turbulent flow of the jet mixer (see Sect. 4.1), with validation of

numerical models for this case (see Sect. 4.2), and also with numerical investigations of the influence of various parameters on competitive-consecutive reactions
in the jet mixer (see Sect. 4.3).

2 Methods of Investigation

2.1 Experimental Setup and Measurement Procedure

Experiments were made in a water closed-circuit channel (see Fig. 1). Soft water
was pumped from tank (1) into tank (2) and then entered through tube (3) with diameter $D = 0.05m$ mixer (4) as a co-flow. The jet of soft water with a premixed
dye (Rhodamine 6G or Rhodamine B) was ejected through nozzle (6) with diameter $d = 0.01m$ from tank (5). In experiments with chemically reacting substances
the co-flow and the jet were the base and acid water solutions. The mixed liquids
were collected in tank (7), neutralized, and drained into the environment. The
mixer was placed into a rectangular glass box filled with water to reduce optical
distortions due to the curvature of the outer tube.

A velocity field was measured by one–component Laser-Doppler-Anemometer
(Flow Light1, Dantec Dynamics). A scalar field was examined by the PLIF
method using two cameras. A 16-bit intensified PI-MAX (Roper Scientific, Inc.)
camera equipped with a Nikkor 50 mm lens and a splitter ring PK-11A provides
the resolution of 300 μm (henceforth, referred to as the coarse LIF). The second
camera (12-bit PCO.qe) equipped with a long distance microscope K2 (Infinity
Photo-Optical Company) and a lens CF-2 provides the resolution of $31\mu m$ within
the measurement window 2.7 x 2.1mm in size (henceforth, referred to as the high
resolved LIF).

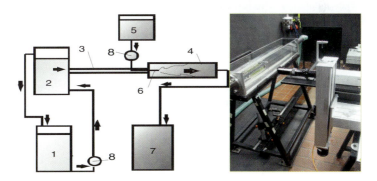

Fig. 1. Sketch of the set up (left) and the test section (right): 1, 2, 5, 7 – tanks; 3 – 6 m long
tube; 4 – test section with the mixer; 6 –nozzle, 8 – pumps.

The x - axis of the coordinate system was aligned along the pipe centreline
downstream of the nozzle exit. The coordinates r were measured from the x -axis.
The point $r/D = 0$ signified that the centre of the measurement window was on

the mixer axis. Measurements were performed at positions ranging from $x/D = 0.1$ to $x/D = 9$ downstream. Over this distance range, at $Re = 10^4$ the Kolmogorov microscale was in the range between 30 and 300 μm, which corresponded to the range of the Batchelor scales between 1 and 10 μm. Therefore, the smallest microstructures of the scalar field were not resolved. An Nd: YAG laser with a repetition rate of 10 Hz and impulse energy stability of $\pm 2\%$ served as an external trigger of the cameras. The thickness of the laser sheet was 0.3-0.5 mm in the coarse LIF and didn't exceed 48 μm within the measurement window in high resolved PLIF measurements. Scalar field development was estimated considering mixture fraction f variations. Mixture fraction distributions were calculated from the emitted Rhodamine 6G intensity I referred to the maximum intensity I_0 determined on the centreline in the first cross section at $x/D = 0.1$. The derivatives of f necessary for further analysis (see Sect. 3) were calculated using the Sobel operator. Details on the experimental setup and data interpretation for the passive scalar study can be found in [14] and [25-27].

Fluorescent dyes – Rhodamine B (RhB) and organic dye – disodium fluorescein ($C_{20}H_{10}O_5Na_2$, uranin) both excited by the Nd: YAG laser were used in experiments with chemical neutralization reaction $HCl + NaOH \rightleftarrows H_2O + NaCl$. Rhodamine B was used to trace the passive scalar, whereas fluorescein was needed to examine the reaction product concentration. The measurement technique was based on the titration method [10] utilizing the sensitivity of uranin fluorescence to the pH value of the solution which changed due to the chemical reaction between the base and the acid. The RhB fluorescence was assumed to be independent of the pH value. Weak concentrations of RhB (0.08mg/l) and uranin (7.4 mg/l) secured linear calibration curves. The emission spectrum of uranin had a maximum at a wavelength of 490 nm, whereas that of RhB at 580nm. Emissions of these dyes measured simultaneously in the same measurement window by two cameras (PCO.qe) were separated by filters. A special calibration procedure was developed to eliminate the mutual influence of the dyes in emission overlapping regions. The product concentration was calculated through the fluorescence light intensity J of $C_{20}H_{10}O_5Na_2$: $Y_p = Y_{p0} \cdot J / J_{max}$, where Y_{p0} and J_{max} are the product concentration and the light intensity for the stoichiometric acid-to-base ratio. Concentrations of the base and the acid were calculated from the mass conservation equation using f and Y_p.

2.2 Numerical Models

The mathematical model is based on the assumption that liquid is incompressible and miscible. The governing equations are the continuity equation, the Navier-Stokes equations, and the transport equation for the mixture fraction $f \in [0,1]$ that is the volume fraction of liquid issued from the nozzle. The governing equations can be read as

$$\frac{\partial \bar{u}_i}{\partial x_i} = 0 \tag{1}$$

$$\frac{\partial \bar{u}_i}{\partial t} + \frac{\partial \bar{u}_j \bar{u}_i}{\partial x_j} = -\frac{1}{\rho}\frac{\partial \bar{p}}{\partial x_i} + \frac{\partial}{\partial x_i}\left[\nu \left(\frac{\partial \bar{u}_i}{\partial x_j} + \frac{\partial \bar{u}_j}{\partial x_i} \right) - \tau_{ij}^t \right] \tag{2}$$

$$\frac{\partial \bar{f}}{\partial t} + \frac{\partial \bar{u}_j \bar{f}}{\partial x_j} = \frac{\partial}{\partial x_j}\left[\frac{\nu}{Sc}\frac{\partial \bar{f}}{\partial x_j} - J_j^t \right], \tag{3}$$

where u_i is the velocity component, p the pressure, ρ the density, and ν the kinematic viscosity. The over-bar symbol stands for Reynolds averaging in the RANS approach and for spatial filtering in the LES. The terms τ_{ij}^t and J_j^t describe the contribution of unresolved flow and scalar structures in momentum (2) and scalar transport (3) equations. Within the RANS approach, τ_{ij}^t is the turbulent stress tensor $\tau_{ij}^t = \overline{u_i' u_j'}$, calculated either from the transport equations or from the Boussinesq assumption. The both approaches require closure models of different complexity. The standard $k-\varepsilon$ model, the SST model, and the Reynolds stress model (RSM) proposed by Launder et al. [15] have been used in the present work. Within the framework of the LES, τ_{ij}^t is the unclosed subgrid stress tensor, $\tau_{ij}^t = \overline{u_i u_j} - \bar{u}_i \bar{u}_j$, modelled through the filtered velocities $\bar{u}_i$ using the dynamic Germano model (henceforth, referred to as the DGM) and the dynamic mixed model (DMM). To prevent the numerical instability when the DMM is applied, a special clipping procedure derived from a rigorous mathematical analysis based on the Taylor series approximation has been adopted [12]. For calculation of the term J_j^t, the gradient diffusion assumption similar to the Boussinesq hypothesis for velocity parameters is adopted in both the RANS and LES. Within the RANS $J_j^t = \nu_t / Sc_t \cdot \partial \bar{f} / \partial x_j$, where ν_t and Sc_t are the turbulent kinematic viscosity and the turbulent Schmidt number, respectively. According to the LES, $J_i^{SGS} = \overline{f u_j} - \bar{f} \bar{u}_j$ is calculated through filtered values of the velocity and the mixture fraction using the dynamic SGS (subgrid scale) models as mentioned above. Simulation of spatially inhomogeneous turbulent flows using the LES requires unsteady turbulent inlet boundary conditions to be specified. The novel inflow generation technique suggested in [13] has been involved. This technique allows generating turbulent inlet velocities with prescribed autocorrelation functions and integral length scales taken from auxiliary LES computations. The wall models are used to avoid high resolution necessary for LES computations in the wall region (see [23]).

Simulation of chemically reacting flows needs additional equations to describe the transport of chemical reagents. Two typical reactions are considered, namely, a simple neutralisation reaction [6] and competitive-consecutive ones. Since the concentration of reagents is low, the thermal effects and the changes in liquid density can be neglected.

The neutralization reaction $A+B \xrightarrow{r_{AB}} P$ is assumed to be fast, irreversible, one-step, two-order and its rate $r_{AB} = K_{AB} Y_A Y_B$ with the constant K_{AB} equal to $10^8 m^3 /(kmol \cdot s)$. The parameters Y_A and Y_B are the concentrations of two reagents A (in co-flow) and B (issued from the nozzle), respectively. For turbulent mixing to be described, the theory of conserved scalars can therefore be adopted to calculate averaged mixture fraction and its variance. The chemical source term in the transport equation for reagent concentration is closed using the Eddy-Dissipation-Concept (hereinafter, the EDC) or the presumed β-PDF of the mixture fraction f [7]. The mixture fraction variance involved to calculate a β-PDF has been defined by solving its transport equation. This equation needs closing the mechanical-to-scalar time ratio. It is determined by its polynomial Re_t dependence valid for liquid with $Sc = 10^3$ and involving the low-Reynolds number effects in the mixing model (see [5] and [18]). The linear dependence for f permits the chemical system to be identified not through three concentrations Y_A, Y_B, and Y_P, but through two magnitudes – mixture fraction f and progress variable Y. For the first of them it is possible to invoke the statistical approaches applied for studying the turbulent motion of inert liquid. The progress variable Y in fact characterizes the concentration of forming reaction products, and the reagent concentration is determined as $Y_A/Y_{A0} = 1 - f - (1 - f_{st})Y$, $Y_B/Y_{B0} = f - f_{st}Y$, and $Y_P/Y_{B0} = f_{st}Y$ [7], where $f_{st} = 1/(1 + Y_{B0}/Y_{A0})$ is a stoichiometric value of the mixture fraction. The final system for calculation of the neutralisation reaction in the jet mixer contains equations (1)-(3) and the transport equation for the progress variable Y.

In the course of competitive-consecutive reactions $A+B \xrightarrow{r_{AB}} R$, $B+R \xrightarrow{r_{BR}} S$, where $r_{\alpha\beta} = -K_{\alpha\beta} Y_\alpha Y_\beta$, a reagent A interacts with a reagent B to form a desired product R that, in turn, reacts with the reagent B, and the second undesired product S is then formed. The second reaction can proceed only after the first has started and depends on the competitions between the reagents A and R to react with the remaining amount of the reagent B. For this chemical system the process of turbulent mixing determines not only a rate of chemical reacting, but can also affect a reaction product yield in the flow. At $K_{AB} > K_{BR}$ the first reaction dominates as compared to the case of $K_{AB} = K_{BR}$. Thereby, the production of the substance S starts to be noticeable at a large content of the reagent R in the mixture. It is assumed here that the first reaction is infinite fast and the second has a finite rate. A mathematical model for competitive-consecutive reactions is based on equations (1)-(3) coupled with the transport equations for normalized concentrations $\overline{Y_B}/\overline{Y}_{B_0}$, $\overline{Y_A}/\overline{Y}_{A_0}$, $\overline{Y_R}/\overline{Y}_{B_0}$, and $\overline{Y}_S/\overline{Y}_{B0}$

$$\frac{\partial \widehat{Y}_\alpha}{\partial t} + \frac{\partial \overline{u}_j \widehat{Y}_\alpha}{\partial x_j} = \frac{\partial}{\partial x_j}\left[\left(\frac{\nu}{\mathrm{Sc}} + \frac{\nu_t}{\mathrm{Sc}_t}\right)\frac{\partial \widehat{Y}_\alpha}{\partial x_j}\right] - \overline{\omega}_\alpha \qquad (4)$$

written following the gradient diffusion assumption. Here $\widehat{Y}_\alpha = \overline{Y}_\alpha / Y_{\alpha 0}$ and $\overline{\omega}_\alpha$ is a corresponding chemical reacting term.

3 Mixing in Jet Mixers

3.1 Performances of Mathematical Models for Simulation of Turbulent Mixing

A series of calculations have been performed to validate different RANS and LES models using software codes: Flowsi, OpenFoam, CFX, and Fluent. A thorough validation of various mathematical models as presented in [23] reveals that the SST model is the most accurate one among the RANS models, and the dynamic mixed model (DMM) is the best choice among the LES models. Distributions of mean values of mixture fraction and velocity calculated using the LES and RANS agree well with LIF and LDV measurements. The rms fluctuations of the mixture fraction and the velocity are in a good agreement with experimental data only for the DMM. Comparison between the LIF and numerical results is exemplified in Fig. 2 for the r-mode in the cross section $x/D = 1.6$ where the strongest influence of the recirculation zone is seen. The Germano model is much worse than the DMM. The superiority of the DMM is both due to intense production of vorticity and due to reproduction of energy backscattering on the starting length of the jet. The most evident advantage of the LES over the URANS modelling is its ability to reproduce unsteady effects in the r-mode. All LES models predict the unsteady behaviour of the flow, whereas asymptotically all URANS ones show a tendency to the steady flow state after some transitional phase at the beginning of simulation. As a result, the opposition-of-phase long-period oscillations revealed in both measurements and LES computations have been proved to be unattainable for URANS ones. The reason for this is the artificial smoothing of small-scale vortices whose collective nonlinear interaction causes unsteady large-scale influences to appear in the r-mode. On the contrary, LES computations capture a back-and-forth motion of a recirculation vortex cluster and opposition-of-phase long-period oscillations.

In the mixing models, consideration of Sc number effects is the problem that is still far from solution. The LES models used in this paper for scalar calculations have been originally developed for the case of $Sc \sim 1$. The variance of the mixture fraction in liquid mixtures should be larger than that in gas mixtures since the diffusion effects governed by the Schmidt number are sufficiently less in liquids than those in gases. This fact, which is beyond doubts, is explicitly used for instance in the multiple-time-scale (MTS) turbulent mixing model proposed in [1]. However the "gas oriented" models have provided quite reasonable results for the scalar variance in liquid mixtures at $Sc \sim 10^3$. This interesting feature is discussed in [14]. It has been shown experimentally that the dominating contribution to the scalar variance is made by large-scale fluctuations corresponding to the energy-containing and inertial-convective subranges of the spectrum. The contribution of fluctuations with scales of a few Batchelor lengths, which have not been taken into account in the SGS LES models, is minor. Obviously, this is the reason why the LES models originally developed for the case of $Sc \sim 1$ provide quite reasonable results for the scalar variance in liquid mixtures.

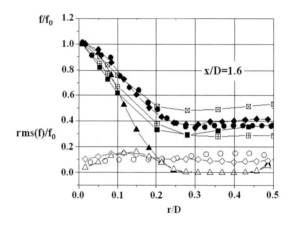

Fig. 2. Profiles of mean mixture fraction and its rms fluctuations at $Re_d = 10^4$, $\dot{V}_D / \dot{V}_d = 1.3$, $D/d = 5$. Mean mixture fraction: • experiment, ■ k-ε model, ◆ LES DMM, ▲ LES DGM, ⊞ URANS RSM, ⊠ SST model; rms fluctuations: ○ experiment, ◊ LES DMM, △ LES DGM.

3.2 Influence of Various Parameters on Mixing

From the dimensional analysis it can be seen that the characteristics of jet mixers depend on the following dimensionless parameters: the diameter ratio D/d, the Reynolds number for the nozzle flow $Re_d = dV_d / \nu$, the Schmidt number(Sc), the flow rate ratio, $\dot{V}_D / \dot{V}_d$, of the co-flow and the injected one from the nozzle, the liquid density ratio ρ_d / ρ_D. In the non-isothermal case, the Prandtl number and the temperature ratio $(T_D - T_d)/T_D$ are additional parameters. Table 1 summarizes our results of both numerical and experimental investigations performed for the both flow regimes. Symbols in Table 1 stand for the influence of increase in these or those parameters on mixing at constant values of others: 0- negligible influence, + weak enhancement, ++ strong enhancement of mixing. The flow rate ratio $\dot{V}_D / \dot{V}_d$ exerts the most influence on mixing. The influence of temperature,

Reynolds and Schmidt numbers, and density ratio on the mixing process in liquid mixtures is negligibly small.

Mixing can be sufficiently enhanced by some active methods of jet control. As seen from Table 1, the flow and mixing can be favourably governed by two following methods of flow control [11]

- Swirling the jet $\Omega = \omega d / U_d$, where ω is the swirl angular velocity. The larger the non-dimensional angular velocity Ω, the shorter is the mixing length.

- Flow rate oscillations $\dot{V}_d / \dot{V}_D = C(1 + A\sin\dfrac{2\pi U_d Sh}{d}t)$, where A is the amplitude and Sh is the Strouhal number of oscillations. If the excitation frequency is close to natural frequencies of vortices generated between the jet and the co-flow, then we have an effect similar to resonance which in this case is observed for Strouhal numbers between 0.2 and 0.4. The mixing length can reduce by 25% in the resonance case.

Table 1. Influence of various parameters on mixing

Parameter	Range	Influence on mixing
$\dot{V}_d / \dot{V}_D$	0.18 - 0.6	++
Re_d	10000 - 15000	0
ρ_d / ρ_D	0.6 - 1.6	0
Sc	0.7 - 1000	0
$\left|T_D - T_d\right| / T_D$	0 - 0.229	0
Swirling	$\Omega = 0 - 1.1$	++
Var $\dot{V}_d / \dot{V}_D$	$Sh = 0...3$ $A = 0...1$	++

3.3 Self-sustained Oscillations in Jet Mixers

One of the most important findings of the present work is the experimental and numerical identification of self-sustained flow oscillations in the r-mode. Topologically, the flow macrostructures in j- and r-modes are quite different. While the j-mode flow is very similar to the free jet, the r-mode flow exhibits two remarkable peculiarities mentioned already in [2]. First, the flow within the recirculation zone is highly unsteady and nearly periodic with a dominating long- period mode. Long-period oscillations with a dominating frequency are typical for the recirculation zone where large-scale vortex clusters are generated. They disappear in areas filled with small-scale vortices and scalar structures at $x/D \geq 2$ and in the vicinity of the mixer centreline. Second, the oscillations at symmetrical points relative to the centreline ($r/D = 0$) of the jet mixer are nearly anti-symmetric. The flow is

self-organized in such a way that the most probable event is a mixing enhancement at one point and at a time the opposite process is seen at the other point and
vice versa. Statistically, this behaviour is competent. These facts are clearly seen
in the time history of the mixture fraction recorded by the PLIF method and from
the analysis of the autocorrelation function of the mixture fraction fluctuations
across the pipe (see Fig. 5 in [25]). Our latest LES computations (see [14]) have
revealed that the reasons for the flow oscillations in the recirculation zone are the
concentrated vortex structures arising due to instability of primary spanwise vortices. Three-dimensional snapshots of vortex and scalar structures explaining the
physical nature of long-period anti-symmetric oscillations are shown in Figs. 3
and 4 in [14]. The interaction between the flow and the vortex structure is schematically illustrated in Fig. 3. Through stretching effects the vortex structures can
become strong and induce flow and scalar flux towards the pipe wall. This leads to
a scalar concentration excess near the pipe wall above the vortex structures and
simultaneously to a scalar deficiency on the opposite side of the pipe. This is the
reason for the phase shift of scalar fluctuations at symmetrical points about the
centreline. When the co-flow velocity U_D is increased the vortices arising in the
primary jet become weaker. They are not able to induce flow motions across and
opposite to the main flow which could be strong enough to overcome the co-flow.
The effects discussed above are weakening and the flow behaviour is changing to
the j-mode.

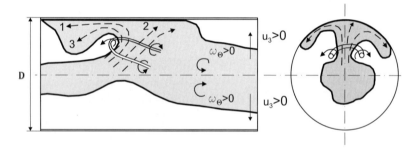

Fig. 3. Schematic illustration of the scalar flux induced by vortex structures

3.4 Structure of the Scalar Field in Jet Mixers

As mentioned above, the large-scale scalar structures of the j- and r-modes are
quite different. The transverse scalar integral scale $L_f(r)$ is growing up to
$x/D \sim 3$ in the both flow modes (see Fig. 4). Within the recirculation zone the
structures become smaller and the integral scale decreases at $x/D > 3$. In the j-
mode the integral scale continues to grow up to $x/D \sim 5$, where the interaction
between the jet and the pipe wall is strong and then diminishes. This fact is reflected in the variation of microstructures of the scalar field whose size is sufficiently smaller than the integral length scale. These structures can therefore be

classified as those of the turbulent scalar field. A snapshot of the mixture fraction field obtained using the high resolved PLIF is presented in Fig. 5. While scalar layers with a thickness of a few dozen microns can be recognized in the j-mode at $x/D = 7$, the scalar field in the r-mode seems to be fully smoothed already at $x/D = 5$. However, as shown in [14], the fine scalar structures have certain similar statistical properties in the both flow modes.

A remarkable feature of the scalar field is the presence of areas with a rapid change in the scalar. The structures called cliffs can easily be seen in Fig. 5. They cause the small-scale intermittency [24]. As a result, the statistics of the scalar difference $\Delta f(r) = f(r) - f(0)$ (the structure function of first order) is non-Gaussian and the intermittency is observed at inertial range scales. The intermittency is most pronounced in the front part of the recirculation zone at $x/D = 1$ and $r/D = 0.25$. During the well mixed stage at $x/D > 3$ in the r-mode the statistics of the scalar difference is almost Gaussian. In the j-mode the intermittency in the centreline is observed up to $x/D = 9$. In the centreline $r/D = 0$ the intermittency is less than at $r/D = 0.25$ for the both flow modes.

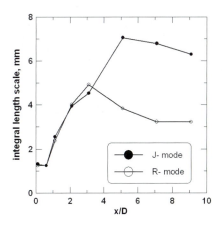

Fig. 4. Distribution of the scalar integral length along the axis

Fig. 5. Snapshot of the scalar distribution within the measurement window. $x/D = 2$, r- mode.

Anisotropy of scalar field at high Sc numbers is particularly important when developing LES SGS models. Small-scale anisotropy is assessed [14] in tests whenever the measured scalar gradient vector $\nabla f(x,r,t)$ shows any preferred orientation within the measurement plane. For this, the pdf distribution of the angle $\vartheta = tan^{-1} \dfrac{\partial f / \partial r}{\partial f / \partial x}$ measured in the (x,r)-plane from the x-axis has been examined. If the underlying fields $\nabla f(x,r,t)$ are fully isotropic, then the distribution of ϑ values should appear uniform. Despite a quite different structure of the macroflow in the *r*-mode and in the *j*-mode the pdf distributions of ϑ are proved to be independent of the flow mode. Along the jet mixer axis the scalar gradient field is very nearly isotropic with a slight preference of the gradient vector to align with the most compressive strain axis and the mean scalar gradient. Therefore, although each of the individual scalar structures exhibits a pronounced anisotropy, the collective statistics of fine scalar structures can be considered as nearly isotropic. The most pronounced anisotropy is detected in the *r*-mode only in the front part of the recirculation zone at $r/D = 0.25$ and $x/D = 1$, whereas at $x/D > 2$ the scalar field becomes nearly isotropic (see Fig. 6). The results show clearly that at $x/D = 1$ the gradient vector of scalar structures corresponding to different dissipation rates has a preferred direction at $\vartheta \approx -50^0$, which is in agreement with the orientation of the most compressive strain axis at $\vartheta_{sa} \approx -46^0$ gained from our LES computations.

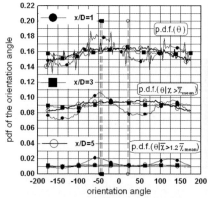

Fig. 6. Distribution of scalar gradient orientation angles at *r/D*=0.25 in the *r*-mode. The vertical dotted lines show the orientation of the most compressive mean strain axis.

Fig. 7 Scalar dissipation rate

$$\chi_{2D} = D_i \left[\left(\frac{\partial f}{\partial x} \right)^2 + \left(\frac{\partial f}{\partial y} \right)^2 \right]$$

in the centreline at *x/D* = 2 in the *r*-mode. The actual size of the windows is 2.74 x 2.075 mm, whereas the corresponding integral length is 4 mm.

3.5 Dissipation Rate Field

The dissipation rate $\chi_{2D} = D_i \nabla f \nabla f (x, y, t)$ obtained from 2D PLIF measurements, where D_i is the scalar diffusivity, is concentrated in thin dissipation layers (see Fig. 7). Three fundamental topologies as introduced in [3] can be identified in the scalar field. They include long regions with many straight and nearly parallel dissipation layers, areas, wherein two such-long regions meet orthogonally, and spiral structures (see Fig. 7).

One of the latest trends in the development of LES SGS models for scalar and velocity fields is the utilization of the multifractal properties of turbulence (see [4]). According to Novikov [20], the fractional coefficient $q_{r,l}$ of the dissipation rate defined by the condition

$$q_{r,l}(h,x) = \chi_r(x_l)/\chi_l(x),$$ (5)

where $\chi_l(x) = l^{-1} \int_{x-l/2}^{x+l/2} \chi(x_l)dx_l$, $-0.5 \leq h = \frac{x_l - x}{l-r} \leq 0.5, r < l$, depends only on the ratio l/r and $|h|$, provided that l and r obey the condition $L_u \gg l > r \gg l_*$ and that in the range between l_* and L_u there are no other characteristic scales influencing the stochastic function $\chi(x)$. Novikov's theorem implies that the function $\chi(x)$ is locally uniform and isotropic at scales smaller than a certain integral length scale L_u and there exists the second characteristic scale l_* determined either by viscosity or by diffusion. A very strong assumption of Novikov's theory is the statistical independence of consecutive fractional coefficients

$$\overline{q_{r,\rho}(h, x+h(l-\rho))\, q_{\rho,l}(h,x)} = 0,$$ (6)

where $r < \rho < l$ at scales between l_* and L_u. The scale similarity postulated by Novikov implies the multifractal nature of turbulence closely related to the concept of cascade turbulence transfer [9].

The analysis of this property [14] is based on considering the multiplier distribution $P(M)$ defined as the pdf of the ratio of dissipation rates calculated on different levels of a hierarchy of parents and children boxes of uniform size. Each parent box with the size p is subdivided into a number of children boxes κ with size $p_l = p/\kappa$, and the ratios of dissipation rates in the original box to those in the children sub-boxes are computed. A histogram of these ratios is then denoted by $P(M)$. According to Novikov's concept, the shape of the distribution $P(M)$ should remain invariant for the inertial range of scales. This fact is used for developing LES SGS models [4]. Unfortunately, the invariance of $P(M)$ has not been supported (see Figs. 19 and 20 in [14]). Since the inertial range has been reliably resolved (see Fig. 6 in [14]) the only possible explanation for the contradiction between our experimental data and Novikov's theory can be the failure of assumption (6). Indeed, the statistical analysis of correlations between successive fractional coefficients (5) reveals a strong statistical dependence among them. A possible explanation for this fact can be found in [9]. As mentioned in [9], there exist elongated coherent objects in each turbulent flow which are difficult to classify into cascade steps. They span a full range of the cascade and some part of the cascade is blocked. This physical mechanism referred to as the blocking effect [9] makes the distributions $P(M)$ not invariant. Therefore, the use of the multifractional concept to develop new SGS models is very questionable.

4 Chemical Reactions in Jet Mixers

4.1 Measurement of Flow with Neutralization Reaction

Measurements were performed with the concentration of HCL Y_{A_0} (4.6 mM/l) and of sodium hydroxide Y_{B_0} (100mM/l) for the stoichiometric ratio $f_{st} = Y_{B0} /(Y_{A0} + Y_{B0}) = 0.956$. A small stoichiometric ratio is favourable since the product cannot be identified at $f > f_{st}$ due to insufficient increase in the pH value. As a result, a zone exists close to the nozzle where measurement of scalar variation is impossible. The smaller the stoichiometric ratio, the shorter is this zone. Soft water was used as solvent to prevent calcium fall-out in the water channel during experiments. Figure 8 shows the radial distributions of the pulsation correlation $\overline{y'_A y'_B}$ in different cross sections of the jet mixer. They are negative and have a peak within the mixing layer. These data can be used for assessment of closure models for the reaction rate. Distributions of the product along the mixer centreline are presented in Fig.9.

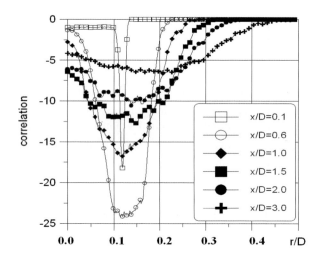

Fig. 8. Radial profiles of the correlation $\overline{y'_A y'_B}$. j-mode: $Re_d = 10^4$, $\dot{V}_D / \dot{V}_d = 5.0$, $D / d = 5$

4.2 RANS and LES Computations of Neutralization Reaction

Figure 9 demonstrates the variations of the averaged concentration of the reaction product along the mixer axis. As seen at the initial stage of the jet development at

$x/D < 0.5$, the reaction product is not formed. In this region the mixture fraction almost does not vary and is close to 1, which corresponds to the fact that only one of the reagents – acid – is present here. Entraining the co-flow liquid by the jet and further mixing causes the mixture fraction to decrease. In doing so, the chemical reaction proceeds and the fraction of the reaction product in the mixture grows (see Fig. 9). If the amount of the co-flow reagent (base) in the jet reaches the level when the stoichiometric reagent ratio is valid, the chemical reaction can be identified. The concentration of the reaction product sharply increases over the range from $x/D = 1.5$ to $x/D = 2$ where it attains a maximum (see Fig. 9). The acid amount drastically decreases due to an excess of the base, and the reaction in fact ceases behind the distance $x/D > 2$. The reaction product behaves like a passive admixture that is mixed with the co-flow liquid. The reaction product concentration decreases and the behavior of $\overline{Y}_P / Y_{A0}$ is identical to the mixture fraction variation in this region. In the region of the intense reaction the RANS models overestimate the quantity $\overline{Y}_P / Y_{A0}$ with reference to the predicted values from experiment and the LES. The LES with a relatively simple closure model for the chemical source term agrees well with measurements of the averaged concentration of the reaction product. The LES provides a better agreement with the measurements than the RANS using similar closure models for the chemical source.

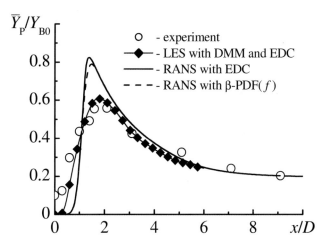

Fig. 9. Averaged product concentration at the mixer axis

4.3 RANS Computation of Competitive-Consecutive Reactions

Six computational runs presented in Table 2 were performed for $d = 0.01m$, $D = 0.05m$, $Sc = 1000$, $Y_{B_0} = 2mol/m^3$, $K_{AB} = 7300m^3/(mol \cdot s)$, $K_{BR} = 1.63m^3/(mol \cdot s)$. The flow rate ratio was $\dot{V}_D / \dot{V}_d = 1.3$ in runs 1-3 (r-mode) and $\dot{V}_D / \dot{V}_d = 5.0$ in runs 4-6. The Reynolds number, Re_d, was consequently increased from 10^4 in runs 1 and 4 to $5 \cdot 10^4$ in runs 2 and 5 and to 10^5 in runs 3 and 6. The turbulence intensity at the inlet was estimated from the formula $Tu = u' / U_{bulk} = 0.16 Re_H^{-1/8}$, where the Reynolds number, $Re_H = U_{bulk} \cdot d_H / v$, was calculated through hydraulic diameters $d_H = d$ for the jet and $d_H = D - d$ for the co-flow. At present only RANS computations were performed using two following approaches to close the chemical reaction rate $\bar{\omega}_{\alpha\beta}$: the *middle kinetics* (MK) approach, which neglects the correlations of concentration fluctuations, $\bar{\omega}_{\alpha\beta} = -K_{\alpha\beta}\bar{Y}_\alpha\bar{Y}_\beta$, and the *Li-Toor closure* using a normal mixture fraction pdf (LT) [17]. The objective of the study was to calculate the selectivity $X_S = 2\bar{Y}_S / (\bar{Y}_R + 2\bar{Y}_S)$. Table 2 cites the X_S values taken at the jet mixer outlet where the chemical reaction and mixing were completed and the radial profile of selectivity was constant in the mixer cross section.

Table 2. Selectivity versus Reynolds number and concentration ratio

Run	Mode	Y_{B_0} / Y_{A_0}	Selectivity X_S	
			MK model	LT model
		1	0.046	0.001
1	r	2	0.659	0.496
		4	0.950	0.644
		1	0.012	0.005
2	r	2	0.388	0.325
		4	0.634	0.482
		1	0.006	0.004
3	r	2	0.255	0.226
		4	0.443	0.363
		1	0.048	0.030
4	j	2	0.076	0.049
		4	0.114	0.079
		1	0.012	0.012
5	j	2	0.019	0.018
		4	0.029	0.029
		1	0.007	0.007
6	j	2	0.010	0.010
		4	0.016	0.016

The following conclusions can be drawn from the analysis of the Table 2 data:

- Increasing the Reynolds number (velocity in the jet mixer) results in growing a desired product yield.

- At $Y_{B_0}/Y_{A_0} > 1$, the j-mode is more efficient than the r-mode with reference to the yield of a desired product.

- Increasing the initial concentration ratio Y_{B_0}/Y_{A_0} leads to a growth of the undesired by-product yield. Decreasing the initial concentration ratio implies the decrease in the amount of B available for the reaction, so that B is consumed earlier than the second reaction begins to proceed.

- Taking into account the fluctuation correlations $\overline{y'_\alpha y'_\beta}$ by the Li- Toor closure decreases the selectivity X_S. Differences between two MK and LT models attain 30%. Only for the j-mode they tend to each other when the Reynolds number is increased.

5 Conclusions

Co-axial jet mixers are well developed and widespread in many fields of chemical industry. The parameters most influencing the mixing processes in these mixers are the diameter and flow rate ratios. The effect of other parameters is insignificant. Mixing can be sufficiently enhanced by nozzle flow rate oscillations and jet swirling. Depending on the flow rate ratio, two different flow modes are observed, namely, the j-mode (jet-like) and the r-mode (with recirculation zone). In the j-mode the structure of the flow is similar to that of the free jet. In the r-mode a strong flow separation at the pipe walls results in a recirculation zone behind the nozzle. Coherent vortex structures with dominant stream-wise components of vorticity cause oscillations containing a dominating long-period mode. PLIF measurements and numerical LES computations have revealed a phase shift of fluctuations at symmetrical points about the pipe centreline. The LES analysis allows one to explain this phenomenon and to propose a schematic scenario of interaction between vortex structures and a scalar field illustrating the appearance of the phase-shifted oscillations (see Fig. 3). The presence of such oscillations is very critical for numerical methods. Whereas the URANS models have proved to be accurate enough to predict the mean values of velocity and scalar, the unsteady effects in the r-mode are not modelled by the URANS. The reason is the artificial smoothing of small-scale vortices whose collective nonlinear interaction results in unsteady large-scale effects to appear in the r-mode. On the contrary, LES computations capture a back-and-forth motion of the recirculation vortex cluster and the opposition-of-phase long-period oscillations. The best choice among the LES SGS models is the dynamic mixed model. For the flow under consideration the most contribution to the scalar variance is made by large-scale motions, whereas the contribution of fine scales smaller than typical inertial range scales is negligible. This is the reason why the LES models originally developed for the case of Sc~1 provide quite reasonable results for the scalar variance in liquid mixtures at Sc~10^3.

Fine structures of the scalar field are topologically similar in the both flow modes. In the r-mode the fine structures become smaller behind the recirculation zone and are difficult to recognize visually at $x/D=5$. On the contrary, in the j-mode the fine structures can be clearly identified up to $x/D=7$. Despite a big difference in the structure of the macroflow and in the dynamics of fine structures, the typical statistical properties of the turbulent scalar fields are very similar for the both flow modes. The scalar field is very intermittent with the areas of rapid change in the scalar. These structures called the cliffs cause small-scale intermittency that is strongly dependent on the flow mode. The intermittency is most pronounced in the front part of the recirculation zone and becomes weaker in the centreline and downstream. At the well mixed stage the scalar field has Gaussian statistics. Each of the particular scalar structures exhibits pronounced anisotropy. However, the collective statistics of fine scalar structures can be considered as nearly isotropic. The orientation of the scalar gradient is more or less uniform in space with a slight preference to align with the direction of the most compressive mean strain axis. The scalar dissipation rate is concentrated in thin dissipation layers having the form of long regions consisting of many straight and nearly parallel dissipation layers, spiral structures, and saddle-type structures. The study of multiplier distributions confirms the fact that the dissipation rate is distributed very unevenly in space. However, Novikov's ideas of the cascade model have not been supported in our measurements.

Reacting flows in the jet mixer have been investigated using the LES, the RANS and simultaneous PLIF measurements of product concentration and mixture fraction. As for the neutralization reaction, the useable RANS models overestimate the generation rate of a reaction product and are in a poor agreement with the LES results and the experimental data on the flow region with intense chemical reacting. The jet mixer is an efficient reactor design to suppress the yield of undesired by-product in competitive-consecutive chemical reactions. Such reactions have been studied numerically using RANS models. It has been shown that the yield of desired product and the reactor efficiency increase when either the Reynolds number (velocity in the jet mixer) is increased or the initial reagent concentration ratio Y_{B_0}/Y_{A_0} is decreased. At $Y_{B_0}/Y_{A_0}>1$, the j-mode is more efficient than the r-mode with reference to the yield of desired product.

Acknowledgment

The authors acknowledge gratefully the support of the German Research Foundation (DFG) through the program SPP 1141. LES computations have been performed on an IBM pSeries 690 Supercomputer at the North German Alliance for the Advancement of High-Performance Computing (HLRN). A number of results presented in the paper have been obtained in cooperation with Dr. I. Tkatchenko, Dr. S. Jahnke, and Dipl.Ing. J. Turnow whose contribution is highly appreciated.

Nomenclature

d	nozzle diameter	p	pressure
D	pipe diameter	ρ	density
u	velocity	K_{xx}	reaction constant
u'	velocity fluctuations	r_{xx}	reaction rate
u_d, u_D	velocity in nozzle, velocity in pipe	ω_{xx}	chemical source term
$\dot{V}_d$, $\dot{V}_D$	volume flow rate in nozzle, volume flow rate in pipe	Sh	Strouhal number
ν, ν_t	kinematic viscosity, turbulent kinematic viscosity	X	selectivity
Re	Reynolds number	T, ϑ	temperature
Sc, Sc_t	Schmidt number, turbulent Schmidt number	χ	dissipation rate
f, f_{st}	mixture fraction, stoichiometric mixture fraction	τ_{ij}^t	turbulent stress tensor (RANS) or subgrid stress tensor (LES) in momentum equations
Y_{xx}	concentration of reagent xx	J_j^t	subgrid stress flux in scalar transport equations
y'_{xx}	concentration fluctuations of reagent xx	J, I	intensity
Tu	turbulence intensity	Ω	non-dimensional angular velocity
$q_{r,l}$	fractional coefficient of the dissipation rate		

References

1. Baldyga, J., Bourne, J.R.: Turbulent Mixing and Chemical Reactions. John Wiley and Sons, Chichester (1999)
2. Barchilon, M., Curtet, R.: Some details of the structure of an axisymmetric confined jet with backflow. J. Basic Engineering, 777–787 (1964)
3. Buch, K.A., Dahm, W.J.A.: Experimental study of the fine-scale structure of conserved scalar mixing in turbulent shear flows. J. Fluid Mech. 317, 21–71 (1996)
4. Burton, G.C., Dahm, W.J.A., Dowling, D.R., Powell, K.G.: A new multifractal subgrid-scale model for large-eddy simulation (2002) AIAA Paper 2002-0983
5. Chorny, A.D., Kornev, N.V., Hassel, E.: Simulation of the passive scalar mixing in the jet mixer. J. Eng. Phys. and Thermophysics 81(4), 666–681 (2008)

6. Chorny, A.D., Turnow, J., Kornev, N., Hassel, E.: LES versus RANS modelling of turbulent mixing involving chemical reacting in a co-axial jet mixer. In: Proc. Int. Symp. Advances in Comp. Heat Transfer Paper CHT-08-223 (2008)
7. Fox, R.: Computational models for turbulent reacting flows. Cambridge University Press, Cambridge (2003)
8. Henzler, H.J.: Investigations on mixing fluids. Dissertation. RWTH Aachen (1978)
9. Jimenez, J.: Intermittency and cascades. J. Fluid Mech. 409, 99–120 (2000)
10. Koochesfahani, M.M., Dimotakis, P.E.: Mixing and chemical reactions in a turbulent liquid mixing layer. J. Fluid Mech. 170, 83–112 (1986)
11. Kornev, N., Tkatchenko, I., Zhdanov, V., Hassel, E., Jahnke, S.: Simulation and measurement of flow phenomena in a co-axial jet mixer. In: Humphrey, J.A.C., et al. (eds.) Proc. 4th International Symposium Turbulence and Shear Flow Phenomena, Williamsburg VA USA, vol. 2, pp. 723–728 (2005)
12. Kornev, N., Tkatchenko, I., Hassel, E.: A simple clipping procedure for the dynamic mixed model based on Taylor series approximation. Commun. Numer. Meth. Eng. 22(1), 55–61 (2006)
13. Kornev, N., Hassel, E.: Synthesis of homogeneous anisotropic divergence free turbulent fields with prescribed second-order statistics by vortex dipoles. Physics of Fluids 19, 068101 (2007)
14. Kornev, N., Zhdanov, V., Hassel, E.: Study of scalar macro- and micro-structures in a confined jet. Int. J. Heat and Fluid Flow 29(3), 665–674 (2008)
15. Launder, B., Reece, G., Rodi, W.: Progress on the developments of a Reynold-stress turbulence closure. J. Fluid Mech. 68, 537–566 (1975)
16. Lima, M., Palma, J.: Mixing in co-axial confined jets of large velocity ratio. In: Proc. 10th Int. Symp. Application of Laser Technique in Fluid Mechanics, Lisboa (2002)
17. Li, K.T., Toor, H.L.: Turbulent Reactive Mixing with a Series-Parallel Reaction: Effect of Mixing on Yield. AIChE J. 32(8), 1312–1320 (1986)
18. Liu, Y., Fox, R.O.: CFD Predictions for chemical processing in a confined impinging-jets reactor. AIChE J. 52(2), 731–744 (2005)
19. Mortensen, M., Orciuch, W., Bouaifi, M., Andersson, B.: Mixing of a jet in a pipe. Trans. IChemE 81A, 1–7 (2003)
20. Novikov, E.A.: Intermittency and scale similarity of the turbulent flow structure. Applied Mathematics and Mechanics 35(2), 266–277 (1971)
21. Rehab, B., Villermaux, E., Hopfinger, E.: Flow regimes of large-velocity-ratio coaxial jet. J. Fluid Mech. 345, 357–381 (1997)
22. Tebel, K.H., May, H.O.: Der Freistrahlreaktor - Ein effektives Reaktordesign zur Unterdrückung von Selektivitätsverlusten durch schnelle, unerwünschte Folgereaktionen. Chem. Ing. Tech. 60(11) (1988)
23. Tkatchenko, I., Kornev, N., Jahnke, S., Steffen, G., Hassel, E.: Performances of LES and RANS models for simulation of complex flows in a coaxial jet mixer. Flow Turbulence and Combustion 78, 111–127 (2007)
24. Warhaft, Z.: Passive scalars in turbulent flows. Annual Review of Fluid Mechanics 32, 203–240 (2000)
25. Zhdanov, V., Kornev, N., Hassel, E., Chorny, A.: Mixing of confined coaxial flows. Intl. J. Heat and Mass Transfer 49, 3942–3956 (2006)
26. Zhdanov, V., Kornev, N., Hassel, E.: The influence of the mixer geometry on the scalar field formation. In: Egbers, C., et al. (eds.) Lasermethoden in der Strömungsmesstechnik, 13 Fachtagung, Cottbus: Brandenburgische Technische Universität, Germany, September 6-8, pp. 37-1–37-8 (2005)
27. Zhdanov, V., Kornev, N., Hassel, E.: Identification of the microstructure in an axial jet mixer by LIF. In: 14 Fachtagung Lasermethoden in der Strömungsmesstechnik, GALA, Braunschweig, September 5-7, pp. 32.1–32.8 (2006)

Formulation and Validation of an LES Model for Ternary Mixing and Reaction Based on Joint Presumed Discrete Distributions

Frank Victor Fischer and Wolfgang Polifke

Fachgebiet für Thermodynamik, Technische Universität München, Boltzmannstraße 15, 85748 Garching

Abstract. A method for large eddy simulation (LES) of ternary mixing and reaction in turbulent flows is presented and partly validated against results of direct numerical simulation (DNS). The subgrid-scale mixing state is characterized by joint presumed discrete distributions (jPDDs), i.e. discrete particle ensembles which approximate multivariate filtered density functions. Biased mixing models are employed to generate particle ensembles with prescribed first and second order moments of the mixture fractions. Filtered reaction rates, which are computed as an average over the discrete distributions, can in this way be parameterized by the first and second order moments. It is emphasized that the covariance, which can have a significant impact on the filtered reaction rate, is taken into account. A closure for the moments transport equations is formulated in the LES context, including a non-equilibrium model for the subgrid scale scalar dissipation and the subgrid scale scalar cross-dissipation rates.

For validation purposes, DNS results for the mixture fraction distribution in a co-annular jet-in-crossflow configuration have been filtered on a coarser grid, as it would be suitable for LES computations. The DNS mixture state is then compared against particle ensembles generated with the mixing model. In terms of filtered reaction rates, good agreement has been found over a wide range of first and second order moments.

1 Introduction

In both chemical engineering and combustion, mixing plays a crucial role in assessing and understanding the performance of reacting flow systems, as mixing processes bring the reactants into direct molecular contact, which is required for a chemical reaction to actually take place. For turbulent flows, Large-Eddy-Simulation (LES) has been shown to be a suitable tool for the numerical simulation of such systems, as large-scale process are resolved directly. However, processes occurring on smaller scales (i.e. subgrid-scale) need to be modeled. It is generally assumed that the subgrid-scale flow and mixing processes are more universal in nature than the resolved processes, and thus permit the use of simpler, less costly and more robust models than corresponding Reynolds-averaged approaches.

Jaberi and Colucci [13, 14] show that for LES of turbulent mixing and reaction, three closure problems occur: The subgrid-scale stresses, the subgrid-scale scalar fluxes and the subgrid-scale unmixedness. For the two former closure problems, established models are used in this work: The subgrid-scale stress is modeled by an eddy-viscosity (i.e. Smagorinsky-type) approach along with the dynamic procedure as introduced by Germano et al [9] and the subgrid-scale scalar fluxes are modeled by an eddy-diffusivity approach as suggested by Eidson [7], where a dynamic procedure is employed as well.

Thus, modelling the subgrid-scale unmixedness in order to achieve closure for the chemical source term in reacting flows is the focus of the present work. Consider the transport equation for a mass fraction f of a species:

$$\frac{D\overline{f}}{Dt} = D\frac{\partial^2 \overline{f}}{\partial x_i^2} - \frac{\partial \tau_{if}}{\partial x_i} + \overline{\omega}, \tag{1}$$

where τ_{if} is the subgrid-scale scalar flux as described above and $\overline{\omega}$ is the filtered reaction rate. It has been shown that the performance of the subgrid-scale models, namely their accurate modelling of the subgrid-scale mixing is paramount when accurate prediction of the (filtered) reaction rates is required.

Filtered Density Functions (FDFs) as suggested by Fox and others can be used to describe the mixing state within each cell. Several alternatives exist to model such FDFs:

1. In order to minimize computational effort, functions of presumed shape can be employed. This approach has been proposed, for example by Gutheil and Bockhorn [11, 10] and others using, for example, clipped Gaussian or β-functions.
2. For high accuracy a transported PDF, which involves the tracking of a very large number of particles, is used. Mixing models are used to implement particle-particle interaction. However, this approach poses a large computational overhead on the simulation compared to the non-reacting and non-mixing LES of the velocity and pressure fields.

The present work combines ideas from both approaches, as it employs multi-variate ensembles of discrete particles to represent Filtered Density Functions (as in transported PDFs), but generates those ensembles using transport equations for statistical moments (as in presumed PDF). Those FDFs provide a closure for the chemical source term and are used to model the subgrid-scale mixing state of each cell. The particle ensembles (or *distributions*) are hereby defined by their first and second order statistical moments. This so called joint probability discrete distribution (or jPDD) approach has been suggested previously for the Reynolds-Averaged-Navier-Stokes (RANS) context [1, 2, 3, 5]. Brandt et al have employed the jPDD approach successfully for combustion processes. In this work, the jPDD approach is refined, adapted for LES and validated using DNS data.

The particle ensembles are generated using particle-based mixing models [6, 12, 18, 16] that have been modified so that they can create *joint probability discrete distributions*. The mixing model have been modified so that, given a set of *target values* for first and second order moments, it generates a particle ensemble which has the desired moments. It is emphasized that the mixing model obtains no other parameter from the flow field than this set of moments, and as such the coupling of the CFD with the mixing model is fundamentally different from that in the transported PDF approach, where there is a close linkage between the mixing model and the flow field. This causes the present model to be unable to account for effects like the mixing frequency, the movement of particles in space, etc which are accounted for in transported PDF approaches.

The first and second-order moments are determined by a set of transport equations in the LES solver. As shown later, these transport equation require closure by modelling e. g. the scalar variance production and scalar variance dissipation terms. While it has been suggested (e.g. [17]) to assume local equilibrium for the production and dissipation of scalar variance, this idea is not pursued here, as scalar variance is not decaying while using this approach (which actually implies that no subgrid-scale mixing occurs).

In order to describe mixing of multiple streams, called *ternary mixing* in the case of three streams, each distribution depicts multiple scalar fractions that can represent mixture fraction. As such, there are multiple statistical moments for each cell: The first-order moment represents the mean values for each mass fraction, while the second-order moment represent the variance *and* the co-variance between any two of such fractions.

As demonstrated by Brandt et al. [1, 2, 4] and in Sect. 5, neglecting the co-variance in the ternary mixing case can lead to significant errors in computing the filtered reaction rate. Therefore, the model developed in this work includes a transport equation for the subgrid-scale scalar covariance, including models for the scalar cross-dissipation rate (which is the co-variance-equivalent to the scalar dissipation rate in the variance transport equation). Furthermore, the mixing models used to generate jPDDs are extended so that multi-variate distributions of arbitrary co-variances can be generated.

The present work validates the proposed LES approach by comparing the distributions and the corresponding reaction rates of a model reaction to DNS data. The benchmark case comprises a coannular double jet-in-crossflow configuration, where the jet itself consists of two distinct streams, one central jet surrounded by a second, annular jet. There are three streams altogether (the main flow, plus the two jets), so that the mixing state can be described by two scalar mixture fractions, which need to add up to unity:

$$\sum f_i = 1. \qquad (2)$$

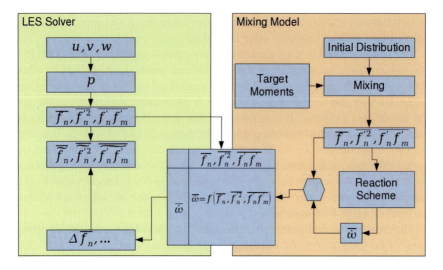

Fig. 1. Workflowing during the LES of tenary mixing

Thus, the LES needs five additional transport equations: Two for the means of the scalars $\overline{f}_i$, two for the variances of the scalar $\overline{f_i'^2}$ and one for the co-variance of between the two scalars $\overline{f_1' f_2'}$. The LES solver then uses the mixing model to generate a distribution with those five moments for each cell. In order to speed up computations, the distributions (or their associated mixture fractions) can be stored in a look-up table during or prior to the computation. The entire workflow is shown in figure 1.

The objectives of this work have consequently been:

1. Extend the jPDD approach by Brandt et al. [1, 2, 4] and adapt it for LES, i.e. implement the transport equations for the moments, including all required subgrid-scale models for the source terms.
2. Extend the mixing model so that jPDDs with arbitrary co-variance can be generated.
3. Perform DNS and prepare data for *a priori* validation of the model.
4. Validate the filtered reaction rates computed from jPDDs against those obtained from DNS data.

2 Biased, Multi-variate Mixing Modes

Formulation of a biased, multi-variate mixing model is used to generate joint particle distributions which possess prescribed statical moments, i.e. means, variances and covariances as requested by the LES solver for each cell and each timestep. The generation of the distributions is performed indepedently of the solver (and can even be offloaded to a preprocessing step with the results being stored in a lookup table). As such, the number of mixing steps or the intensity of the single mixing process (called a *mixing event*) are of subordinate importance, since no parameters (like for example the mixing frequency) of the flow are required for obtaining the mixing parameter.

As mentioned, particle-interaction models have been used in this work. These models display the molecular mixing by moving (in state space) the particles from their positions within the initial distribution towards the mean value, which is conserved. One of the most basic features of a mixing model is to relax the variance of the distribution. Under the influence of a mixing model, any distribution will evolve to a single Dirac-Delta peak (i.e. no variance). The IEM [6] and Modified Curl [12] models both meet these requirements. Other mixing models like the EMST suggested by Subramaniam and Pope [18] or the PSP suggested by Mayer and Jenny [16] meet additional requirements that may be formulated for mixing models like the localness in scalar space, correct dependence on scalar length scales and correct dependence on Reynolds, Schmidt and Damköhler Numbers [16].

For the present work, distributions have been created using a variant of the Modified Curl model, extended to support multiple mass fractions using indepedent mixing operations for all mass fractions. Futhermore, the model has been adapted to generate arbitrary covariances, this feature will be described in Sect. 2.2. The mixing model uses an initial distribution that already possesses the required means while presenting the maximum possible variance as shown in figure 2. This can be achieved by moving all particles to the extremes of the state space, which, in terms of an FDF, represents Dirac-Delta peaks. The Modified Curl model then randomly selects pairs of particles $f^{(p)}$ and $f^{(q)}$ from the distribution and their compositions

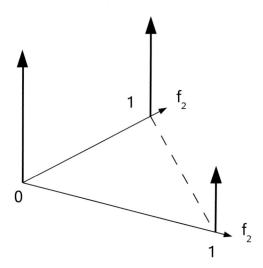

Fig. 2. Initial distribution for the mixing model. Dashed line indicates the boundary of the physically accessible state-space: The sum of mixture fractions can never exceed unity.

are changed over a mixing event. Since in this work every particle carries multiple mixture fractions $f_i, i = 1, 2$ with $\sum_{i=1}^{2} f_i \leq 1$, these mixture fractions are changed as

$$f_{i,n+1}^{(p)} = f_{i,n}^{(p)} + \frac{1}{2}a\left(f_{i,n}^{(p)} - f_{i,n}^{(q)}\right), \tag{3}$$

$$f_{i,n+1}^{(q)} = f_{i,n}^{(q)} + \frac{1}{2}a\left(f_{i,n}^{(q)} - f_{i,n}^{(p)}\right), \tag{4}$$

where (p) and (q) denote the two chosen particles, n and $n+1$ denote the state of the particle before and after the process and a is a randomly chosen number (called the mixing parameter) which is distributed uniformly between zero and unity.

2.1 Variance Adjustment

The variance V of a random variable X with the expected value $\mu = E(X)$ can be written as

$$V(X) = E\left[(X - \mu)^2\right], \tag{5}$$

In analogy to Subramaniam and Pope [18] it is assumed that the scalar variance decays as:

$$V(n) = Ve^{-Cn}, \tag{6}$$

where V_0 represents the initial variance of the distribution, C denotes a constant that is a function of the mixing parameter in equation 3 and n is the number of mixing processes. If the mixing parameter a in equations (3) and (4) is distributed in the same fashion for each mixture fraction i, it is obviously not possible to obtain distributions that have greatly differing variances for both mass fractions, unless the initial distribution already possesses such variances. Thus, a different mixing parameter a_i is needs to be assigned for each mass fraction:

$$f_{1,n+1}^{(p)} = f_{1,n}^{(p)} + \frac{1}{2}a_1\left(f_{1,n}^{(p)} - f_{1,n}^{(q)}\right), \tag{7}$$

$$f_{1,n+1}^{(q)} = f_{1,n}^{(q)} + \frac{1}{2}a_1\left(f_{1,n}^{(q)} - f_{1,n}^{(p)}\right), \tag{8}$$

$$f_{2,n+1}^{(p)} = f_{2,n}^{(p)} + \frac{1}{2}a_2\left(f_{2,n}^{(p)} - f_{2,n}^{(q)}\right), \tag{9}$$

$$f_{2,n+1}^{(q)} = f_{2,n}^{(q)} + \frac{1}{2}a_2\left(f_{2,n}^{(q)} - f_{2,n}^{(p)}\right). \tag{10}$$

Here, i indicates the which mass fraction is just modified.

If different mixing parameters a_i are chosen, the decay law (6) for the two mass fractions can be written as

$$V_{1;\,\text{target}} = V_{1;\,\text{initial}} e^{-C_1 n}, \tag{11}$$

$$V_{2;\,\text{target}} = V_{2;\,\text{initial}} e^{-C_2 n}. \tag{12}$$

It can been seen that if the initial and target variances are known, the number of mixing processes is canceling out and the only remaining unknown from equations (11) and (12) is the quotient of the mixing parameters C_1/C_2

$$\frac{a_1}{a_2} \propto \frac{C_1}{C_2} = \ln\left(\frac{V_{1;\,\text{target}} V_{2;\,\text{inital}}}{V_{2;\,\text{target}} V_{1;\,\text{initial}}}\right). \tag{13}$$

It should be noted for stability reasons that a_i cannot exceed unity, which imposes a minimal number of steps required to reach the given target values for the variances.

Thus, when two particles are selected for mixing, one (random) mixing parameter a_i is generated for *each* of the mixture fractions i, but the maximum mixing parameter that can be generated for each mixture fraction have the relation shown in equation (13). This process is called the *biasing* of the random number generator. A typical variance evolution over the mixing processes is shown in Figure 3. It can be seen that while both mass fractions commence at similar variances, the desired terminal variances have been reached after the same number of mixing processes. The deviation from the perfect e^{-x} law by both variances can be explained by the statistical nature of the selection of the mixing parameter and particles.

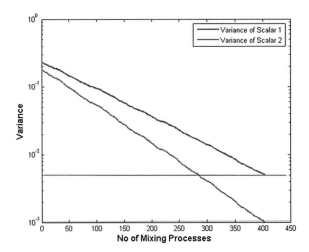

Fig. 3. Variance decay over mixing processes for the Modified Curl Model over a total of 400 mixing processes with 1600 particles. Both target variances are met after the same number of mixing events.

2.2 Covariance

Distributions can have different covariances or correlations, with the covariance C of two random variables X and Y is defined as

$$C(X,Y) = E\left[(X - \mu)(Y - v)\right], \tag{14}$$

with $\mu = E(X)$ and $v = E(Y)$ being the expected values of X and Y. The correlation $\rho_{X,Y}$ of X and Y can be interpreted as the normalized covariance (i. e. the correlation is always between -1 and 1).

$$\rho_{X,Y} = \frac{C(X,Y)}{\sqrt{V(X)V(Y)}}. \tag{15}$$

As opposed to distributions with a lower absolute value of the correlation, which are more or less circular (as can be seen in figure 4a), the ones with higher correlations (absolute value) have a more elliptic shape (figure 4b). The slope of the semi–major axis of the ellipsis, with respect to the f_1, f_2–coordinate system, corresponds to the correlation of the distribution. This fact is made use of in the further *biasing* of the mixing process, with the objective of generating jPDDs with certain values of covariance. As seen from figure 9, neglecting the covariance can result in significant errors in the reaction rate.

In order to obtain a distribution with a certain correlation, the mechanism used to bias the mixing parameter a_i in equations (3) and (4) is also employed for the correlation: As seen earlier, correlated functions are of elliptic nature, the correla-

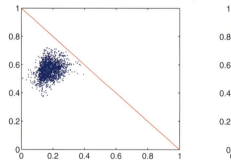

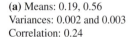
(a) Means: 0.19, 0.56
Variances: 0.002 and 0.003
Correlation: 0.24

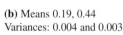
(b) Means 0.19, 0.44
Variances: 0.004 and 0.003
Correlation: -0.48

Fig. 4. Distributions with different correlations taken from DNS. Left: distribution with a low (absolute value) correlation, which is more circular in nature. Right: distribution with a higher correlation, rather elliptic.

tion itself can be interpreted as the slope of the semi-major axis of the ellipis (the *correlation slope*)[1]. Hence, if the line that connects two mixing particles is parallel to the correlation slope, the mixing parameter will be selected so that these particles will mix slowly, while those partciles whose connecting line is perpendicular to the correlation slope will mix fastest (cf. figure 5). This is another factor which is required to bias the pseudo-random number generator to obtain the mixing parameter for each mixing event. Because of the statistical nature of the mixing process, the development of the covariance is also statistical. Due to this, and the fact that the initial distribution already has a significant correlation, it is insufficient to just adjust the mixing process along the correlation slope, as the target correlations will usually not be met or overshot. The following iterative scheme is applied for the correlation to meet the final correlation after the correct number of mixing events:

1. Start with a *reference slope* that is parallel to correlation slope.
2. Assess the correlation after a number of mixing events so that all particles have mixed once.
3. Modify the reference slope. If the correlation is building up too fast with respect to the required number of mixing events (See Sect. 2.1), reduce the reference slope with respect to the correlation slope, if the correlation is growing too slow, increase the reference slope.

Using this scheme, it can be seen to that the target correlation is obtained after the correct number of mixing events. An example for the evolution of the covariance by using this method can be seen in figure 6. The large-scale fluctuations of the correlation are the influence of the reference slope, while the small-scale fluctuations are due to the statistical nature of the mixing. At each mixing process, the reference slope was updated to meet the required correlation.

Compared to the variance, where the biasing of the mixing parameter is performed on a global level, that is, it is selected for the mixing process as a whole depending on initial and final variance, the co-variance performs a further adjustment of the mixing parameter for each mixing event. As such, the variances and the correlation can be adjust independently from each other.

3 Transport Equations

The transport equation for a conserved, diffusive and passive scalar ϕ reads [8]:

$$\frac{D\phi}{Dt} = \Gamma \frac{\partial^2 \phi}{\partial x_j \partial x_j}. \tag{16}$$

Filtering equation (16) for the LES approach and using a conventional eddy-diffusivity approach for the subgrid-scale transport as suggested by Eidson [7], or its dynamic counterpart yields a transport equation for filtered scalar

---

[1] The correlation slope is measured as the angle between the semi-major axis and the X-axis.

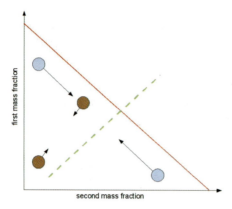

Fig. 5. Principle of covariance based mixing: As the blue particles are perpendicular to the reference slope (dashed line), they mix more rapidly as the brown particicles, which are more in-line to the reference slope (mixing intensity indicated by arrow length)

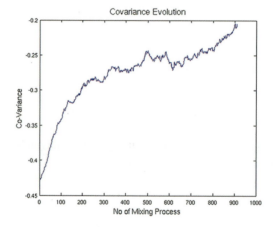

Fig. 6. Evolution of the covariance of a given distribution. The initial covariance is determined by the initial Dirac–Delta peaks, the targeted covariance is -0.2.

$$\frac{D\bar{\phi}}{Dt} = \frac{\partial}{\partial x_j}\left((\Gamma + \Gamma_T)\frac{\partial\bar{\phi}}{\partial x_j}\right). \tag{17}$$

Here, Γ_T is the subgrid-scale diffusivity from the model.

The jPDDs outlined in Sect 2 are characterized by the first and second order moments. Transport Equations are formulated for those quantities in this work in the incompressible LES context. According to Jimenez [15], the transport equation for the subgrid-scale scalar variance can be written as

$$\frac{\partial \overline{V}_f}{\partial t} + \frac{\partial \overline{u_i V_f}}{\partial x_i} = \frac{\partial}{\partial x_i}\left((D+D_T)\frac{\partial \overline{V}_f}{\partial x_i}\right) - 2D\overline{\frac{\partial f}{\partial x_i}\frac{\partial f}{\partial x_i}} + 2(D+D_T)\frac{\partial \overline{f}}{\partial x_i}\frac{\partial \overline{f}}{\partial x_i}, \quad (18)$$

assuming that the LES filters fulfil the following restriction: $\overline{\overline{f}} = \overline{f}$. For the box filters which are used typically used in non-spectral LES solvers, this is not necessarily true, so a small error will be induced by the filter operation. The eddy-diffusivity D_T (according to [7]) relates to the eddy-viscosity like

$$D_T = \frac{\nu_T}{Sc_T}, \quad (19)$$

with ν_T as the eddy-viscosity from the Smagorinsky Model. Using a dynamic procedure by Germano thus closes (18) with respect to D_T. However (18) still has an unclosed term,

$$2D\overline{\frac{\partial f}{\partial x_i}\frac{\partial f}{\partial x_i}}. \quad (20)$$

This term, known as the *scalar dissipation rate*, requires another closure approach.

3.1 Closure for the Scalar Dissipation Rate

It has been suggested [17] to assume that the scalar dissipation and the variance production are in equilibrium. However, this assumption will conserve variance, whereas mixing is a process that is known to reduce variance [16]. As such, Jimenez et al [15] have suggested a non-equilibrium approach that models the scalar dissipation rate using the mechanical-to-scalar time-scale ratio:

$$\overline{\chi} = C\left(V_f\right)\overline{\tau_f}, \quad (21)$$

by approximating the characteristic mixing time as the ratio between kinetic energy $\overline{\kappa}$ and the kinetic energy dissipation $\overline{\varepsilon}$:

$$\overline{\tau_f} = \frac{\overline{\kappa}}{\overline{\varepsilon}}. \quad (22)$$

Fox [8] shows that at Schmidt numbers of around unity the mechanical-to-scalar time-scale ratio will be approximately constant for a given Taylor-scale Reynolds number.

Thus the scalar dissipation rate could be modelled as:

$$\overline{\chi} = \frac{\nu + C_S \overline{\Delta}^2 |\overline{S}|}{Sc\, C_I \overline{\Delta}^2}\left(\overline{f^2} - \overline{f}^2\right). \quad (23)$$

Finally, the transport equation for the scalar variance reads:

$$\frac{\partial \overline{V}_f}{\partial t} + \frac{\partial \overline{u_i V_f}}{\partial x_i} = \frac{\partial}{\partial x_i}\left((D+D_T))\frac{\partial \overline{V}_f}{\partial x_i}\right) - \tag{24}$$

$$2D\frac{v+C_S\overline{\Delta}^2|\overline{S}|}{Sc\,C_I\overline{\Delta}^2}\left(\overline{f^2}-\overline{f}^2\right) + 2\,(D+D_T)\frac{\partial \overline{f}}{\partial x_i}\frac{\partial \overline{f}}{\partial x_i}. \tag{25}$$

Using a dynamic prodecure for C_I and C_S fully closes this equation.

3.2 Closure for the Covariance

As seen in Sect. 2.2, the co-variance can be regarded as a generalization of the variance. Thus, generalizing the transport equation for the variance (18) yields a transport equation for the co-variance,

$$\frac{\partial \overline{C}_{f12}}{\partial t} + \frac{\partial \overline{u_i C_{f12}}}{\partial x_i} = \frac{\partial}{\partial x_i}\left((D+D_T))\frac{\partial \overline{C}_{f12}}{\partial x_i}\right) - 2D\overline{\frac{\partial f_1}{\partial x_i}\frac{\partial f_2}{\partial x_i}} + 2\,(D+D_T)\frac{\partial \overline{f}_1}{\partial x_i}\frac{\partial \overline{f}_2}{\partial x_i}. \tag{26}$$

Here a new term appears,

$$2D\overline{\frac{\partial f_1}{\partial x_i}\frac{\partial f_2}{\partial x_i}}, \tag{27}$$

called the *scalar cross-dissipation rate*. The scalar cross dissipation rate has a similar effect on the covariance as the scalar dissipation rate has on the variance. However, as (27) is not quadratic (compared to (23)), the scalar cross-dissipation rate does not necessarily reduce the covariance.

Using the same approach as in Sect. 3.1 the resulting model for the scalar cross-dissipation rate reads:

$$\overline{\chi} = \frac{v+C_S\overline{\Delta}^2|\overline{S}|}{Sc\,C_I\overline{\Delta}^2}\left(\overline{f_1 f_2}-\overline{f}_1\,\overline{f}_2\right). \tag{28}$$

4 DNS Simulations for a Priori Validation

DNS simulations of a turbulent co-annular jet-in-crossflow have been performed in order to analyse several items:

- The quality of the generated jPDDs by the mixing models.
- The closures for the transport equations for variance and covariance
- The dependence of the reaction rate upon additional flow parameters.

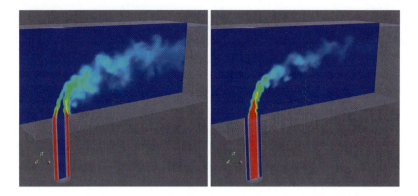

Fig. 7. DNS of a co-annular jet in crossflow. Indicates the scalar fraction of the annular jet on the left side, and the scalar fraction of the central jet on the right side. Main flow from left to right.

It is emphasized that the reaction rate is *solely* used a quality indicator; no actual conversion from educts to products is performed.

The Schmidt number of the flow is unity, so that the Batchelor-lengthscale is identical to the Kolmogorov-lengthscale. The resolution of the DNS is then selected to be sufficient to eliminate scalar subgrid-scale fluctuations, thus the DNS cells contain no scalar variance or co-variance. Transport equation for two mixture fractions have been solved. One mixture fraction is set to unity for the central jet, the other for the annular jet.

In order to compare the data of the distributions generated by the mixing models with the DNS, a spatial and temporal filtering scheme is applied to the DNS results. A test-grid, which has a significantly lower resolution than the DNS grid, is superimposed onto the domain. As each of these test cells contains a number of DNS cells, each of those test cells have a distribution of scalar values of their own. Also, statistical moments can then be computed for those test cells. However, the number of scalar samples within one of the test cells is still low. To overcome this drawback, test cells are grouped together. All cells in the domain that have similar[2] first and second order moments are combined into a single distribution. All those distributions with similar moments are also grouped together in the temporal dimension.

In this scheme, five parameters are used to group cells together: two mean values, two variances and the correlation. In order to assess the accuracy of the presented model, additional parameters have been taken into account:

- Scalar dissipation rate of either scalar
- Scalar cross-dissipation rate
- Any combination thereof

---

[2] As no two distributions have *exactly* the same moments, a certain tolerance has been added to grouping the cells together.

It should be noted that using up to eight parameters (two means, two variances, two scalar dissipation rates, the correlation and the cross dissipation rate) requires vast amounts of cells and timesteps in order to yield distributions with a large number of samples, especially at the extremes of the composition space.

In order to assess the quality of the model, three individual parts need to be analysed:

1. The distributions obtained from the mixing models need to display the actual microscopic behaviour of the flow as described by the DNS.
2. The transport equations need to approximate the moment with high accuracy. Even low errors, especially for the mean values, deteriorate the overall quality of the model greatly.
3. The reaction scheme needs to simulate the actual reaction process precisely.

Each of these items is discussed within this section. For assessing the quality of the model, an filtered reaction rate (based on a fast Arrhenius-type reaction) is computed for the cells. Two different reaction rates are then compared:

1. The sum of the reaction rates taken from all DNS cells that constitute the test-cell.
2. The mixing model generates a distribution using the moments known for the test-cell.

These two reaction rates are then compared to indicate the performance of the model.

5 Results

This section will first discuss the quality of the distributions that have been generated, and then study the dependency of the filtered reaction rates on various parameters.

5.1 Distributions

As described before, the first part of the results for the validation of the model copes with the accuracy of the distributions themselves. Distributions that have been obtained from the mixing models are compared against those extracted from DNS by the method described in section 4. Figure 8 shows this comparison, where the left column shows distributions that have been obtained from the mixing model, and the right column shows those extracted from DNS data. Each row represents a sample set of first and second order moments.

Generally it can be seen that the actual distributions found in a filtered DNS computation and those generated by the model match reasonably well for a variety of first and second order moments. The most noticeable difference can be seen in the

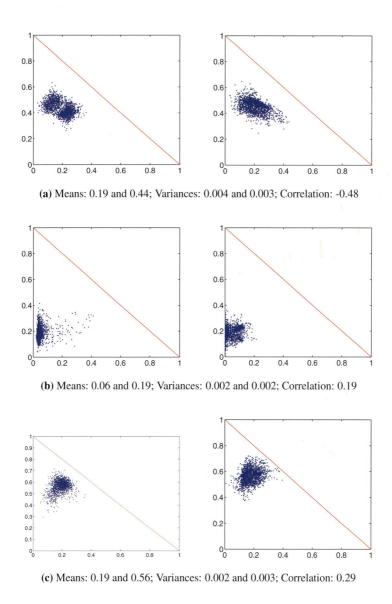

(a) Means: 0.19 and 0.44; Variances: 0.004 and 0.003; Correlation: -0.48

(b) Means: 0.06 and 0.19; Variances: 0.002 and 0.002; Correlation: 0.19

(c) Means: 0.19 and 0.56; Variances: 0.002 and 0.003; Correlation: 0.29

Fig. 8. Comparison between distributions. Distributions generated by mixing models are displayed in the left column, those taken from the DNS data are shown in the right column. The distributions in one row have the same statistical moments. One scalar fraction depicted on the the X–axis, the other on the Y–axis. Red lines indicate the boundary of the physical space, so that the sum of scalar fractions does not exceed unity.

set 8b, where the model generates a number of artifacts with a high value of the first scalar (the X-axis). In the same set it can also be seen that while the DNS shows a number of points having a zero or near-zero value of the first scalar, these points are missing altogether in the distribution produced by the model. As described before, a mixing model moves all particles towards the mean, dragging them away from the zero-value of the first scalar with each mixing event, thus leaving the gap between the Y axis and the points with the lowest X values. Since the requested mean of the first scalar is rather low, the model requires a number of points with high values of the first scalar to offset the lack points with low values; this fact leads to the artifacts (few points with values far from the mean of 0.06) that can be seen in the distribution.

In Figure 9, the significance of the covariance can be seen. The mixing model is run twice, beginning with the same initial distribution. However, for one run (blue line), the mixing model has been set to establish a correlation of -0.4, while during the second run (green line), a correlation of +0.4 has been produced. After each mixing cycle, the reaction of a sample reaction (in this case, the oxidation of methane) has been computed using an Arrhenius term. As the mixing takes place, the variance decays for both distributions, however the reaction rates after a given number of mixing events differs greatly, although the means are identical and the variances are similar within statistical noise. After a large number of mixing events, the variance is converging towards zero, and the covariance plays a subsequently subordinate role.

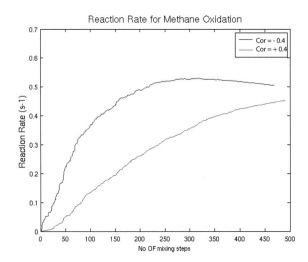

Fig. 9. Example reaction rate over the number of mixing processes. Means are conserved, Variances decay from left to right. Large differences can be seen between distributions of different correlations. Blue line: negative correlation of -0.4; Green line: positive correlation of +0.4.

5.2 Reaction Rates

In order to assess the quality of the model, reaction rates are obtained for each cell in the grid, for both the LES and the DNS. As according to Sect. 4, the composition of a DNS cell can be assumed as homogeneous, while the LES cells need convolution

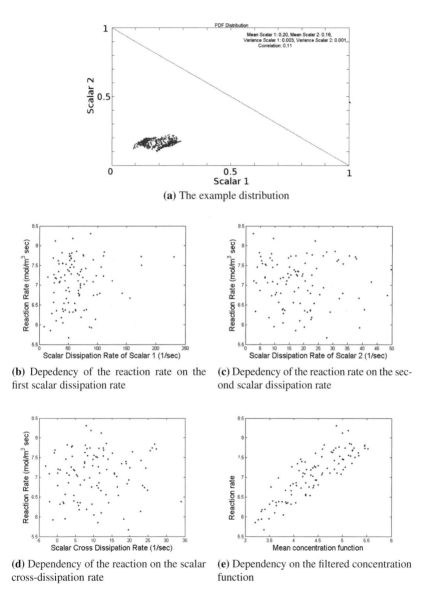

(a) The example distribution

(b) Depedency of the reaction rate on the first scalar dissipation rate

(c) Depedency of the reaction rate on the second scalar dissipation rate

(d) Dependency of the reaction on the scalar cross-dissipation rate

(e) Dependency on the filtered concentration function

Fig. 10. Reaction rate dependency for a given distribution

over the distribution from the model. For an FDF that is represented by particle ensembles, this is the average over the individual reaction rates:

$$\overline{\dot{\omega}}\left(\overline{f_1},\overline{f_2},\overline{f_1'^2},\overline{f_2'^2},\overline{f_1'f_2'}\right) = \frac{1}{N}\sum_{i=1}^{N_{part}} \dot{\omega}_i(f_1,f_2). \tag{29}$$

The individual reaction rates $\dot{\omega}_i(f_1,f_2)$ can, for example, be obtained using a conventional Arrhenius-like approach, however the model allows any other scheme to be used that computes reaction rates from concentrations. In Sect. 4, it has been explained that numerous test-cells (which are coarser than the actual DNS cells), that have similar first and second order moments, are integrated into a single distribution in order to make the results statistically more meaningful. In Figure 10, the individual reaction rates for each of those test cells within a sample distribution (the distribution itself is seen in Fig. 10a) is shown. In order to collocate the test-cells, a certain tolerance was applied with respect to the means and variances to allow grouping (as no two cells have absolutely identical moments). Computing a mean filtered concentration function M with

$$M = f_1^{[\alpha]} f_2^{[\beta]}, \tag{30}$$

the concentrations and the Arrhenius-exponents, and comparing this Mean concentration function against the filtered reaction rate, as shown in Figure 10e, reveals that the deviation of the means of the individual test cells is still paramount to the reaction rate, as a clear dependence can be seen. Figs. 10b and 10c show the dependency of the reaction rate on the scalar dissipation rates of the first and second scalar, respectively. No dependency between those parameters can be seen. Also the reaction rate does not seem to depend directly on the scalar cross dissipation rate 10d.

6 Conclusion

Biased, multi-variate particle-based mixing models have been used to generate discrete subgrid-scale particle ensembles to represent mixture fractions for turbulent reacting LES. Used in conjunction with transport equations for first and second order statistical moments for mixture fractions they provide an accurate and computationally inexpensive model for mixing and reaction in turbulent flows. It should be noted however, that the transport equations require modelling of several terms such as the subgrid-scale scalar flux, the scalar dissipation rates and the scalar cross-dissipation rate. Should multiple mass fractions occur in a simulation, mixing models can be extended to generate such distributions and additional transport equations can be used, especially for the co-variance (correlation) between any two mass fractions, to increase accuracy. Using only the first and second order moments appears to be sufficient to obtain an acceptable accuracy, and no direct dependence of the reaction on the scalar dissipation rates and scalar cross-dissipation rates has been found.

However it should be noted that the dissipation rates do of course influence the reaction rates as they appear as a source term in the transport equation for the scalar variance, and covariance.

However, further assessment of the model is required, especially its performance under various conditions, like different geometries and different reaction schemes. Furthermore, the influence of the higher-order moments on the reaction rates needs to be studied in depth.

Acknowledgements

The authors are indebted to Michael Manhart and Florian Schwertfirm of the Technische Universität München, Fachgebiet Hydromechanik, for supplying and supporting the DNS solver "MGLET".

References

1. Brandt, M.: Beschreibung der Selbstzündung in turbulenter Strömung unter Einbeziehung ternärer Mischvorgänge. Ph.D. thesis, Technische Universität München (2005)
2. Brandt, M., Gharaibah, E., Polifke, W.: Modellierung von Mischung und Reaktion in turbulenten Mehrphasenströmungen mittels Verteilungsfunktion. Chemie Ingenieur Technik 76, 46–51 (2004)
3. Brandt, M., Polifke, W.: Tabulation of mean reaction rates from multivariate, correlated distributions with a monte carlo model. In: 9th Int. Conference on Numerical Combustion, p. 2. SIAM, Sorrento (2002)
4. Brandt, M., Polifke, W., Flohr, P.: Approximation of joint PDFs by discrete distributions generated with Monte-Carlo methods. Combustion Theory and Modelling 10(4), 535–558 (2006)
5. Brandt, M., Polifke, W., Ivancic, B., Flohr, P., Paikert, B.: Auto-ignition in a gas turbine burner at elevated temperature. In: Proc. of ASME Turbo Expo 2003 Power for Land, Sea and Air, ASME, Atlanta, Georgia, USA, p. 11 (2003), No. 2003-GT-38224
6. Dopazo, C., Obrien, E.E.: Approach to autoignition of a turbulent mixture. Acta Astronautica 1(9-10), 1239–1266 (1974)
7. Eidson, T.M.: Numerical simulation of the turbulent rayleigh-benard problem using subgrid modelling. Journal of Fluid Mechanics 158, 245–268 (1985)
8. Fox, R.O.: Computational Models for Turbulent Reacting Flows. Cambridge (2003)
9. Germano, M., Piomelly, U., Moin, P., Cabot, W.H.: A dynamic subgrid-scale eddy viscosity model. Physics of Fluids 3, 1760–1765 (1991)
10. Gutheil, E.: Multivariate PDF closure applied to oxidation of CO in a turbulent flow. In: Kuhl, A.L., Leyer, J.-C., Borisov, A.A., Sirignano, W.A. (eds.) Dynamics of Deflagrations and Reactive Systems. Progress in Astronautics and Aeronautics (1991)

11. Gutheil, E., Bockhorn, H.: The effect of multi-dimensional PDFs on the turbulent reaction rate in turbulent reactive flows at moderate Damköhler numbers. PhysicoChemical Hydrodynamics 9(3/4), 525–535 (1987)
12. Janicka, J., Kolbe, W., Kollmann, W.: Closure of the transport–equation for the probability denstiy function of turbulent scalar fields. Journal of Non-Equilibrium Thermodynamics 4(1), 47–66 (1979)
13. Jaberi, F.A., Colucci, P.J.: Large Eddy Simulation of heat and mass transport in turbulent flows. Part 1: Velocity field. Int. J. Heat Mass Transf. 46(10), 1811–1825 (2003)
14. Jaberi, F.A., Colucci, P.J.: Large Eddy Simulation of heat and mass transport in turbulent flows. Part 2: Scalar field. Int. J. Heat Mass Transf. 46(10), 1827–1840 (2003)
15. Jimenez, C., Ducros, F., Cuenot, B., Bedat, B.: Subgrid scale variance and dissipation of a scalar field in large eddy simulations. Physics of Fluids 13, 1748–1754 (2001)
16. Meyer, D.W., Jenny, P.: A mixing model for turbulent flows based on parameterized scalar profiles. Physics of Fluids 18(3), 035105 (2006), http://link.aip.org/link/?PHF/18/035105/1
17. Pierce, C., Moin, P.: A dynamic model for subgrid-scale variance and dissipation rate of a conserved scalar. Physics of Fluids 10, 3041–3044 (1998)
18. Subramaniam, S., Pope, S.B.: A mixing model for turbulent reactive flows based on euclidean minimum spanning trees. Combustion and Flame 115(4), 487–514 (1998)

Mixing Analysis and Optimization in Jet Mixer Systems by Means of Large Eddy Simulation

Pradeep Pantangi[1], Ying Huai[2], and Amsini Sadiki[1]

[1] Department of Mechanical Engineering, Institute for Energy and Powerplant Technology,
Technical University Darmstadt, Petersenstraße 30, D-64287 Darmstadt, Germany
sadiki@ekt.tu-darmstadt.de
[2] 701 Group, Dalian Institute of Chemical Physics Chinese Academy of Sciences
457 Zhongshan Road, 116023 Dalian, China

Abstract. To achieve the analysis, control and optimization of mixing processes by means of Large Eddy Simulation (LES) technique, an advanced subgrid scale (SGS) scalar model package for the description of turbulent mixing in gaseous and liquid flows is developed and validated. This aims to strongly improve the prediction accuracy of the mixing field quantities prior to any mixing modification studies. Both non-reacting and reacting systems are considered. To cover different reaction regimes, the mixing processes with chemical reaction in jet mixer systems under investigation are described in terms of mixture fraction and two reaction progress variables. To assess the accuracy of the SGS model package and focussed on the high Schmidt-number phenomena, the results of LES are compared with experimental data and other previous simulation results for both a non-reacting jet in channel flow configuration and a confined impinging jets reactor (CIJR) featuring a parallel reaction system. The mixing and reaction processes are analyzed. Especially for the CIJR, the influence of operating conditions (active modification) on mixing properties is evaluated and a quality of measure, that is a prerequisite for mixing control and optimization in turbulent reacting flows by means of passive modification, is highlighted.

Keywords: Large Eddy Simulations, SGS scalar model package, Schmidt number-dependency, jet mixer systems, chemical reactions, mixing analysis and optimization.

1 Introduction

Turbulent mixing systems, as they are often encountered in many engineering applications, involve many phenomena and processes, such as turbulence, mass and heat transfer, chemical reactions, etc., which may strongly interact. In order to modify the mixing processes on the basis of essential parameters, two different basic approaches can be applied. The first is to change the internal conditions such as time-dependent perturbations or flow boundary conditions (active modification). Another is by varying the external conditions, for example, the geometric properties (passive modification). In both cases the characterization of the mixing, the appropriate quality of measure and the accuracy of the prediction of the effects of different operating conditions on the mixing properties are challenging.

For complex configurations of technical importance in which experimental investigations are difficult to be accomplished, the required need for comprehensive knowledge of phenomena can well be achieved only by solving the equations governing the processes involved in the frame of Computational Fluid Mixing (CFM), a subset of the well known Computational Fluid Dynamics (CFD) [1, 2]. Thereby an accurate prediction strongly depends on the availability of adequate and efficient physics-preserving mathematical models and techniques. Reynolds Averaging based Numerical Simulation (here, RANS) tends to wipe out most of the important characteristics of a time-dependent solution and appears inappropriate in dealing with intrinsically unsteady phenomena [3]; Direct Numerical Simulation (DNS) resolves all turbulent scale structures and remains still computationally unrealistic for turbulent flows of high Reynolds number and Schmidt number [2 – 7]. Even though a rapid development of computers and applications-oriented numerical methods are notable in the last years, a need to maintain adequate resolution of the small scales requires that the high-Sc number DNS-calculations be performed only at low Reynolds numbers without particular technical or practical interest [1 -- 21].

Having in mind the dominant role of large flow structures in governing turbulent mixing processes, Large Eddy Simulation (LES) emerges as a suitable tool for such flow systems and is carried out in this work. Classical LES allows to compute the resolved large scales structures whilst the non resolved small scales structures are modelled by means of so called subgrid scale (SGS) models. Although some issues for the LES technique must still be clarified as pointed out in [4 – 7] LES for non-reacting flows has reached a level of development making its accepted advantages useful for reacting flow simulations. SGS turbulence/mixing models are needed to close the SGS stress tensor and scalar flux vector and to provide the appropriate amount of energy exchange, backscatter, and dissipation. Because chemical reactions in turbulent reacting flows occur essentially at the smallest scales of the sub-filter level, a SGS model is also required for the chemical reaction rate [4, 8].

Comprehensive reviews of existing SGS models for the SGS stress tensor may be found in [4 – 6]. In the LES, the combination of accurate discretisation and integration schemes with large numerical efficiency plays a central role. These issues have already been reported elsewhere (see in [7]) and will not be discussed here. Nevertheless, it is worth mentioning that the need for realiable and accurate predictions require compromise on numerical accuracy, the range of scales resolved, and testing for numerical and physical accuracy, where computer power is limited [4, 8]. While this task is easy in RANS-context, this is more difficult for LES. The final aspect of the numerical algorithm is the enforcement of boundary conditions. A reliable predictability using LES could be improved when appropriate choice of boundary and inflow conditions is made. For these special issues, the reader may refer to [9, 10].

For the SGS scalar flux vector, the most studies applied the linear eddy diffusivity model even though, it is well known that this simple model has a limited validity in turbulent reacting flows [11, 12, 19, 20, 21]. Especially this model is not valid for high Schmidt number fluid transport processes [13 -- 15]. Different model approaches of the SGS scalar flux have been therefore proposed and evaluated in the past [12, 13, 16]. As pointed out by Huai et al. [14, 15], they all experience some weaknesses in accurately predicting mixing characterized by high Schmidt number effects. Toward the development and assessment of SGS model for the filtered chemical reaction rate, most of previous research works focused on chemical reaction systems related to combustion, especially dealing with gaseous flows. A recent comprehensive review has been provided in [8].

For turbulent liquid flow systems that are mostly encountered in chemical reaction and process engineering applications, the investigations using LES are very few. A well-known issue is that side reactions produce undesired byproducts, which reduce the reaction yield of the main product and later on complicate its downstream purification [1, 2]. In order to be able to manipulate the selectivity of specific reactions by changing mixing conditions that govern the course of reactions a rational description of the chemical reaction rate is essential.

In [17] Michioka et al. proposed a SGS model for the filtered chemical reaction rate and developed LES of non-premixed turbulent liquid flow with a moderately fast chemistry. Their model based on a presumed (beta) PDF model and an algebraic expression for the SGS conditional expectation. This conditional expectation model has been deduced from DNS filtered data of a stationary isotropic turbulence with a second order chemical reaction. Thereby the SGS scalar variance has been determined according to a scale similarity approach. No particular emphasis of the SGS scalar flux effect has been made, rather they used the linear eddy diffusivity model known for its inaccuracy in such applications. In a complex multi-jets configuration, Olbricht, Janicka and Sadiki [18] carried out a mixing analysis based on a transported (Eulerian) PDF method to clarify the validity of a (beta) presumed PDF for complex mixing processes. It turned out that the presumed approach is not really a good approximation for these processes. It is worth mentioning that techniques based on conditional expectation or on solutions of the transported PDF equation are time consuming and not yet appropriate for design and optimization tasks of engineering or industrial importance [8].

To meet the urgent need for reliable predictive method to aid mixing safety studies and the design/optimization of practical high Reynolds number mixing systems, it is essential that turbulent SGS models for scalar in CFD be able to address accurately major effects at low computational cost. Hence Huai et al. [14, 15] reported LES results in which an adequate model for the SGS scalar flux vector has been applied. This methodology was validated in different configurations of various complexity involving gaseous and liquid flows, respectively [14, 15, 23, 24].

In this present work, an advanced SGS scalar model package is presented to develop Large Eddy Simulation (LES) of both turbulent gaseous and liquid flows. It includes a SGS scalar flux model for the filtered scalar flux vector, a corresponding SGS scalar variance model and related scalar dissipation rate as well as a SGS model for the filtered chemical reaction rate of the involved reacting scalars.

Dealing with chemical reactions a parallel reacting scheme is considered in configurations under study. To well cover different reaction regimes, the configurations may be described in terms of the mixture fraction and two reaction progress variables. According to [22, 25] that concluded that the scale similarity model with a constant model parameter reproduced correctly the locations of the unresolved reaction rate, a dynamic scale similarity approach is applied in this work to close the chemical source terms in the reaction-progress variable equations. This eliminates the uncertainties associated with tunable model parameters and allows to well address the problem of extrapolation from larger scales [26].

The paper will run as follows. In the next section (section 2), the basic governing equations used to describe the flow and scalar field are briefly recalled within the LES framework. In particular the SGS scalar model package is shortly outlined. In section 3, the numerical and experimental setups are provided. Two configurations featuring high Schmidt number effects are chosen to assess the SGS scalar model package. The results of LES are then presented and discussed for a non-reacting jet in channel flow configuration in comparison with both experimental data from [27] and predictions using other models. As specific optimisation issues of mixer jets configurations have already been reported in [42], this aspect will not be repeated here. Next, a confined impinging jets reactor (CIJR) as experimentally investigated in [28] is simulated to appraise how the present SGS model package is applicable for a practical turbulent reacting liquid flow. The flow and mixing results are analyzed. In particular, the influence of operating conditions (active modification) on mixing properties of the CIJR is evaluated and a quality of measure that is a prerequisite for a mixing optimisation in turbulent reacting flows by means of passive modification is highlighted. We devote section 4 to conclusions.

2   Mathematical Models and Numerical Procedure

To separate the large from small-scale structures in LES, filtering operations are applied to the governing equations, which are the momentum equation (2) along with the continuity equation (1) used to describe the motion of low Mach number Newtonian fluids. In addition, the change of mixture fraction f caused by the turbulent convection and diffusion of a passive (or conserved) scalar is given by the transport equation (3).

$$\frac{\partial \bar{\rho}}{\partial t} + \frac{\partial \bar{\rho}\bar{u}_i}{\partial x_i} = 0,$$ (1)

$$\frac{\partial}{\partial t}(\bar{\rho}\bar{u}_i) + \frac{\partial}{\partial x_j}(\bar{\rho}\bar{u}_i\bar{u}_j) = \frac{\partial}{\partial x_j}\left[\bar{\rho}\bar{\nu}\left(\frac{\partial \bar{u}_i}{\partial x_j} + \frac{\partial \bar{u}_j}{\partial x_i}\right) - \frac{2}{3}\bar{\rho}\bar{\nu}\frac{\partial \bar{u}_k}{\partial x_k}\delta_{ij} - \bar{\rho}\tau_{ij}^{sgs}\right] - \frac{\partial \bar{p}}{\partial x_i} + \bar{\rho}$$ (2)

$$\frac{\partial}{\partial t}\bar{\rho}\bar{f} + \frac{\partial}{\partial x_i}\left(\bar{\rho}\bar{u}_i\bar{f}\right) = \frac{\partial}{\partial x_i}\left(\bar{\rho}\bar{D}_f\frac{\partial \bar{f}}{\partial x_i}\right) - \frac{\partial}{\partial x_i}\left(\bar{\rho}J_i^{sgs}\right).$$ (3)

In equations (1)-(3) the quantity ui (i=1, 2, 3) denotes the velocity components at xi direction, ρ the density, p the hydrostatic pressure and δ_{ij} the Kronecker delta. The quantity ν is the molecular viscosity and Df the molecular diffusivity coefficient.

In a rapid reaction case, the chemical timescale is far smaller than that of the turbulent diffusion. Since such a small time step cannot be set even for the fastest supercomputer, a conserved scalar approach found therefore usual applications. Assuming the equilibrium chemistry, the concentrations of all species shall be related to the conserved scalar as prescribed in [2] or [17]. To take into account different chemical reaction regimes, the introduction of reaction-progress variables is useful. In this case additional transport equations for the involved reacting scalars or reaction-progress variables are needed:

$$\frac{\partial}{\partial t}\bar{\rho}\bar{Y}_\alpha + \frac{\partial}{\partial x_i}\left(\bar{\rho}\bar{u}_i\bar{Y}_\alpha\right) = \frac{\partial}{\partial x_i}\left(\bar{\rho}D\frac{\partial \bar{Y}_\alpha}{\partial x_i}\right) - \frac{\partial}{\partial x_i}\left(\bar{\rho}J_i^{sgs}\right) + \bar{S}_\alpha, \quad \alpha = \{1,2,...\} \qquad (4)$$

where $\bar{Y}_\alpha$ is the filtered concentration of the reaction progress variable α. The quantity D denotes the molecular diffusivity coefficient. The equations (1)-(4) govern the evolution of the large, energy-carrying, scales of flow and mixing field denoted by an over-bar. In flow and scalar field, the effect of the small scales appears through the SGS stress tensor and the SGS scalar flux vector,

$$\tau_{ij}^{SGS} = \overline{u_i u_j} - \bar{u}_i\bar{u}_j \,, \qquad (5)$$

$$z_{\alpha i}^{sgs} = \overline{u_i Y_\alpha} - \bar{u}_i\bar{Y}_\alpha \,, \qquad z \equiv (f, Y_\alpha) \qquad (6)$$

respectively. The last term, $\bar{S}_\alpha$, in equation (4) is the filtered chemical reaction rate. Together with the quantities (5) and (6) it must be modelled in order to obtain a closed system of equations (1) - (4). Due to observed poor predictions of scalar properties achieved by the use of SGS scalar flux models exiting in the literature (see in [13, 16]), especially in dealing with mixing processes characterized by high-Schmidt numbers [15], a new model package for SGS scalar transport description without chemical reactions was proposed in [14, 15]. This is now being extended [23, 24] by a corresponding SGS scalar variance and an associated scalar dissipation rate model as well as a chemical reaction rate model. Since we do not focus especially on the flow field, we rather apply the simplest SGS model for the stress tensor that is compatible with the SGS scalar flux models according to the second law of thermodynamics [29]. Thus we use the Smagorinsky model (scalar eddy viscosity model) with a standard dynamic procedure for determining the model coefficient to close the subgrid scale stress-tensor, τ_{ij}^{SGS} [4 – 22].

2.1 Model Package for SGS Scalar Transport Description

The proposed model package consists of physics-preserving submodels required to accurately characterize different scalar transport processes involved. It includes advanced SGS models for the filtered scalar flux vector, for its corresponding scalar variance along with scalar dissipation and for the chemical reaction rate.

2.1.1 Proposed SGS Scalar Flux Model

The modelling is based on the second law of thermodynamics in conjunction with the invariant theory. From this formalism an explicit anisotropy resolving algebraic SGS scalar flux model emerged, such that the irreversibility requirements of the second law of thermodynamics are automatically fulfilled by the suggested parameterization. The new model combines in its cubic form three terms as given in equation (7).

$$J_i^{SGS} = -D_{ed}\frac{\partial \overline{z}}{\partial x_i} + D_{dev}(T_{SGS}\tau_{ij}^{SGS(dev)}\frac{\partial \overline{z}}{\partial x_j}) +$$

$$+ D_\lambda(T_{SGS}\tau_{ij}^{SGS}\frac{\partial \overline{z}}{\partial x_k}\frac{\partial \overline{z}}{\partial x_k})\frac{\partial \overline{z}}{\partial x_i} = D_{ij}^{SGS}\frac{\partial \overline{z}}{\partial x_j}, \quad \text{with} \quad z \equiv (f, Y_\alpha) \tag{7}$$

where D(-) are the model coefficients. T_{SGS} is the SGS turbulent characteristic time. This model obviously involves a nonlinear tensor of diffusivity

$$D_{ij}^{SGS} = -D_{ed}\delta_{ij} + D_{dev}T_{SGS}\tau_{ij}^{SGS(dev)} + D_\lambda T_{SGS}\tau_{ij}^{SGS}\frac{\partial \overline{z}}{\partial x_k}\frac{\partial \overline{z}}{\partial x_k} \tag{8}$$

in terms of the quadratic of the gradient of the filtered scalar field. In Eq (7) the model coefficients depend on the invariants of $\frac{\partial \overline{z}}{\partial x_i}$ and τ_{ij}^{SGS} as well as on the grids size, Δ. As known from [29], these coefficients cannot be determined independently, so that a dynamical procedure is suitable for their determination. This will be done in this work.

The new model (7) combines obviously the conventional linear eddy diffusivity model with two additional terms and goes out the commonly known mixed models [12, 16]. The first additional term expresses the influence of the SGS flow fluctuations acting on the filtered scalar gradient. This couples the (deviatoric) SGS stress tensor and the gradient of the filtered scalar field. The second term involves the production of scalar variance (proportional to the quadratic of scalar gradients) giving rise to a nonlinear diffusivity tensor in terms of scalar gradients (see Eq. 8). Both terms include transport properties of the fluid incorporated into the vortices through the SGS flow fluctuations, the turbulent Reynolds number and the molecular Schmidt number making the model sensitive to sub-Kolmogorov scales. The Sc- and Ret-number dependency of this model may clearly appear when the SGS characteristic time, T_{SGS}, is related to the SGS scalar mixing time, T_{SGS}^s, through the mechanical-to-scalar time ratio as

$$R = T_{SGS}/T_{SGS}^s \sim (2 + \frac{1}{Sc})/(3 + Re_t^{-1/2}\ln Sc) \tag{9}$$

It is also clear that the last term in (8) may be related to the production of the scalar variance (or to the scalar dissipation once a local equilibrium assumption is assumed in the scalar variance transport equation [30]). This term may be of special importance in the case of reacting mixing processes [36].

According to the modeling level used for the deviatoric part of the SGS stress tensor (isotropic, non-linear, anisotropic, similarity, etc.) this model may lead to various specific models that have been proposed in the literature. The model in (7) can then be considered as a generalized expression.

For conserved scalar case, we restrict ourselves in this paper to Smagorinsky type model for the flow field and to linear terms in scalar gradient in (7). The simplest model case is then derived by expressing the SGS time scale in terms of quantities of equation (9). Practically, this can also be achieved in terms of the filter size and the SGS viscosity defined within the Smagorinsky model. Using the expression of the Smagorinsky model, the model (7) thus reduces to

$$J_i^{SGS} = -D_{ed}\frac{\partial \overline{z}}{\partial x_i} + D_{dev}(T_{SGS}\tau_{ij}^{SGS(dev)}\frac{\partial \overline{z}}{\partial x_j})$$

$$= -D_{ed}\frac{\partial \overline{f}}{\partial x_i} + D_{an}\Delta^2 \overline{S}_{ij}\frac{\partial \overline{f}}{\partial x_j} = D_{ij}^{Smag}\frac{\partial \overline{f}}{\partial x_j}; \quad (z \equiv f), \qquad (10)$$

$$D_{ij}^{Smag} = -D_{ed}\delta_{ij} + D_{an}\Delta^2 \overline{S}_{ij}; \quad D_{ed} = \frac{\nu_t}{\sigma_t}; \quad \sigma_t \equiv (Sc_t, Pr_t)$$

where D(-) are model coefficients. D_{ij}^{Smag} expresses the reduced eddy diffusivity tensor. The simplest form (10b) resembles to so-called "classical mixed models" (nonlinear closure plus scalar eddy diffusivity) used in the literature. It extends the expression in [13] by including the first term, the conventional eddy diffusivity part. In fact, Peng et al. [13] modified the non-linear model by Clark et al. (see in [12]) by introducing a tensor of diffusivity being linear to the resolved strain-rate tensor that does not involve the eddy diffusivity coefficient. It was shown that the model by Peng et al. could reproduce reasonably some results as compared with DNS data. However, other important features (e.g. counter-gradient effects) related to simple configurations could not be well captured.

The SGS scalar flux model (10) has been successfully validated in different configurations with both gaseous streams and liquid jets, like jet in cross flow (air streams) [31, 32], mixing layer (water streams) [14, 33], jet in channel flow (water streams) [14, 27] and in coaxial jet mixer [34, 35].

Based on the proved performance in non-reacting mixing flows the new model is now applied to investigate high Schmidt number dominated mixing processes in reacting environments. Complex algebraic models for the scalar flux vector, that include chemical contributions, have been derived in [36] from a RANS-based transport equation of the scalar flux vector. Within the transported PDF approach the chemical source appears closed so that its effect on the scalar flux can be treated exactly. As this latter method suffers from too high computational costs, algebraic model as proposed in (7) is useful. For reacting scalars, we restrict ourselves to the following reduced form

$$J_i^{SGS} = -(D_{ed} + \lambda_1 \frac{\partial \bar{Y}_\alpha}{\partial x_k} \frac{\partial \bar{Y}_\alpha}{\partial x_k}) \frac{\partial \bar{Y}_\alpha}{\partial x_i} + D_{dev} T_{SGS} \tau_{ij}^{SGS(dev)} \frac{\partial \bar{Y}_\alpha}{\partial x_j} = D_{ij}^{SGS(\lambda)} \frac{\partial \bar{Y}_\alpha}{\partial x_j}$$

(11)

$$\text{with } D_{ij}^{SGS(\lambda)} = -(D_{ed} \delta_{ij} + \lambda_1 \frac{\partial \bar{Y}_\alpha}{\partial x_k} \frac{\partial \bar{Y}_\alpha}{\partial x_k}) \delta_{ij} + D_{dev} T_{SGS} \tau_{ij}^{SGS(dev)} \text{ and } (z \equiv Y)$$

With respect to (8) the parameter λ_1 in (11) includes the SGS stress tensor and the coefficient $D_\lambda T_{SGS}$ and will also be determined dynamically. The expression of the SGS scalar flux vector (11) will be used in equation (4) while (10b) is applied in equation (3).

2.1.2 Closure for the Chemical Reaction Rate

When a second-order, irreversible, and isothermal reaction ($A + B \rightarrow P$) is considered, the filtered reaction source term in (4) may be expressed as $\bar{S} = Da\overline{Y_A Y_B}$ where Da is the Damköhler number. In a rapid reaction case, equation (3) will be used. The filtered concentration values including SGS mixing should be computed from the SGS PDF of the conserved scalar, in particular from a presumed assumption. In a moderately fast or slow reaction case, the timescale of the chemical reaction may be equivalent to that of the turbulent diffusion. The filtered reaction source term can be decomposed into two terms, as:

$$\bar{S} = Da\overline{Y_A Y_B} = \left(\bar{Y}_A \bar{Y}_B + \overline{Y_A' Y_B'} \right)$$

(12)

where Y_A' is the concentration fluctuation of the chemical species, A, at subgrid scale level. Different approaches can be used for the calculation of (12). For review, please refer to [8]. According to [18] a presumed PDF based approach is not introduced. Rather the turbulence-chemistry interaction is included in the frame of a scale similarity based model following [22] that demonstrated that the scale similarity filtered reaction rate model provided the best agreement with the DNS for product formation in comparison to laminar flamelet model. See also [25, 26]. In particular we follow the dynamic scale similarity approach in contrast to [17], [22] and [25]. By assuming scale similarity between resolved and sub-resolved grid scales, the subgrid covariance in (12) can be modelled as proportional to the resolved covariance at larger scales. It can then be expressed as

$$\overline{Y_A' Y_B'} = \overline{Y_A Y_B} - \bar{Y}_A \bar{Y}_B \sim C_{sim}^\Delta \left(\widetilde{\bar{Y}_A \bar{Y}_B} - \widetilde{\bar{Y}}_A \widetilde{\bar{Y}}_B \right)$$

(13)

where the overtilde represents spatial filtering at scale 2Δ and C_{sim}^Δ is the similarity coefficient. It is dynamically determined by

$$\Upsilon = \widetilde{\bar{Y}_A \bar{Y}_B} - \widetilde{\bar{Y}}_A \widetilde{\bar{Y}}_B$$

$$C_{sim}^\Delta = \frac{\langle \Upsilon X \rangle}{\langle XX \rangle};$$

$$X = \left(\widehat{\widetilde{\bar{Y}_A \bar{Y}_B}} - \widehat{\widetilde{\bar{Y}}_A \widetilde{\bar{Y}}_B} \right) - \left(\widetilde{\widetilde{\bar{Y}_A \bar{Y}_B}} - \widetilde{\widetilde{\bar{Y}}_A \widetilde{\bar{Y}}_B} \right)$$

(14)

This dynamic procedure eliminates the uncertainties associated with tunable model parameters and allows to well address the issue of extrapolation from larger scales. In (14) the brackets represent the averaging operation. While a scale invariance assumption has been postulated in deriving the expression in (14), a scale dependent consideration is being implemented and will be reported in future works.

The reaction systems under investigation in section 3 consist of a parallel reacting system [28] that can be schematically indicated as follows:

$$A + B \xrightarrow{k_1} P_1,$$
$$A + C \xrightarrow{k_2} P_2 + P_3 + (A), \tag{15}$$

where A, B and C are reactants. The first reaction is instantaneous and gives the desired product P_1 whilst the second one is a finite rate reaction that gives the undesired products P_2 and P_3. As the second reaction is catalytic with respect to A, there is no net consumption of this reactant. This system can be described in terms of mixture fraction (see eq. (3)) and two reaction-progress variables, Y_1 for the first reaction and Y_2 for the second one (see eq. (4)), respectively. A consistent relationship between the reaction-progress variables and the species concentrations involved in (15) is formulated according to [2] as

$$\Gamma_A = A_0\left[f - f_{s1}Y_1\right]; \quad \Gamma_B = B_0\left[1 - f - (1 - f_{s1})Y_1\right];$$
$$\Gamma_C = C_0\left[1 - f - (1 - f_{s2})Y_2\right]; \quad \Gamma_{P_1} = B_0\left[(1 - f_{s1})Y_1\right]; \tag{16}$$
$$\Gamma_{P_2} = C_0\left[(1 - f_{s2})Y_2\right]; \quad f_{s1} \equiv \frac{B_0}{B_0 + A_0}; \quad f_{s2} \equiv \frac{C_0}{C_0 + A_0}$$

In these expressions A_0, B_0 and C_0 are the start concentration. As the first reaction is instantaneous, the variables that describe the system are only two (f and Y_2), the quantity Y_1 being immediately determined by the local equilibrium yielding

$$Y_{1\infty} = \min\left(\frac{f}{f_{s1}}, \frac{1 - f}{1 - f_{s1}}\right). \tag{17}$$

In equations (4) for the reaction progress variables the source terms are calculated by

$$\overline{S}_{1,2} = \frac{k_{1,2}}{B_0, C_0 f_{s1,2}} \overline{\Gamma_A \Gamma_{B,C}} = \frac{k_{1,2}}{B_0, C_0 f_{s1,2}}\left(\overline{\Gamma}_A \overline{\Gamma}_{B,C} + \overline{\gamma_A \gamma_{B,C}}\right) \tag{18}$$

according to (12) and (13).

2.1.3 SGS Scalar Variance Model and Scalar Dissipation Rate
In most papers the scalar variance is formulated by means of a scale similarity model and an independent model for the scalar dissipation. As pointed out by Jimenez et al. [30] this can result in situations in which the variance evolves in a

way incompatible with the local scalar dissipation (see [38]). The scalar variance and dissipation cannot be independently modelled without taking into account the fact that they are intimately related through the mixing. Therefore we follow [30] and extend their equation of variance by applying the SGS scalar flux models (10b) for the conserved scalar and (7) for the reaction-progress variables. In both cases the scalar dissipation is postulated in the resulting equations according to [30]. For specific details, the reader may refer to this reference.

2.2 Numerical Procedure

All the filtered continuity, momentum, conserved scalar and reaction progress variable equations were implemented into the FASTEST-3D CFD code. It features geometry-flexible block-structured, boundary fitted grids with collocated variable arrangement. Second-order central schemes are used for spatial discretization except for the convective term in the scalar transport equation. Here, a flux-limiter with TVD (total variation diminishing) properties is employed to ensure bounded solutions for the mixture fraction. Pressure-velocity coupling is achieved via a SIMPLE similar procedure. As time integration scheme the second-order implicit Crank-Nicolson method is used. The resulting set of linear equations is solved iteratively using a SIP-solver. The code is parallelized based on domain decomposition using the *MPI* message passing library (e.g. [14, 15, 22, 24]).

3 Configurations, Results and Discussion

To assess the prediction ability of the proposed SGS model package two configurations are investigated, a non-reacting (water) jet in channel (water) flow and a confined impinging jet reactor featuring a parallel reacting system. The reader interested in results for mixing systems with gaseous streams may refer to [14, 15, 23, 24, 31].

3.1 Non-reacting Jet in Channel Flow (Water)

3.1.1 Configuration

The reference geometry consists of a straight duct with a square cross section of *40*40 mm* as experimentally investigated in [27]. The jet, emerging from the midpoint of one side wall and perpendicular to this wall, is created from a pipe with an inner diameter of $D=4$ mm. Both channel and jet consists of clean water. The water temperature is *26°C*. The Reynolds number $Re = 33750$ is based on hydraulic diameter and bulk mean velocity U_{cf} of the duct flow. The velocity ratio $R=2$.

The coordinate system is centered at the jet axis at the entrance to the duct, x-axis being in the direction of duct flow and z-axis in the direction of the jet axis. Experimental data are available in the x-z plane centered in the duct, yielding the U and W components of velocity in x and z directions, respectively. Values of the mean mixture fraction and corresponding fluctuations are also provided.

To make a compromise between the real CV requirement for mixing process simulation using LES (see in [15]) and the available computational capacity the total control volumes of *415788* cells have been found to be acceptable.

The inflow velocity of the jet is simply a constant bulk velocity because the narrow diameter of jet makes the turbulent inflow condition difficult to implement. To overcome the shortage of laminar inflow, relative long jet geometry should be simulated so that the jet flow can self develop into turbulence before injecting into the channel. The inflow condition of the channel flow uses a polynomial fit method to describe the shape of the inlet velocity profile. In particular a fourth order polynomial function was used to better fit the experimental inflow velocity field.

3.1.2 Results and Discussions

Prior to any quantitative validation, let us give some qualitative properties of the flow and mixing. Figure 1 presents contour plots of the mean streamwise velocity U/U_{cb} both from experimental and numerical results. The comparison shows that the numerical results can predict a similar flow structure as revealed in experiments. Thereby the region of reverse flow is denoted by doted areas downstream of the jet trajectory which is indicated by diamond symbols. For a turbulent jet exhausting into a uniform free stream, Zcan and Larsen (see in [27]) found that the reverse flow region extends all the way down to the jet wall where it covers the range 0.5< x/D <3 for R=3.3 and 0.5<x/D<2.2 for R=1.3. In contrast, for the confined channel flow of the present study, reverse flow near the jet exit is observed in a rather small region around the trailing edge of the jet hole. Side wall effects and the resulting pressure distribution, may be responsible for the absence of a large reverse flow region near the jet wall. It may be hypothesized that the duct flow, which is deflected laterally away from centre plane upstream of the jet, is forced to turn back along side walls of the duct downstream of the jet exit. The subsequent acceleration of the cross flow may cause a reduction in the size of the reverse flow region.

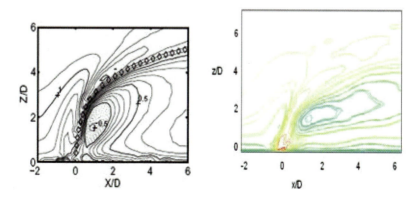

Fig. 1. Contour lines of the streamwise mean velocity $U=U_{cb}$ for jet in Channel Flow: Experiment (left); LES (right)

In order to achieve a fair assessment for the SGS scalar flux models for this high Schmidt number case, simulation results with different SGS scalar flux models as listed in Table 1 are compared to experimental data.

Table 1. Different SGS scalar flux models used for the jet in cross flow configuration

SGS model	Case 1	Case 2	Case 3	Case 4
SGS stress	Dynamic	Dynamic	Dynamic	Dynamic
SGS Scalar Flux	Eddy Diffusivity	Dynamic	Scale Similarity	Anisotropy

Huai et al. demonstrated in [24] how the anisotropy model appreciably improves the prediction of the mean mixture fraction. Concentrated on the fluctuation quantity, the results obtained by different models are presented in Figure 2. The eddy diffusivity model with dynamic procedure reproduces the experimental observations poorly in comparison to the scale similarity model. In contrast, the anisotropy model retrieves clearly better both the position and the maximum values of scalar fluctuations.

3.1.3 Towards Optimisation

The high performance that was also be proved for gaseous mixing processes (see in [15, 23, 24]) qualifies the new SGS scalar package for reliable analysis and optimisation studies of mixing systems. For that purpose appropriate measure parameters are generally introduced to quantify the mixing processes. In [14] the authors considered the so-called "spatial mixing deficiency" and "temporal mixing deficiency" based on the behavior of the scalar variance in space and in time as well as a mixedness parameter to quantify the effects of passive modifications on the mixing mechanism in a jet in cross flow configuration. In [31] they quantified the effects of the variation of the angles of the jet on the mixing mechanism from [32] by means of a mixedness parameter. Further high quality trade-off studies have been carried out in [42]. Thereby alternative design configurations to that studied in [40] have been analyzed. The aim was to optimise the mixer jet configuration in [40] by coupling the LES-CFD and a gradient approach based optimization algorithm [43]. The "spatial mixing deficiency" has been used as objective function depending on the impinging jet angle and respecting the constraint conditions defined in [40]. To avoid repetition, we recommend the reader to refer to [42] for details and results.

In the case of reacting systems, chemical reactions are essentially molecular-level processes. Only mixing on that level can directly influence their course, mixing mechanisms on larger scales have an indirect influence by changing the environment for local mixing. Accordingly, it is possible to modify, even to enhance selectivity of chemical reactions by modifying the mixing conditions of reactants in the reactor and then to strongly influence the yield and purity of end product.

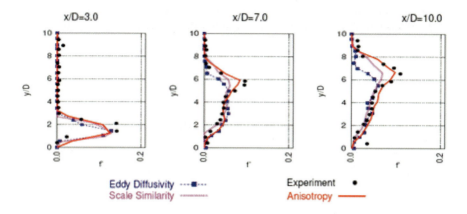

Fig. 2. Comparison of radial profiles of mixture fraction fluctuation predictions (*f'*) (obtained by the use of different SGS scalar flux models) with experimental data for the jet in channel flow configuration.

As most mixing processes are not running at optimal performance because the mixing has been either neglected or designed incorrectly [1], a well physically founded understanding of mixing processes and, in turn, a reliable mixing analysis and reaction control are prerequisite for meeting the demand of efficient mixing. A configuration which includes chemical reactions is subject of the next section.

3.2 Confined Impinging-Jets Reactor (CIJR)

3.2.1 Configuration and Properties

The geometry in figure 3 corresponds to that used in [28]. This consists of a small cylindrical chamber, where an impinging plane is originated by two high-velocity, coaxial liquid inlet streams that collide and produce mixing times on the order of milliseconds. The authors especially investigated the effect of reactor geometry and operating conditions on parallel reaction schemes as schematically indicated by eq. (15) and derived a scaling relationship in terms of some geometric and operating parameters. Thereby the fast reaction is the neutralisation of sodium hydroxide and the slower, second reaction is the acid catalyze hydrolysis of dimethoxypropane (DMP) to form one mole of acetone and two moles of methanol:

$$H^{+}+OH^{-} \xrightarrow{k_1} H_2O,$$
$$H^{+}+CH_3C(OCH_3)_2CH_3 \xrightarrow{k_2} CH_3COCH_3+2CH_3OH+H^{+}$$

Since the first reaction is very rapid, when excess B (see eq. 15) is present the second reaction will take place under conditions where mixing is slow compared to its reaction rate. The conversion of C is thus a sensitive measure of the extent of mixing in the reactor. As quality of measure a mixing efficiency has been used by considering the conversion of the second reaction in the form:

$$X=\text{Conversion} = 1 - \frac{\overline{\Gamma}_C}{\Gamma_{C,m}} \tag{19}$$

where $\Gamma_{C,m}$ is the average concentration after mixing as if no reaction had occurred. $\overline{\Gamma}_C$ is the volume-averaged value at the outlet of the filtered concentration C. Turbulent reacting flows are commonly classified using a turbulent and a Kolmogorov Damköhler numbers [11]. In the context of LES where only eddies smaller that the filter size are parameterized a SGS Damköhler number can also be introduced (see in [26]). Because this latter quantity is not experimentally provided in [28] we focus, for further purpose, on the first one defined by the ratio between mixing t_m and reaction time t_r

$$Da = \frac{t_m}{t_r} \quad \text{with} \quad t_r = \frac{1}{k_2 \Gamma_{A,m}} \tag{20}$$

as used in experiments [28] where the mixing time has been approximated by $t_m = 0.17u^{-3/2}$ (u representing the velocity fluctuation).

Numerically, the CIJR in figure 4 has been already investigated in [39] who adopted a RANS-based CFD technique to simulate the mixing and reaction ongoing in the configuration labelled 500-Y2X (see Table 2). Gavi et al. [37] also applied a RANS-based CFD technique by means of the FLUENT 6.1.22 code where the attention was devoted to the sensitivity of prediction with regard to various turbulence models. They described mixing and reaction at different inlet reactant concentrations, different jet Reynolds numbers and two different reactor geometries (configurations labelled 500A-Y2X and 1000A-Y2X, see Table 2). Recently Marchisio [41] reported simulations by means of LES in which the SGS scalar flux vector has been approximated by the well known linear eddy diffusivity model. Both contributions followed the Direct Quadrature Method of Moments (DQMOM) approach.

Even though these works demonstrated the ability of CFD to reproduce satisfactorily experimental data for a CIJR, it turned out that the predictions of mixing quantities, especially of the DMP conversion, are strongly dependent from the turbulence model used within the operating conditions under consideration. Furthermore the use of a linear eddy diffusivity model for describing the SGS scalar flux vector weakens among others the LES results in [41]. Toward a reliable control and optimisation purposes the prediction accuracy should be improved. Huai et al. [24] investigated the configuration labelled 500A-Y2X (see conf. 1 in Table 2) using LES by applying the new SGS model package. An encouraging agreement with experimental data with respect to distributions of species A, B and DMP has been reported. In the present work LES technique is applied to the configurations labelled 1000A-Y2X (see conf. 2-5 in Table 3). The simulation results are compared to experiments and numerical results obtained in [37] and [41]. All these configurations are summarized in Table 3.

The configuration under study (see cases conf. 2-5 in Table 3) is characterized by a diameter of the impinging jets of 1mm. Having in mind the quantities D, H, Z and δ as the chamber diameter, height, length, and the outlet diameter, respectively, their scaled values are D=4.76d, H=0.8D, Z=1.2D and δ =2d (see figure 4). This configuration is numerically represented in all cases by a computational grid of about 1.25 million CV. The inflow conditions were taken according to experimental data.

Table 2. Configurations investigated for consideration of the influence of geometry and operating conditions

Cases: Conf.	Mixing Head Jet	d, mm	Chamber Multiples, D	Outlet Type	Outlet multiples	Outlet δ, mm	Re
1	500A-Y2X	0.5	4.76	Conical	2	1.0	400
2	1000A-Y2X	1.0	4.76	Conical	2	2.0	704
3	1000A-Y2X	1.0	4.76	Conical	2	2.0	1006
4	1000A-Y2X	1.0	4.76	Conical	2	2.0	1892
5	1000A-Y2X	1.0	4.76	Conical	2	2.0	2696

Non-premixed feed conditions are used. According to (15), if one stream carries A, in concentration A_0 and the other feed stream carries $B+C$ in concentration $B_0 + C_0$ (see figure 4), the mixture fraction is set to $f=1$ where $\Gamma_A = A_0$ and $f=0$ where $\Gamma_B + \Gamma_C = B_0 + C_0$. The reactor is operated in continuous mode with mass flow rates according to experiments as given in see figure 3.

3.2.2 Results and Discussions: Mixing Analysis

Prior to any analysis of mixing let us focus first on the flow field. Using the time series recorded throughout the simulation to compute power spectra, the 3D power spectrum obtained at a point (monitoring point P) located at the central plane is shown in figure 4. The characteristic -5/3 slope of the inertial sub-range is clearly recovered and no specific frequency is exhibited by the turbulent flow. Contours plots for the velocity magnitude at two different Reynolds numbers (Re=704 and 2696) shown in figure 5a,b (top) confirm that the two feed streams collide in the centre of the cylindrical chamber forming an impinging plane where turbulence is rapidly produced and dissipated. In fact, at this impinging plane the kinetic energy that becomes higher with increasing jet Reynolds numbers as result of higher velocity magnitude is quickly converted into turbulent kinetic energy. Thus the local

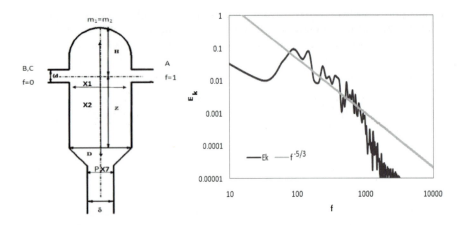

Fig. 3. Schematic drawing of the Impinging jet reactor

Fig. 4. 3D Energy spectrum obtained from the LES in monitoring point P at Re=704

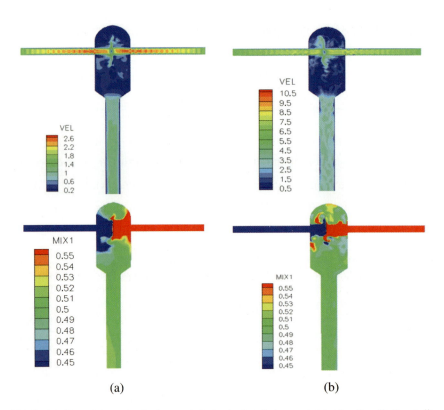

(a)   (b)

Fig. 5. (a). Instantaneous velocity magnitude (top) and mean mixture fraction (bottom) distribution obtained for Re=704. **(b).** Instantaneous velocity magnitude (top) and mean mixture fraction (bottom) distribution obtained for Re=2696.

magnitude of turbulence becomes higher with increasing jet Reynolds numbers while the turbulent fluid motions cover a wider portion of the reactor. This yields in turn an increasing mixing at the macro-scale level as the (mean) mixture fraction becomes more uniform with increasing Reynolds numbers (figure 5a,b (bottom)).

Furthermore, as higher Reynolds numbers imply higher turbulent kinetic energy and higher turbulent dissipation rate, it is expected that higher micro-mixing rates and therefore faster dissipation of the micro-scale variance will occur enabling mixing at the macro- and micro-scale levels to become more efficient. This is confirmed by the contour plots of the reaction progress variable Y2 (here note RPV2) for the second reaction in figure 6. Because this quantity can be seen as an index of mixing efficiency, it reduces from 0.26 to 0.085 with increasing Reynolds numbers from 704 to 2696 demonstrating that the mixing at all scales becomes faster with increasing Reynolds numbers. Let us mention that the micro-scale mixture fraction variance is mainly generated by the filtered mixture fraction gradients and reaches its maximum values on the impinging plane where most of the gradients are present. This fact has also been observed in [37].

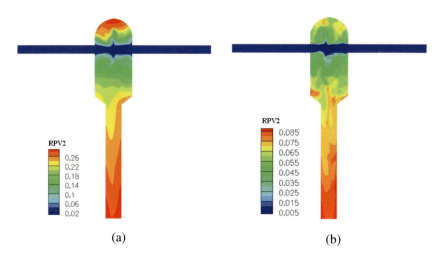

Fig. 6. (a). Instantaneous contours plots for the filtered reaction progress variable of the second reaction (RPV2) obtained for Re=704. **(b).** Instantaneous contours plots for the filtered reaction progress variable of the second reaction (RPV2) obtained for Re=2696.

3.2.3 Results and Discussions: Reaction Analysis and Dependency on Damköhler Number

Focussed now on the conversion of DMP as measure of the process output, figure 7a displays the conversion as function of the jet Reynolds numbers (i.e., different global mixing times) at an initial reactant concentration (i.e., a characteristic reaction time of t_r=14,7 ms). One can clearly observe that the conversion of the second reaction decreases by increasing the jet Reynolds numbers. In other words, as mixing improves the conversion of the second reaction is reduced.

The agreement achieved between experimental values and simulation results is quite impressive in comparison with those obtained in [37] and [41] shown in figure 7b. In [41] the authors combined the DQMOM and LES where the SGS scalar flux vector has been approximated by a linear eddy diffusivity model, well known for its weakness. The result in the present work proves that the model package proposed can credibly be used for derivation or verification of scale-up criterion. In this regard it is worth noted that the Reynolds numbers have been shown in [28] and [37] to be not appropriate scaling variable, as data relative to the different geometries in Table 2 are not well correlated.

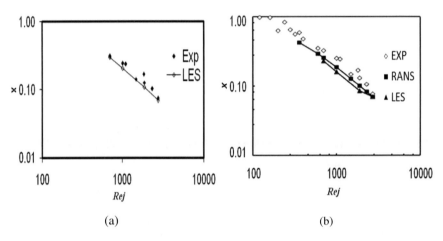

(a)   (b)

Fig. 7. (a). Conversion of the second reaction as function of the Reynolds numbers (present LES). **(b).** Conversion of the second reaction as function of the Reynolds numbers as obtained in [41].

Following [28], one can examine the turbulence scaling model by plotting the conversion (18) versus the Damköhler number (19). A comparison between experimental values for the conversion (18) and simulation predictions obtained for different Reynolds numbers and the initial reactant concentration corresponding to $t_r = 14,7$ ms reported versus the Damköhler number (19) is shown in figure 8a. It turns out that a direct relationship between these two quantities exists for the parallel reaction. Clearly it can be noticed that when the mixing is much faster than the second chemical reaction (lower Da), conversion values are very low. The flow is almost well mixed and, therefore the turbulent scales do not influence much the chemical transformations (see high Re in figure 6b). In contrary, when mixing is much slower than the chemical reaction (higher Da) conversion becomes higher. The reactants are not well mixed resulting in a enhancement of the chemical transformations (see low Re in figure 6a).

This insight may help to accurately determine the scale-up behaviour of confined impinging jet and the functional form of the mixing process at different scale levels. For this purpose additional computations of a number of mixing heads of different geometry have to be carried out. This task is out of the scope of the present work.

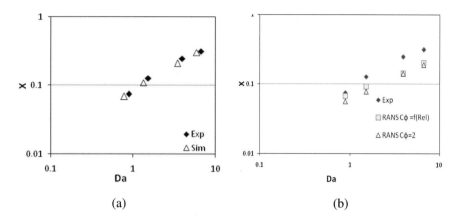

(a)   (b)

Fig. 8. (a). Conversion of the second reaction as function of Damköhler number (present LES. **(b).** Conversion of the second reaction as function of Damköhler number obtained in [37].

4 Conclusions

In this work an advanced subgrid scale (SGS) scalar model package for the description of turbulent mixing in gaseous and liquid flows has been proposed and validated in both non-reacting and reacting systems. It includes advanced SGS models for the filtered scalar flux vector, for its corresponding scalar variance along with scalar dissipation and for the chemical reaction rate. To cover different reaction regimes, the mixing processes with chemical reaction in jet mixer systems were described in terms of mixture fraction and two reaction progress variables.

To assess the accuracy of the SGS model package focus has been put on high Schmidt-number dominated mixing systems: a non-reacting jet in channel flow configuration and a confined impinging jets reactor (CIJR) featuring a parallel reaction system. The model package achieved an overall improvement of prediction of mixing properties in all scale levels. Especially for the CIJR, mixng and reaction have been investigated. The influence of operating conditions (active modification) on mixing properties has been evaluated and a quality of measure based on the conversion of the second reaction has been successfully studied.

Through a quite impressive agreement with experimental values that has been accomplished in all cases the proposed SGS scalar model package strongly improves the prediction accuracy of the mixing field quantities enabling a reliable mixing modification and optimization studies as already reported in [42] for non-reacting mixer jet configurations.

Acknowledgments

The authors express their gratitude to the Deutsche Forschungsgemeinschaft (DFG-SPP1141) that supported the research reported in this paper.

References

1. Baldyga, J., Bourne, J.R.: Turbulent mixing and chemical reactions. John Wiley & Sons, Chichester (1999)
2. Fox, R.O.: Computational models for turbulent reacting flows (2003)
3. Sadiki, A., Maltsev, A., Wegner, B., Flemming, F., Kempf, A., Janicka, J.: Unsteady methods (URANS and LES) for simulation of combustion systems. Int. J. Thermal Sciences 45, 760–773 (2006)
4. Pope, S.B.: Turbulent Flows. Cambridge University Press, Cambridge (2000)
5. Sagaut, P.: Large Eddy simulation for incompressible flows. Springer, Berlin (2001)
6. Germano, M., Piomelli, U., Moin, P., Cabot, W.: A dynamic subgrid scale eddy viscosity model. Physics of Fluids A 3, 1760–1765 (1991)
7. Meyers, J., Geurts, B., Sagaut, P.: Quality and Reliability of Large Eddy Simulatios. Ercoftac Series, vol. 12. Springer, Heidelberg (2008)
8. Janicka, J., Sadiki, A.: Large eddy simulation of turbulent combustion systems. Proceedings of the Combustion Institute 30(1), 537–547 (2005)
9. Klein, M., Sadiki, A., Janicka, J.: A digital filter based generation of inflow data for spatially developing direct numerical or large eddy simulations. Journal of Computational Physics 186(2, 10), 652–665 (2003)
10. Veloudis, I., Yang, Z., McGuirk, J.J., Page, G.J., Spencer, A.: Novel Implementation and Assessment of a Digital Filter Based Approach for the Generation of LES Inlet Conditions. Flow, Turbulence and Combustion 79(1) (July 2007)
11. Peters, N.: Turbulent Combustion. Cambridge University Press, Cambridge (2000)
12. Kang, H.S., Memeveau, C.: Passive scalar anisotropy in a heated turbulent wake: new observations and implications for large-eddy simulations. J. Fluid Mech. 442, 161–170 (2001)
13. Peng, S.H., Davidson, L.: On a subgrid-scale heat flux model for large eddy simulation of turbulent thermal flow. International Journal of Heat Mass Transfer 45, 1393–1405 (2002)
14. Huai, Y., Sadiki, A.: Investigation of heat and mass transfer in a jet in channel flow configuration using LES. In: Proceedings of the 4th International Conference of Heat and Mass Transfer, Paris, vol. 1, pp. 223–228 (2005)
15. Huai, Y.: Large Eddy Simulation in the Scalar field, Doctoral thesis (2005)
16. Jaberi, F.A., Colucci, P.J.: Large eddy simulation of heat and mass transport in turbulent flows. part 2: scalar field. International Journal of Heat and Mass Transfer 46(10), 1826–1840 (2003)
17. Michioka, T., Komori, S.: Large-Eddy Simulation of a Turbulent Reacting Liquid Flow. AIChE Journal 50(11) (2004)
18. Olbricht, C., Hahn, F., Sadiki, A., Janicka, J.: Analysis of subgrid scale mixing using a hybrid LES-Monte-Carlo PDF method. International Journal of Heat and Fluid Flow 28(6), 1215–1226 (2007)

19. Huai, Y., Sadiki, A., Pfadler, S., Löffler, M., Beyrau, F., Leipertz, A., Dinkelacker, F.: Experimental Assessment of Scalar Flux Models for Large Eddy Simulations of Non-Reacting Flows. In: 4th ICHT, Dubrovnik (2006)
20. Pfadler, S., Kerl, J., Beyrau, F., Leipertz, A., Sadiki, A., Scheuerlein, J., Dinkelacker, F.: Direct evaluation of the subgrid scale scalar flux in turbulent premixed flames with conditioned dual-plane stereo PIV. In: Proceedings of the Combustion Institute (2008) (in Press)
21. Chorny, A., Turnow, J., Kornev, N., Hassel, E.: LES versus RANS modeling of turbulent mixing involving chemical reacting in a co-axial jet mixer. In: ICHMT International Symposium on Advances in Computational Heat Transfer, Marrakech, Morocco, May 11-16 (2008)
22. DesJardin, P.E., Frankel, S.H.: Large eddy simulation of a nonpremixed reacting jet: Application and assessment of subgrid-scale combustion models. Phys. Fluids 10, 2298–2314 (1998)
23. Huai, Y., Björg, K., Sadiki, A., Jakirlic, S.: Large Eddy Simulations of Passive-scalar Mixing using a New Tensorial Eddy Diffusivity based SGS-Modeling. In: European Turbulence Conference, Portugal (2007)
24. Huai, Y., Sadiki, A.: Large eddy simulations of mixing processes in turbulent liquid flows with chemical reactions. In: Turbulent Shear Flow Phenomena, München, vol. 5, pp. 1137–1142 (2007)
25. Knikker, R., Veynante, D., Meneveau, C.: A priori testing of a similarity model for large eddy simulations of turbulent premixed combustion. In: Proceedings of the Combustion Institute, pp. 2105–2111 (2002)
26. Vinuesa, J.F., Porté-Agel, F.: Dynamic Models for the Subgrid-Scale Mixing of Reactants in Atmospheric Turbulent Reacting Flows. Journal of the Atmospheric Sciences 65(5), 1692–1699 (2008)
27. Meyer, K.E., Ozcan, O., Larsen, P.S., Gjelstrup, P., Westergaard, G.H.: Point and planar LIF for velocity-concentration correlation in a jet in cross flow. In: Proceedings of FEDSM (2001)
28. Johnson, B.K., Prud'homme, R.K.: Chemical Processing and Micromixing in confined Impinging Jets. AIChE Journal 49(9), 2264–2282 (2003)
29. Sadiki, A.: Extended Thermodynamics as Modeling Tool of Turbulence in Fluid Flows. In: Trends in Applications of Mathematics to Mechanics, pp. 451–462. Shaker Verlag, Aachen (2005)
30. Jiménez, C., Ducros, F., Cuenot, B., Bédat, B.: Subgrid scale variance and dissipation of a scalar field in large eddy simulations. Phys. Fluids 13, 1748–1754 (2001)
31. Wegner, B., Huai, Y., Sadiki, A.: Comparative study of turbulent mixing in jet in cross-flow configurations using LES. International Journal of Heat and Fluid Flow 25(5), 767–775 (2004)
32. Andreopoulos, J., Rodi, W.: Experimental investigation of jets in a crossflow. J. Fluid Mech. 138, 93–127 (1984)
33. Hjertager, L.K., Hjertager, B.H., Deen, N.G., Solberg, T.: Measurement of turbulent mixing in a confined wake flow using combined PIV And PLIF. Can. J. Chem. Eng. 81(6), 1149–1158 (2002)
34. Tkatchenko, I., Kornev, N., Jahnke, S., Steffen, G., Hassel, E.: Performances of LES and RANS Models for Simulation of Complex Flows in a Coaxial Jet Mixer. Flow, Turbulence and Combustion 78(2), 111–127 (2007)

35. Sadiki, A., Huai, Y.: Assessment of an Explicit Anisotropy-resolving Algebraic Scalar Flux SGS Model for LES of Turbulent Mixing Processes. Int. J. Heat and Fluid Flow (submitted)
36. Adumitroaie, V., Taulbee, D.B., Givi, P.: Explicit algebraic scalar-flux models for turbulent reacting flows. AIChE Journal 43(8), 1935–1946 (1997)
37. Gavi Movichinio, E.G., Marchisio, D., Barresi, A.: CFD modelling and scale-up of Confined Impinging Jet Reactors. Chemical Engineering Science 62(8), 2228–2241 (2007)
38. Prière Gravere, C., Gicquel, L.Y.M., Kaufmann, P., Krebs, W., Poinsot, T.: Large eddy simulation predictions of mixing enhancement for jets in cross-flows. Journal of Turbulence 5(5) (2004)
39. Liu, Y., Fox, R.O.: CFD Predictions for chemical processing in a confined Impinging-Jets reactor. AIChE Journal 52(2), 731–744 (2006)
40. Unger, D.R., Muzzio, F.J.: Laser-induced fluorescence technique for the quantification of mixing in impinging jets. AIChE Journal 45(12), 2477–2486 (1999)
41. Marchisio, D.: Large Eddy Simulation of mixing and reaction in a Confined Impinging Jets Reactor. Computers and Chemical Engineering (2008)
42. Huai, Y., Sadiki, A.: Analysis and optimization of turbulent mixing with large eddy simulation. In: ASME 2nd Joint U.S.-European Fluids Engineering Summer Meeting, FEDSM 2006-98416, Miami (2006)
43. Campos, F., Weston, S., Scumacher, T.: Automatic Optimisation of CFD Engineering Designs. In: Automated Design and Optimisation Techniques Using CFD 2006, IMechE, London, UK (2006)

Experimental Investigation of a Static Mixer for Validation of Numerical Simulations

Andreas Lehwald[1], Stefan Leschka[1,2], Dominique Thévenin[1], and Katharina Zähringer[1]

[1] Lab. of Fluid Dynamics and Technical Flows, University of Magdeburg "Otto von Guericke", Universitätsplatz 2, 39106 Magdeburg, Germany
[2] Hydraulic & Coastal Engineering, DHI-WASY GmbH, Syke, Germany

Abstract. The mixing behavior around a SMX-type static mixer segment has been investigated experimentally by using Particle Image Velocimetry (PIV) and Planar Laser-Induced Fluorescence (PLIF) simultaneously. The newly conceived experimental rig allows quantifying mixing, by measuring velocity fields and concentration fields for different Reynolds numbers along vertical and horizontal sections directly behind the mixer. A spectral analysis of the simultaneous PIV and PLIF measurements has been carried out, in order to determine the characteristic frequencies induced by the mixer and relationships between velocity and concentration. Furthermore, the mixing efficiency has been quantified by considering the segregation index. In order to quantify micro-mixing a chemical reaction, involving a fluorescent tracer, has been chosen and imaged simultaneously with the macro-mixing, characterized by another tracer. Theoretical micro-mixing models have been tested by considering their accuracy in reproducing experimental data and first comparisons between experimental and numerical velocity fields are presented. The experimental data is documented in a data base, freely accessible through Internet, to allow a quantitative validation of numerical codes.

1 Introduction

In the past the mixing behavior of static mixers has been investigated in a global manner for various mixer configurations, but usually without relying on non-intrusive optical diagnostics. For example Pahl and Muschelknautz (1982) have compared 12 different static mixer types, but considering only global properties, like pressure drop and overall mixing efficiency.

Recently, static mixers have been also characterized using Planar Laser-Induced Fluorescence (PLIF) to determine the concentration fields and the concentration variance, a characteristic parameter to quantify macro-mixing. A few generic configurations (Wadley and Dawson 2005) and commercial static mixers (Pust *et al.* 2006) have been investigated in this manner. This is an essential step, but an isolated PLIF measurement cannot deliver information concerning the flow structure.

In the present project Particle Image Velocimetry (PIV) and PLIF have been combined and used simultaneously to quantify experimentally the fluid behavior

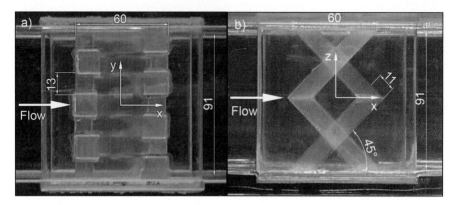

Fig. 1. SMX-type static mixer considered in this study, a) top view: horizontal (x-y) plane, b) side view: vertical (x-z) plane

around a static mixer, in particular to determine the flow and mixing conditions behind an isolated mixer segment. The considered configuration is a SMX-type static mixer (fig. 1). These extensive experimental measurements are additionally employed to improve numerical mixing models and to validate computational simulations relying on Large-Eddy Simulations (LES), carried out in a companion project at the University of Heidelberg. Such LES computations should ultimately be able to characterize in an accurate manner the properties of industrial static mixers. First results of such validations are also presented here.

Both reacting and non-reacting mixing flows have been considered. Since our test rig is not a small model installation, a chemical reaction had to be chosen, which involved only affordable reactants, easy to handle and not poisonous for the waste water. Several reaction systems and tracer dyes have been analyzed for this purpose. The final choice is presented here, as well as first experimental results of the simultaneous characterization of macro- and micro-mixing.

The specific issue of micro-mixing (Fox 2003) has also been considered in a theoretical manner, in order to identify appropriate models for numerical calculations.

2 Experimental Set-Up and Measurement Methods

A gravity-driven flow channel installation is used to obtain laminar conditions at the inlet of the mixer (fig. 2). For that, a pump delivers the fluid from a lower tank to an upper tank, whose water level lies roughly 10 m above the channel axis. A diffuser is installed, which is used after the down-pipe to adapt the cross-section from circular to square. This square shape has been chosen for the whole measurement section in order to increase the quality of all optical measurements by limiting refraction, laser beam divergence and image distortion. After the diffuser, a 2.45 m long inflow section leads the fluid to the static mixer. The mixer consists of seven rectangular lamellas with an individual width of 13 mm, as illustrated in figure 1. Four lamellas are turned against the flow; between them, three are turned in the direction of the flow. The straight outflow section behind the mixer also has a length of 2.45 m. Inflow, outflow and mixer consist of acrylic glass and have a cross section of 91 x 91 mm². A flow meter is installed at the end of the test section. The required flow-rates are adjusted using a precision control valve. From there, the fluid goes to a damper. For simultaneous PLIF measurements, this water contains the fluorescence dye and is sent to a purification process, so that the dye does not re-circulate in the installation (open loop). A schematic view of the experimental set-up developed for the present investigation is shown in figure 2. Further details can be found in Leschka *et al.* (2006).

For PIV measurements tracer particles are injected into the upper tank (fig. 2). Spherical hollow glass balls (Type 110 P8 CP00) with a mean diameter of $d_{50} = 10.2$ μm ($d_{10} = 2.9$ μm, $d_{90} = 21.6$ μm), a mean density of 1.1 g/cm³ and a bulk volume between 2.22 l/kg and 2.85 l/kg have been used. Depending on the fluid velocities, the tracer concentration is adapted to obtain 30 to 40 particles in a 32 x 32 pixel interrogation area of the PIV image. This results in a typical volume fraction of $V_p/V_f = 6.86 \cdot 10^{-5}$, low enough to avoid any influence of the particles on the flow.

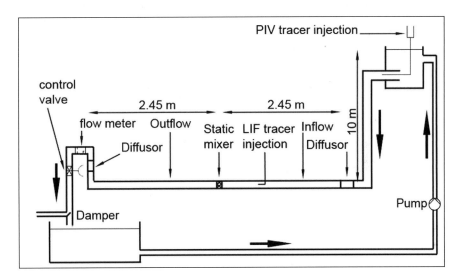

Fig. 2. Schematic view of the experimental set-up

For macro-mixing measurements using PIV and PLIF simultaneously, the fluorescence tracer (99% pure Rhodamine 6G) is injected along the centerline, 39 mm upstream of the mixer segment, with the same velocity as the main fluid and with a Rhodamine concentration of 20 μg/l. The PLIF images have been acquired using an intensified CCD camera (LaVision NanoStar, 1280 x 1024 pixels) during the first of two PIV laser pulses (Nd:YAG, 532 nm, 80 mJ per pulse), while the CCD camera (LaVision Imager Intense, 1600 x 1186 pixels) used for PIV takes double images, used to calculate the two-dimensional velocity fields. To separate PIV and

PLIF signals, the cameras are equipped with two different filters. For recording PIV images a laser-line filter BP532/10 nm and for recording PLIF images a high-pass filter LP580 nm have been used, respectively.

The outlet of the central LIF tracer injection system has a square section of 25 mm². Before entering the test-section, the injected mixture of water and fluorescing dye flows through a thermostat (Haake Phoenix 2 P1-C25P), which is used to control the temperature of the injected fluid and keep the difference with the main fluid temperature within the channel below 0.5 K. The temperatures typically lie around 17 °C in all experiments presented here. This very accurate temperature control was observed to be absolutely necessary, in order to avoid a rapid stratification between the injected mixture and the main flow, due to buoyancy.

The static over-pressure in the test-section is 0.872 bar. The Reynolds numbers are always calculated using the hydraulic diameter of the channel and the mean streamwise velocities.

The resulting values of the Reynolds number Re together with the corresponding total flow rate and the flow rates Q_1 of main fluid and Q_2 of the injected mixture, their ratio and the mean streamwise velocities are summarized in table 1 for the two main cases (Re = 562 and Re = 1000) discussed in this paper.

Static mixers are usually employed for highly viscous fluids and thus very low Reynolds numbers. The present experiments use water as a main fluid but keep the analogy by considering the same typical Reynolds numbers (similarity conditions), thus leading to very low velocities.

Table 1. Reynolds numbers, flow rates and resulting mean axial velocities

Re [-]	Q [l/h]	Q_1 [l/h]	Q_2 [l/h]	$\overline{u}_x$ [mm/s]	$\dfrac{Q_2}{Q_1+Q_2}$ [-]
562	184.8	184.0	0.826	6.2	$4.5 \cdot 10^{-3}$
1000	328.9	327.2	1.714	11.0	$5.2 \cdot 10^{-3}$

The PIV data have been post-processed and evaluated using an adaptive multi pass cross-correlation with decreasing size over interrogation areas of 128 x 128, 64 x 64 and 32 x 32 pixels, each with 50 % overlapping. Furthermore, a Gaussian low-pass filter has been applied. In the interrogation area, values lying within the range of 120 % of the second highest value have been accepted.

For PLIF a concentration calibration is necessary. In the literature different specifications can be found concerning the linear part of the fluorescence behavior of Rhodamine 6G. Therefore several tests have been conducted in order to find an appropriate range for the present optical and flow conditions. As recommended by Law and Wang (2000), a concentration range of 0 µg/l to 20 µg/l has been finally found to be suitable for these experiments. The Nd:YAG-laser has been systematically combined with an energy monitor, which allows to correct laser energy fluctuations during post-processing. This correction is necessary to obtain accurate quantitative concentrations from PLIF. Post-processing masks have been defined to exclude non-evaluable areas due to locally insufficiently good optical conditions, resulting from the lamellas, shadows of mounting screws, etc. (see also fig. 1). In addition, background images have been acquired. In this manner negative effects like background scattering (reflection, diffraction and refraction) can be eliminated.

The dyes and experimental methods used for simultaneous macro- and micromixing investigations are described in the section dealing with that subject, since several preliminary tests were necessary and have shown interesting results, which are presented later.

3 Results

3.1 Simultaneous PIV-PLIF Measurements

The results presented in figure 3 exemplify instantaneous simultaneous measurements involving PIV and PLIF directly behind the mixer ($x = 50$ mm), in the central horizontal and vertical plane at Re = 562 and Re = 1000. White zones in the images correspond to masked areas, in which measurements are not possible.

Reynolds number 562 is the reference case for this study. Averaging the velocity measurements over 300 images, velocity ranges of 0.3 mm/s up to 33 mm/s in the horizontal plane and of 0.4 mm/s to 22 mm/s in the vertical plane are obtained. The maximum velocity is measured, as expected, in-between the mixer lamellas (lowest hydraulic section).

At Re = 1000 the flow shows more vertical structures and the velocity magnitudes range between 0.2 mm/s and 57 mm/s in the horizontal plane and between 0.3 mm/s and 42 mm/s in the vertical plane.

The differences observed between horizontal and vertical measurements are not surprising, since the single-stage mixer structure is not at all isotropic (fig. 1), with all lamellas positioned in the vertical direction. For such low Reynolds numbers, the influence of the mixer geometry is expected to be very high, leading to different results in the horizontal and vertical plane. This difference will decrease when increasing the Reynolds number or when considering realistic mixer geometry with several stages, turned one to the other at 90°, and thus increasing flow isotropy.

The concentration measurements clearly show at Re = 562 small-scale, coherent vortex structures both in the horizontal and in the vertical plane. The measurements for the higher Reynolds number do not present such clear vortex structures any more. Due to the small quantity of dye injected into the main flow and to the higher dynamics of the flow, a large part of the injected fluid is already well-mixed when the fluid leaves the mixer segment. Qualitatively, a high level of mixing is in particular seen in the horizontal plane just behind the mixer, since the influence of the lamellas is maximal in this plane. In the vertical plane, elongated vortex filaments are observed. As a whole, mixing homogeneity increases with the Reynolds number, as expected. This will be quantified in what follows.

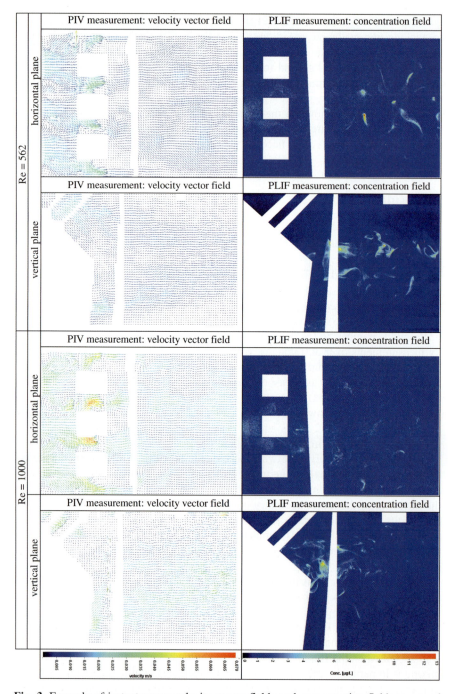

Fig. 3. Example of instantaneous velocity vector fields and concentration fields measured simultaneously, directly behind the static mixer at Reynolds number 562 (top) and 1000 (bottom) for the central horizontal and vertical middle planes

Behind the mixer the flow velocities and concentrations have been analyzed extensively using Non-equidistant Fast-Fourier Transformation (NFFT) in order to identify the characteristic frequencies f of the structures induced by the static mixer, both in the velocity and in the concentration fields. During the simultaneous PIV/PLIF measurements, the obtained acquisition frequency is limited to maximum 1.6 Hz and is furthermore not constant, due to the limited and different data transfer-rates of the cameras. As a consequence, a classical FFT analysis is not possible any more. For the present post-processing the efficient Lomb-Scargle-Algorithm for NFFT (Press *et al.*, 1995), has been used.

The frequency analysis of the PIV data delivers clear peaks and harmonics for Re = 562 and Re = 1000 in the horizontal as well as in the vertical planes (fig. 4).

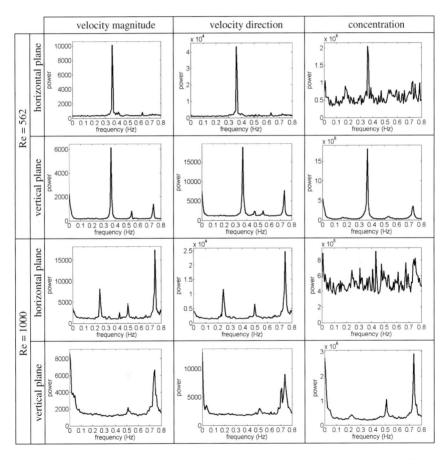

Fig. 4. NFFT for velocity vector fields considering separately vector magnitude (left) and vector direction (middle), as well as NFFT of concentration fields (right). All measurements have been carried out directly behind the mixer at Reynolds number 562 (top) and 1000 (bottom) for horizontal and vertical planes.

The PLIF frequency analysis shows a lower signal-to-noise ratio in the horizontal plane, especially for Re = 1000. For Re = 562 the peak frequencies for velocity magnitude, velocity direction and concentration are all equal to 0.36 Hz. This means that the structures induced by the mixer repeat after 2.78 s, as can also be seen by looking directly at time-sequences. For Re = 1000, a peak frequency of 0.75 Hz (period of 1.33 s) has been found, unfortunately quite close to the Shannon-Nyquist frequency. For both planes and both values of Re the variations of the velocity field correspond very well to the fluctuations of the concentration field.

For the quantification of the mixing efficiency the classical segregation index $I_S = \dfrac{\sigma^2}{\sigma_{max}^2}$, given by Danckwerts (1952) and based on the normalized concentration variance, has been calculated along the channel axis. The variance of the concentration $c(x)$ of the injected dye is calculated by $\sigma^2 = \dfrac{1}{A}\int_A [c(x)-\mu]^2 \cdot dA$, with the expectation (mean) value $\mu = \dfrac{1}{A}\int_A c(x)\cdot dA$. The maximum computed variance σ_{max}^2, used as a reference value for normalization, is computed by considering independently all individual results (global maximum). For this treatment the measurement area has been divided in regular stripes along the streamwise direction. In each stripe, the "local", mean concentration value and variance can be determined. After having determined the global maximum of the variance, a segregation index can be calculated as a function of the streamwise coordinate for each instantaneous PLIF image. The resulting segregation indices have then been averaged over 300 images. The final, average values are shown in figure 5. The fastest mixing is, as expected, obtained at the higher Reynolds number, as can clearly be seen from this figure. The difference between the vertical and the horizontal plane can again be explained by the non-isotropic geometry of the single-element mixer.

For a further quantification of the mixing efficiency, the mixing length L_{mix} can be defined as the distance from the mixer outlet ($x = 0$ mm) to the first location after which a given threshold value of the segregation index, usually 5%, is not exceeded any more. Considering the mean velocities $\overline{u}_x$ given in table 1, a characteristic mixing time t_{mix} can then be deduced as $t_{mix} = L_{mix}/\overline{u}_x$.

For the conditions considered here this analysis leads to a characteristic mixing distance of 58 mm and to a corresponding mixing time of 9.3 s for Re = 562. For Re = 1000, $L_{mix} = 38$ mm and $t_{mix} = 3.5$ s are found. These are the limiting (slowest) values, obtained in the horizontal plane. In the vertical plane this criterion leads to a slightly faster mixing process, which is not obvious from a visual, qualitative analysis of the images (see fig.3). This quantitative analysis shows that the mixing is roughly three times faster when doubling the Reynolds number.

Thanks to such simultaneous PIV/PLIF measurements it is possible to characterize quantitatively and in a non-intrusive manner the flow and mixing conditions behind mixers, in particular to determine characteristic flow frequencies, relationships between velocity and concentration fields and to quantify mixing efficiency.

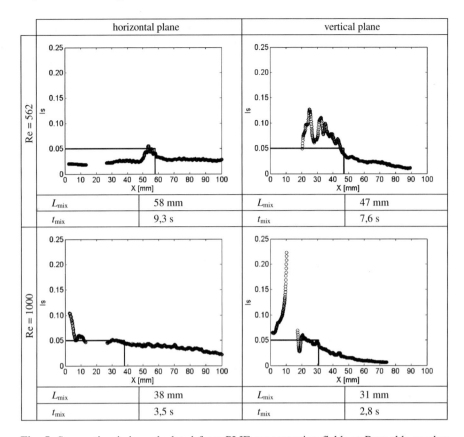

Fig. 5. Segregation index calculated from PLIF concentration fields at Reynolds number 562 (top) and 1000 (bottom) for horizontal and vertical planes. Resulting characteristic mixing length L_{mix} and mixing time t_{mix} for Re = 562 and Re = 1000.

In the near future similar simultaneous PIV and PLIF measurements will be realized for up to four consecutive mixer segments, each of them turned by $\pi/2$ compared to the previous one.

3.2 Experimental Characterization of Macro- and Micro-Mixing

The mixing of chemically reacting species in a liquid-liquid system can be used to quantify simultaneously micro-mixing and macro-mixing.

Macro-mixing considers the mixing at large scales, up to the size of the apparatus. Micro-mixing is associated specifically with small-scale mixing, down to the molecular scale (Hjertager Osenbroch, 2004).

In order to separate macro-mixing and micro-mixing, a preliminary study has demonstrated that fluorescein disodium salt (Uranine) can be used as a tracer of a neutralization reaction (Lehwald *et al.*, 2008). This fluorescence dye changes its fluorescence intensity depending on the local pH-value, which can be used to track micro-mixing. Figure 6 shows the fluorescence intensity I as a function of the pH-value for a range of the pH between 3.5 and 8.

For these results small transparent tanks have been filled with calibrated solutions. Then, 300 fluorescence images have been taken, with the same laser and camera described before. These have been averaged and the mean value of the image has been extracted, before normalizing by the maximum, attained at pH=8.

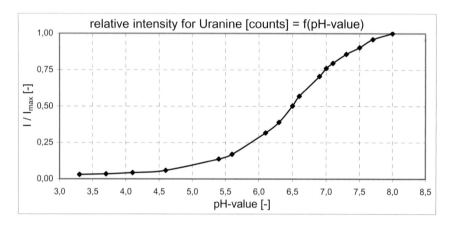

Fig. 6. Fluorescence intensity of fluorescein disodium salt (Uranine) as a function of the pH-value

The experimental concept is now to mix a water/hydrochloride acid (HCl) solution with fluorescein disodium salt and inject it into the main fluid, which is now an alkaline fluorescein disodium salt solution. The resulting acid-base reaction changes the local pH-value and as a consequence the fluorescence properties of the fluorescein disodium salt.

Using equal concentrations of fluorescein disodium salt in both liquids, this local pH modification can be quantified by PLIF and is a direct marker of the very fast acid-base chemical reaction and thus of micro-mixing.

For correlation between micro-mixing and macro-mixing, a second dye is required for characterizing the macro-mixing, that can be stimulated at the same wavelength (532 nm) and differentiated from the first one. This dye fluorescence intensity has furthermore to be independent of the local pH-value.

Figure 7 shows the fluorescence emission spectra of different fluorescent dyes. These spectra have been acquired using an Acton Research spectrograph combined with the intensified NanoStar camera described before. The dye solution has been filled in small transparent tanks and fluorescence has been stimulated at 532 nm.

In the low wavelength range (445 - 555 nm) Uranine has to be detected in order to quantify micro-mixing. The dye tracer employed for macro-mixing thus must emit in a higher wavelength range (for instance 700 - 710 nm), in order to be able to separate its signal from the Uranine signal by a filter.

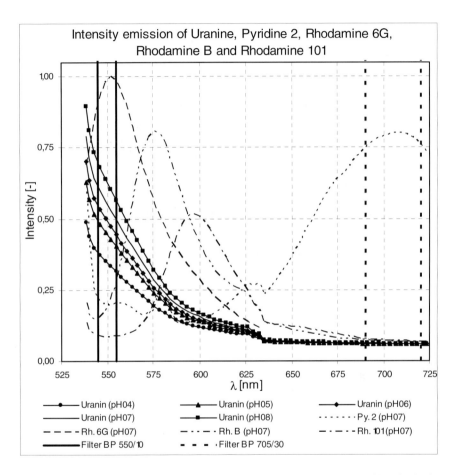

Fig. 7. Fluorescence intensity emission of Uranine, Pyridine 2, Rhodamine 6G, Rhodamine B and Rhodamine 101 as a function of the pH

As can be seen in figure 7, Pyridine 2 turned out to be the most suitable macro-mixing tracer for this experimental combination. It can be introduced into the injection flow and can be separated optically from the Uranine emission by a filter. Its fluorescence intensity does not depend noticeably on the pH-value. Simultaneously quantifying micro- and macro-mixing is thus possible with this combination (Uranine: micro-mixing; Pyridine 2: macro-mixing).

The following images (fig. 8) show very first results of comparisons between micro- and macro-mixing in the x-z-plane just behind the mixer outlet. They have been acquired using two PLIF cameras, identical to the one described previously. These first results are very encouraging and show that it should be possible to investigate simultaneously and in a quantitative manner micro-mixing and macro-mixing using this tracer combination. Further details can be found in a dedicated publication.

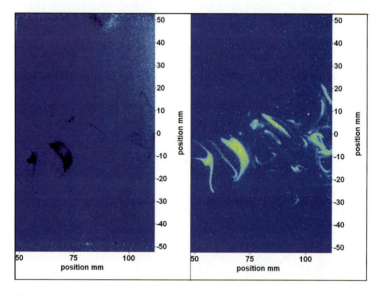

Fig. 8. PLIF recordings that could be used to quantify micro-mixing (left: Uranine) and macro-mixing (right: Pyridine 2)

Under such conditions, PIV can again be employed simultaneously with PLIF, in order to analyze the velocity field. In this manner, it will be possible to determine correlations between the velocity field, macro-mixing and micro-mixing. This is the subject of our future work.

3.3 First Comparisons with Numerical Simulations

As explained previously, these experimental results should also be used for the validation of companion numerical simulations. Corresponding results are found in the chapter of our partners (IWR, Univ. of Heidelberg, research group of G. Wittum) in the present book. In order to simulate the mixer, Large-Eddy Simulations (LES) employing the simulation code UG, developed at the Univ. of Heidelberg, have been employed. One example of an early comparison between PIV measurements and LES computations obtained on an adaptive grid with 4 levels of refinement is shown in Fig. 9. In this figure, the time-averaged axial velocity (x-direction) is compared, showing a very good agreement, both qualitatively and quantitatively. Further, more recent results can be found in the chapter written by our partners.

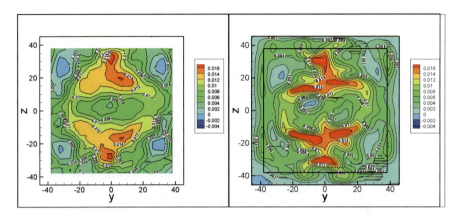

Fig. 9. Experimentally measured average velocity component in the *x*-direction (left), compared with similar results obtained by LES (right). The same scale is used for both figures. Only the part of the numerical results contained within the solid black lines has been measured. All values in m/s.

3.4 Modeling Numerically Micro-Mixing

The properties and quality of a product processed by a static mixer are of course highly dependent on the specific mixing conditions encountered in the mixer, and are to a large extent controlled by micro-mixing when fast chemical reactions are involved. In order to check this issue for future numerical simulations of our static mixer, a precipitation (also called reactive crystallization) process has been considered first, since it involves globally the same physical phenomena as those of interest for static mixer applications in reacting flows. Furthermore, it should be very sensitive to the employed micro-mixing model, since precipitation reactions are usually quite fast. More specifically, the precipitation of barium sulphate ($BaSO_4$) has been numerically investigated considering two different tubular reactors involving as a whole four different Configurations (Bałdyga & Orciuch, 2001; Marchisio *et al.*, 2002). Five different micro-mixing models associated with an increasing level of complexity have been implemented in the industrial CFD code FLUENT® 6.3. Steady-state simulations of the two considered experimental setups based on Reynolds-averaging have been carried out. The particle properties predicted by the simulations have been then compared to experimental results from the literature and the resulting error has been computed.

The micro-mixing models tested in this project are ME (multi-environment, using either 2 or 3 three different environments), Eng (engulfment), and DQMOM-IEM (direct quadrature method of moments – interaction by exchange with the mean, using again either 2 or 3 different environments). All have been formulated here in an Eulerian framework and compared to the predicted results without taking into account micro-mixing at all. All further details concerning the model equations, the practical implementation, and all numerical issues are available in a dedicated publication (Öncül et al., 2008), containing also further references to the employed formulations.

In Figure 10, a comparison with the Configuration A of Marchisio et al. (2002) is proposed. In this configuration the concentration of the reactant in the inner pipe is kept constant at 0.034 kmol m^{-3} whereas that of the other reactant is equal to α·0.034 kmol m^{-3}. The concentration ratio, α, takes values between 0.1 and 3.0. Figure 10 shows the mean particle diameter d_{43} at reactor outlet vs. α. The experimental data including the error bars are those presented in Marchisio et al. (2002). The numerical average at reactor outlet is based on the local mass flowrate and should thus be directly comparable to the experimental measurements. As can be seen both qualitatively and quantitatively, the agreement is globally satisfactory, but the different micro-mixing models show a quite different behaviour as a function of α, i.e. when varying supersaturation and therefore changing the characteristic chemical time of the system.

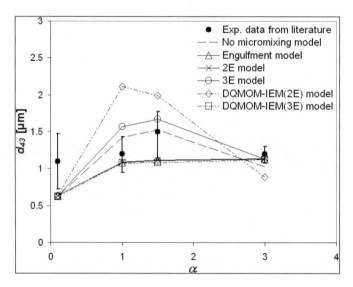

Fig. 10. Mean crystal size profiles averaged at reactor outlet in Configuration A of Marchisio et al. (2002) according to our numerical calculations in comparison with the experimental data.

At lowest supersaturation ratio (i.e. $\alpha = 0.1$) the influence of the micromixing model is almost negligible due to the lack of competition between the mixing and reaction mechanisms. When computing the relevant timescales, it becomes obvious that micromixing plays a much more significant role for $\alpha = 1.0$. The averaged errors associated with the different micro-mixing models considering globally all four different configurations are shown in table 2, together with the total requested computing time, normalized by that needed without any micro-mixing model.

Table 2. Averaged relative error between numerical results and experimental data and increase in computing time compared to the simulation without any micro-mixing model

Model\Relative error (%)	$Err_{mean}(\%)$	Computing time increase factor (compared to no micro-mixing)
1E model (no micro-mixing)	29.4	1 (reference)
Eng model	25.6	1.33
2E model	25.0	1.33
3E model	32.7	2.1
DQMOM-IEM(2E) model	48.7	19.3
DQMOM-IEM(3E) model	24.3	24.0

According to these results and to the complementary information contained in Öncül *et al.* (2008), it can be concluded that although the influence of the micro-mixing model is generally clearly visible, it might be negligible for example at higher Reynolds numbers or lower supersaturation ratios. Moreover, the predictions of the employed closures may show similarities with each other and/or may show drastic deviations from the experimental data. For two models, the computations without any micro-mixing model lie even closer to the experimental values! Thus, it is not an easy task to come up with a clear "best approach". Nevertheless, the 2E model represents globally the best compromise for describing micro-mixing in the present configurations, when both the overall accuracy of the obtained results and the computational times are taken into account. DQMOM-IEM(3E) appears to be even more accurate, but requires a much larger computing effort. It should therefore only be considered when accuracy is an essential issue. No model leads to a perfect agreement with the experiments for all considered configurations. It is not absolutely clear if the discrepancy results from numerical approximations, inappropriate physical models, or even experimental uncertainties…Further investigations, and probably further model improvements, will be needed to progress toward a model with acceptable numerical costs and a more general validity. In the meantime, using either 2E or DQMOM-IEM(3E) is recommended. The comparisons will be of course repeated for our static mixer configuration, as soon as quantitative experimental results allowing a separation between micro- and macro-mixing based on simultaneous PLIF measurements, as described in Section 3.2, will be available.

4   Conclusions

Thanks to simultaneous PIV/PLIF measurements it has been possible to characterize quantitatively and in a non-intrusive manner the mixing process in a SMX-type static mixer. In particular characteristic flow frequencies, relations between velocity and concentration fields and mixing efficiency have been quantified. In the near future, similar simultaneous PIV and PLIF measurements will be realized for up to four consecutive mixer segments.

At present, this single segment is employed to quantify simultaneously micro-mixing and macro-mixing by using at the same time PLIF of Uranine (micro-mixing) and Pyridine 2 (macro-mixing). First studies have shown that this combination should be usable at an acceptable cost in our set-up. In principle, it is again possible to combine with PIV in order to get a complete description of the mixing process. These measurement results will be used to check the accuracy of different micro-mixing models available in the literature, needed for a later numerical simulation of this static mixer.

All results of the experimental measurements are and will be made freely available for the scientific community through a data-base accessible via Internet under http://www.uni-magdeburg.de/isut/LSS. This database contains all the data needed for comparisons and validations of numerical simulations: the boundary conditions, the coordinates, the velocity components and the concentration values. This data-base will be continuously extended as soon as new results are available.

Acknowledgments

A companion project takes place at IWR, University of Heidelberg (A. Hauser, V. Aizinger, research group of G. Wittum), concerning the numerical simulation of this static mixer using Large-Eddy Simulations. Interesting discussions with H. Nobach (Max Planck Institute for Dynamics and Self-Organization, Göttingen, Germany) concerning the frequency analysis of non-equidistant time-signals are gratefully acknowledged.

References

Bałdyga, J., Orciuch, W.: Barium Sulphate Precipitation in a Pipe - An experimental study and CFD modelling. Chem. Eng. Sci. 56, 2435–2444 (2001)

Danckwerths, P.V.: The definition and measurement of some characteristics of mixtures. Appl. Sci. Res. A3, 279–296 (1952)

Fox, R.O.: Computational Models for Turbulent Reacting Flows. Cambridge University Press, Cambridge (2003)

Hjertager Osenbroch, L.K.: Experimental and Computational Study of Mixing and Fast Chemical Reactions in Turbulent Liquid Flows. Ph.D. thesis, Aalborg University Esbjerg, Denmark (2004)

Law, A.W.K., Wang, H.: Measurement of mixing process with combined Digital Particle Image Velocimetry and Planar Laser Induced Fluorescence. Exp. Therm. Fluid Sci. 22, 213–229 (2000)

Lehwald, A., Leschka, S., Zähringer, K., Thévenin, D.: Fluid dynamics and mixing behavior of a static mixer using simultaneously Particle Image Velocimetry and Planar Laser-Induced Fluorescence measurements. In: 14th Int. Symp. on Applications of Laser Techniques to Fluid Mechanics, Lisbon, Portugal (2008)

Leschka, S., Thévenin, D., Zähringer, K.: Fluid velocity measurements around a static mixer using Laser-Doppler Anemometry and Particle Image Velocimetry. In: Lajos, T., Vad, J. (eds.) Proceedings of the Conference on Modelling Fluid Flow, Budapest, vol. 1, pp. 639–646 (2006) ISBN 963 06 0361 6

Marchisio, D.L., Barresi, A.A., Garbero, M.: Nucleation, growth, and agglomeration in Barium Sulfate turbulent precipitation. AIChE J. 48, 2039–2050 (2002)

Öncül, A.A., Janiga, G., Thévenin, D.: Comparison of various micromixing approaches for CFD simulation of barium sulphate precipitation in tubular reactors. Ind. Eng. Chem. Res. 48, 999–1007 (2009)

Pahl, M.H., Muschelknautz, E.: Static mixers and their applications. Int. Chem. Eng. 22, 197–205 (1982)

Press, W., Teukolsky, S., Vetterling, W., Flannery, B.: Numerical Recipes in FORTRAN. Cambridge University Press, Cambridge (1995)

Pust, O., Strand, T., Mathys, P., Rütti, A.: Quantification of laminar mixing performance using Laser-Induced Fluorescence. In: 13th International Symposium on Applications of Laser Techniques to Fluid Mechanics, Lisbon (2006)

Wadley, R., Dawson, M.K.: LIF measurements of blending in static mixers in the turbulent and transitional flow regimes. Chem. Eng. Sci. 60, 2469–2478 (2005)

Simulation of Flow and Transport in a Static Mixer Using Adaptive and Higher Order Numerical Methods

Vadym Aizinger[1], Andreas Hauser[2], and Gabriel Wittum[1]

[1] Goethe-Zentrum für Wissenschaftliches Rechnen (G-CSC), Goethe-Universität Frankfurt am Main, Kettenhofweg 139, 60325 Frankfurt am Main
[2] Siemens Forschungszentrum, Günther-Scharowsky-Straße 21, 91058 Erlangen

Abstract. We simulate flow and mixing in a Sulzer static mixer with a complex geometry utilizing adaptive mesh refinement strategies for a node based Finite Volume method and an arbitrary order Discontinuous Galerkin method. Performance of both schemes is compared and tested using experimental results and some additional benchmark problems.

1 Introduction

The main goal of this project is to develop and test advanced numerical methods for flow and mixing in static mixers and compare the results to experiment and traditional numerical techniques. In particular, we are interested in adaptive mesh refinement and higher order schemes. To achieve this end we add adaptive features and error indicators to our Finite Volume package and implement a Discontinuous Galerkin (DG) solver for the incompressible Navier-Stokes equations.

To verify and test our solvers we conduct a series of numerical simulations for some standard benchmarks and carry out comparisons to experimental results provided by the group of Prof. Thévenin of the University of Magdeburg.

The paper is subdivided as follows. The next section briefly specifies the mathematical model used to simulate the flow. In Section 3, we provide definitions of numerical schemes used in our simulations. Numerical results and comparison to experiments are presented in section 4. We conclude the paper with a brief discussion of the results obtained in the course of the project.

2 Mathematical Model

Our discretization is based on the incompressible Navier-Stocks equations formulated for the case of constant density and augmented as needed by a transport equation.

The momentum conservation equations have the form:

$$\partial_t \underline{u} + \nabla \cdot \left(\underline{u} \times \underline{u} - \nu \nabla \underline{u} \right) + \nabla p = 0, \tag{1}$$

where p is the pressure, $\underline{u} = (u,v,w)$ is the velocity vector, and $V = V_m + V_t$ is the diffusion coefficient computed as the sum of molecular diffusivity V_m and turbulent eddy viscosity V_t.

The continuity (mass conservation) equation is simply

$$\nabla \cdot \underline{u} = 0 , \tag{2}$$

and the transport equation for a passive tracer of concentration c can be written as:

$$\partial_t c + \nabla \cdot \left(\underline{u}c - v_c \nabla c\right) = 0 , \tag{3}$$

where V_c denotes the diffusivity coefficient for c.

3   Discretization

All numerical simulations were carried out using the UG (Unstructured Grids) package developed at the group of Prof. Wittum at the University of Frankfurt [2]. This package provides a powerful numerical tool by combining mesh generators, load balancing subroutines, advanced linear solvers, including multigrid, with an extensive set of Finite Volume and Finite Element discretizations for various applications.

In the course of this project, the incompressible Navier-Stocks solver in the UG – based on a vertex centered Finite Volume (FV) method – was expanded to include several error indicators employed in an adaptive mesh refinement algorithm. A transport equation for species propagation was also added to the Navier-Stokes library.

Our work concerning higher order numerical schemes resulted in a Discontinuous Galerkin (DG) scheme for the incompressible Navier-Stocks equations implemented within UG. Due to the fact that the latter solver had to be developed from scratch it does not yet include all the features present in our Finite Volume Navier-Stokes package (e.g., the LES model or transport equations).

3.1   Finite Volume Method

Details of the Finite Volume discretization implemented in the UG can be found in [8]. The LES model employed by the Finite Volume solver is based on Germano's algorithm [5].

A paper discussing various error indicators and mesh adaptivity strategies is currently in preparation. In this work, adaptive mesh refinement/unrefinement was carried out using the maximum error indicator defined on each element K by

$$\eta_{\max} = \frac{1}{|K|} \sum_{\xi \in K} v_t \, |\operatorname{scv}(\xi) \cap K|, \tag{4}$$

where $scv(\xi)$ is the subcontrol volume corresponding to node ξ of element K, and $|.|$ denotes the measure (volume or area) of K or its parts. Is the value of η_{max} above (below) some given C_{ref} (C_{coa}) then element K is marked for refinement (unrefinement). This very simple heuristic error indicator causes the mesh to be refined in areas of high turbulence as indicated by the turbulent eddy viscosity coefficient η_t.

3.2 Discontinuous Galerkin Method

Our implementation of the Discontinuous Galerkin method supports virtually unlimited orders of approximation (storage and the CPU time are the only constraints) and provides automatic generation of orthogonal basis functions for any standard element shape in 2D and 3D. The DG solver is fully integrated in the UG and is built using the standard UG modules such as time stepping routines, nonlinear and linear solvers, grid transfer mechanisms, etc. Moreover, our implementation includes grid prolongation and restriction routines that enable the DG module to employ multigrid solvers in UG.

Out of a whole variety of DG schemes we chose to implement a generalization of the Oden-Babuška-Baumann method [9] called Nonsymmetric Interior Penalty Galerkin method. However, by simple means of changing a few coefficients – which can be done without recompiling the code – our solver can be turned into the original Oden-Babuška-Baumann method or Interior Penalty or Symmetric Interior Penalty Galerkin [1] method.

The stabilization with respect to the saddle point problem utilized in our DG scheme is similar to the technique proposed by Cockburn, Kanschat and Schoetzau [3] for the Local Discontinuous Galerkin method but includes an additional parameter that helps to improve convergence of the multigrid solver in the presence of an advective term.

In order to define our DG discretization we first introduce some notation. By $\mathcal{T}_h$ we denote a partition of domain Ω in polygonal elements of size h. Furthermore, $\Delta_e, \Delta_w, \Delta_i, \Delta_o$ represent sets of interior, wall, inflow, and outflow element boundaries correspondingly (e.g., $\Delta_i = \left(\bigcup_{\Omega_e \in \mathcal{T}_h} \partial \Omega_e \right) \cap \Gamma_i$).

With each element boundary face Γ we associate a unit normal $\underline{n}$ and use $f^\pm = \lim_{\varepsilon \to 0\pm} f(\underline{x} + \varepsilon \underline{n})$ for the traces on Γ of function f computed from the two elements sharing the boundary Γ. We also define the mean value $\bar{f} = 0.5(f^+ + f^-)$ and the jump $[f] = f^- - f^+$ of f at a point on Γ.

Next, we introduce the approximation spaces for velocity and pressure:

$$V_h = \{\underline{v} \in L^2(\Omega)^d : \underline{v}\big|_{\Omega_e} \in \mathcal{P}^k(\Omega_e)^d, \forall \Omega_e \in \mathcal{T}_h\}, \tag{5}$$

$$P_h = \{p \in L^2(\Omega) : p\big|_{\Omega_e} \in \mathcal{P}^k(\Omega_e), \forall \Omega_e \in \mathcal{T}_h\}, \tag{6}$$

where $\mathcal{P}^k(\Omega_e)$ denotes the space of complete polynomials of degree at most k on element Ω_e, and d is the space dimension. In our implementation of the DG method, the polynomial degree can be chosen individually for every independent unknown without any adverse effects on the stability of the scheme.

Then the approximate solution $(\underline{u}_h, p_h) \in V_h \times P_h$ satisfies the following semi-discrete system

$$
\begin{aligned}
&\sum_{\Omega_e \in T_h} \int_{\Omega_e} \partial_t \underline{u}_h \underline{\varphi}\, d\underline{x} - \sum_{\Omega_e \in T_h} \int_{\Omega_e} \left(\underline{u}_h(\underline{u}_h \cdot \nabla) + p_h \nabla\right) \underline{\varphi}\, d\underline{x} \\
&+ \sum_{\gamma \in \Delta_e} \int_\gamma \left(\underline{u}_h^\uparrow(\underline{u}_h \cdot \underline{n}) + \overline{p}_h \underline{n}\right)[\underline{\varphi}]\, ds + \sum_{\gamma \in \Delta_i} \int_\gamma \left(\underline{u}_i(\underline{u}_i \cdot \underline{n}) + p_h \underline{n}\right)\underline{\varphi}\, ds \\
&+ \sum_{\gamma \in \Delta_o} \int_\gamma \underline{u}_h(\underline{u}_h \cdot \underline{n})\underline{\varphi}\, ds + \sum_{\gamma \in \Delta_w} \int_\gamma p_h \underline{n}\underline{\varphi}\, ds + \sum_{\Omega_e \in T_h} \int_{\Omega_e} v \nabla \underline{u}_h \cdot \nabla \underline{\varphi}\, d\underline{x} \\
&- \sum_{\gamma \in \Delta_e} \int_\gamma v\left(\nabla \overline{\underline{u}}_h \cdot \underline{n}[\underline{\varphi}] - B\overline{\nabla\underline{\varphi}} \cdot \underline{n}[\underline{u}_h] - C[\underline{u}_h][\underline{\varphi}]\right) ds \\
&- \sum_{\gamma \in \Delta_i} \int_\gamma v\left(\nabla \overline{\underline{u}}_h \cdot \underline{n}\underline{\varphi} - B\overline{\nabla\underline{\varphi}} \cdot \underline{n}[\underline{u}_h] - C[\underline{u}_h]\underline{\varphi}\right) ds - \sum_{\gamma \in \Delta_o} \int_\gamma v\left(\nabla \overline{\underline{u}}_h \cdot \underline{n}\underline{\varphi}\right) ds \\
&- \sum_{\gamma \in \Delta_w} \int_\gamma v\left(\nabla \overline{\underline{u}}_h \cdot \underline{n}\underline{\varphi} - B\overline{\nabla\underline{\varphi}} \cdot \underline{n}\underline{u}_h - C\underline{u}_h \underline{\varphi}\right) ds = 0,
\end{aligned}
\tag{7}
$$

$$
\begin{aligned}
&- \sum_{\Omega_e \in T_h} \int_{\Omega_e} \underline{u}_h \cdot \nabla \psi\, d\underline{x} + \sum_{\gamma \in \Delta_e} \int_\gamma \left(\underline{u}_h^\uparrow \cdot \underline{n} - D[p_h]\right)[\psi]\, ds \\
&+ \sum_{\gamma \in \Delta_i} \int_\gamma \underline{u}_i \cdot \underline{n}\psi\, ds + \sum_{\gamma \in \Delta_o} \int_\gamma \left(\underline{u}_h \cdot \underline{n} - Dp_h\right)\psi\, ds = 0
\end{aligned}
\tag{8}
$$

for each test function $(\underline{\varphi}, \psi) \in V_h \times P_h$. Above, $\underline{u}_h^\uparrow$ is the standard upwind velocity, and $\underline{u}_i$ denotes the specified velocity vector at the inflow boundary.

Stabilization parameter B together with penalty parameter C determines the scheme type [1]. Parameter $D = (D_{11} + D_{12}|\underline{u}_h|)h$ provides stability with respect to the saddle point problem. The last term in the definition of D improves convergence of the multigrid scheme, particularly for high Re.

A special mention is due to the basis functions employed in the DG scheme. Popular choices for time dependent DG solvers are bases consisting of orthogonal polynomials which produce diagonal mass matrix. However, construction of orthogonal polynomials on simplices or even more general shapes in 2D and 3D is by no means trivial. One possible solution to this problem are Dubiner bases [4], but this approach works only for a limited number of element shapes and requires computation of 1D Jacobi polynomials.

The arbitrary order basis functions employed in our DG code are computed using a generalization of the well known 'three-term-formula' and can be constructed on any polygonal convex elements.

4 Numerical Results

We compare the results of the Discontinuous Galerkin simulations for different approximation orders to those of the node-centered Finite Volume method. The Finite Volume scheme in the UG has been verified and extensively tested using several standard benchmarks for laminar and turbulent flows [8]. Thus, we employ Finite Volume simulations on fine meshes as reference to verify and evaluate the performance of the Discontinous Galerkin method. The last part of this section is dedicated to the comparison of the numerical results to the experiments conducted by the group of Prof. Thévenin.

4.1 Driven Cavity

The first problem we consider is a standard benchmark simulating flow in a square cavity with the moving top boundary. This setup is well known from the literature [6] and serves to verify our implementation of the Discontinuous Galerkin method. The size of the cavity is 1x1 in dimensionless units, and the flow velocity at the top is $u=1$, $v=0$. At all other boundaries, we have $u=v=0$. The initial mesh (level 0) consisting of four triangles was produced by the intersecting diagonals of the square; finer levels were obtained by uniform refinement of the initial mesh using edge bisection.

We simulate Driven Cavity flow with Re=10000. All results were compared to the Finite Volume solution on level 9 (ca. 1 ml. elements) which served as the reference. As stabilization, we took $B=1$, $C=15/h$, and $D = \left(4 + 0.2 \, | \underline{u}_h \, | \right) h$.

In Figure 1, we present plots of the stream function for the Finite Volume solutions on different grid levels. Figure 2, Figure 3, Figure 4 give the stream function plots for the Discontinuous Galerkin discretization. We note that the higher order DG schemes seem to provide comparable accuracy at lower grid levels (e.g., psi=-0.12 curve was resolved by the cubic DG scheme starting from level 3 whereas, in the Finite Volume plots, it is visible starting from level 7 only). Advantages of the DG method appear to be less pronounced if we try to resolve eddies that are considerably smaller than the element size (see the small eddy in the lower left corner).

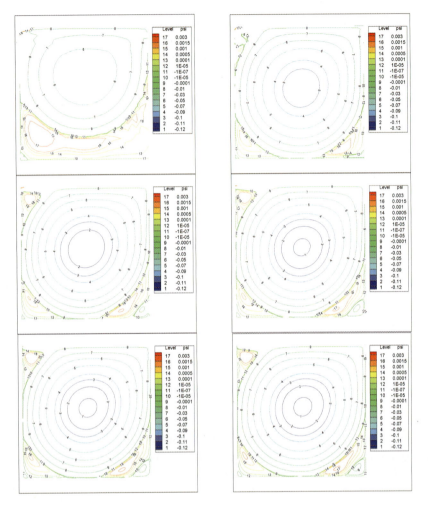

Fig. 1. Finite Volume with Re=10000 for levels 4(top left), 5(top right), 6(middle left), 7(middle right), 8(bottom left), 9(bottom right)

These conclusions are corroborated when we compare u-velocity profiles at x=0.5 (Figure 5) and v-velocity profiles at y=0.5 (Figure 6). Large elements prevent higher order schemes on coarse grids from resolving all the 'kinks' of the velocity profiles, but, in the smoother areas, the DG solutions lie generally closer to the reference line than the Finite Volume results from grid level 5. In particular, the 4th order DG scheme on level 2 (64 elements) is just as accurate in the middle part of the domain as the Finite Volume method on level 5 (4096 elements). However, the comparison of the degrees of freedom is not quite as dramatic: 960 in the DG vs. 2113 in the Finite Volume scheme.

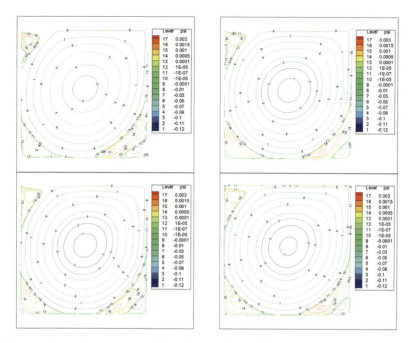

Fig. 2. Linear Discontinuous Galerkin with Re=10000 for levels 4(top left), 5(top right), 6(bottom left), 7(bottom right)

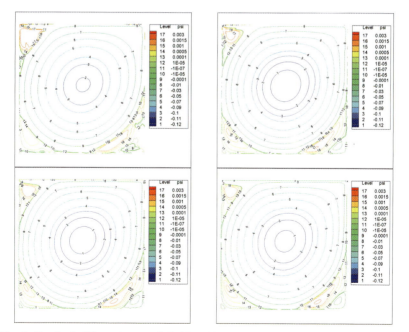

Fig. 3. Quadratic Discontinuous Galerkin with Re=10000 for levels 3(top left), 4 (top right), 5(bottom left), 6(bottom right)

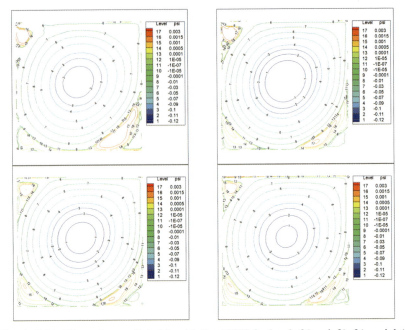

Fig. 4. Cubic Discontinuous Galerkin with Re=10000 for levels 2(top left), 3(top right), 4(bottom left), 5(bottom right)

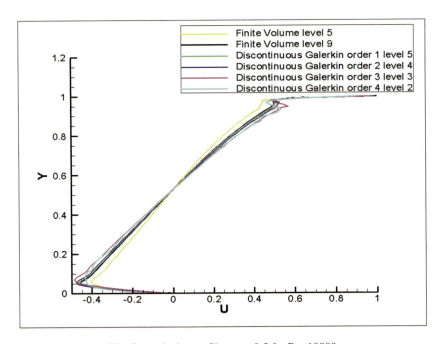

Fig. 5. u-velocity profiles at x=0.5 for Re=10000

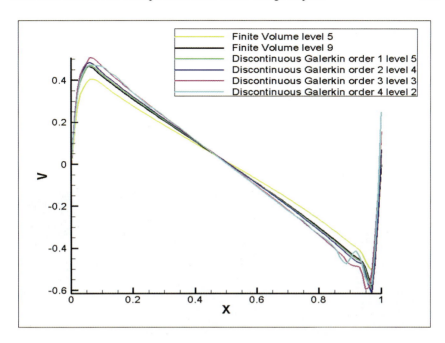

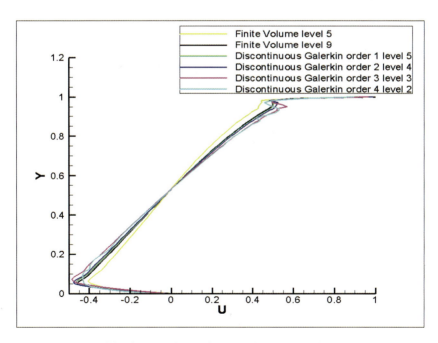

Fig. 6. v-velocity profiles at y=0.5 for Re=10000

4.2 Static Mixer

4.2.1 Problem Description

Details of the experimental setup and a detailed description of the geometry of the static mixer of Sulzer type can be found in the contribution of the group of Prof. Thévenin of the University of Magdeburg to this volume. Figure 7 gives a view of the static mixer in the experimental setup and the CAD geometry used as the input for our discrete model.

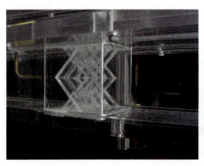 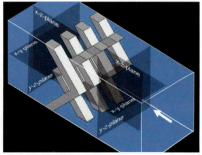

Fig. 7. Static mixer in the experimental setup (left) and as CAD-Geometry (right)

The experiment (and the numerical simulation) deals with a weakly turbulent flow in the mixer. Average axial velocity at the inflow boundary was specified at 6.2×10^{-3} m/s resulting in Re=562. This flow regime allowed us to disregard turbulent effects at the inflow boundary and before the mixer; however, the flow right after the mixer did show some turbulent behavior.

The origin of the coordinate system was placed in the center of the mixing element and the x-axis pointed in the direction of the flow. Experimental measurements were conducted for several cross sections (or fragments of cross-sections) of the flow channel.

In particular, we compared numerical results at the cross sections located at $x_0 = -121.0$, $x_1 = -60.5$, $x_2 = 60.5$, and $x_3 = 121.0$ in the y - z - plane .

4.2.2 Boundary Conditions

At the wall boundary Γ_w, "no flow" Dirichlet boundary conditions were used $\underline{u}\big|_{\Gamma_w} = 0$.

At the outflow boundary Γ_o , we specified the pressure $p\big|_{\Gamma_o} = 0$.

Somewhat more complicated boundary conditions were required at the inflow boundary Γ_i . According to the experimental results the inflow velocity exhibits an unsymmetric profile (comp. to Figure 8(left)) due to the centrifugal acceleration in the pipe leading to the mixer in the experimental setup. After a suitable smoothing postprocessing (see Figure 8(right)) this velocity profile was utilized to provide the inflow boundary conditions for our numerical simulations.

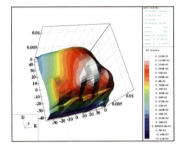 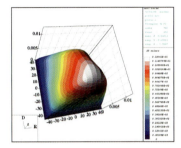

Fig. 8. Axial velocity profiles at the inflow boundary. Measured (left), after smoothing (right).

4.2.3 Grid

In the first step, the mixer geometry – available in the CAD-format – had to be converted to a format readable by the UG, and an initial mesh had to be generated. This was done using the method described in [7].

The initial mesh (level 0) consisting of 3913 tetrahedra is shown in Figure 9. Finer meshes were obtained by uniform or, in the case of an adaptive algorithm,

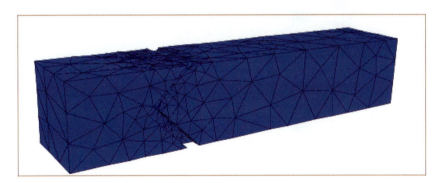

Fig. 9. Initial mesh with 3913 Tetrahedra

selective refinement of lower level meshes. Table 1 gives an overview of different grid levels with the corresponding numbers of degrees of freedom (DOF) for the Finite Volumes and Discontinuous Galerkin methods as well as the minimum (H_{min}) and maximum (H_{max}) element sizes.

Table 1. Mesh info for different refinement levels

Level	Nodes	Elements	DOF FV	DOF DG linear	DOF DG quadratic	H_{min}	H_{max}
0	1086	3913	4344	62608	1.56e+5	2.74	68.6
1	6975	31304	27900	5.e+5	1.25e+6	1.84	34.3
2	48766	2.5e+5	1.95e+5	4.e+6	1.e+7	0.83	17.2
3	3.6e+5	2.e+6	1.4e+6	3.2e+7	8.e+7	0.41	8.58
4	2.8e+6	1.6e+7	1.1e+7	2.6e+8	6.4e+8	0.20	4.29

4.2.4 Finite Volume Method Results

The results in this subsection were obtained from the Finite Volume simulations. Figure 10 gives snapshots of the axial velocity in slices $z=0$ and $y=0$.

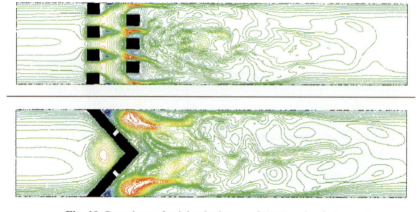

Fig. 10. Snapshots of axial velocity at $z=0$ (top) and $y=0$ (bottom)

Laminar character of the flow before the mixer is easily discernable. One can also see the speed up of the flow in the mixing element due to smaller cross section. After the mixing element the flow converges into two unstationary jets.

In Figure 11 we plotted the corresponding distribution of the turbulent eddy viscosity. Interestingly enough, the areas with the highest turbulent eddy viscosity do not coincide with the areas of highest velocity or greatest velocity gradients.

Fig. 11. Snapshot of the turbulent eddy viscosity at $z=0$

The next series of plots (Figure 12) illustrates the axial velocity profiles in different cross sections before and after the mixing element.

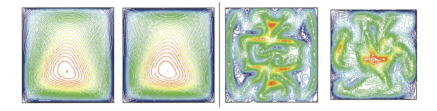

Fig. 12. Snapshots of the axial velocity at x_0, x_1, x_2, x_3 (from left to right)

Almost no changes are noticeable in the inflow profile before the mixer, thus suggesting a good approximation given by the inflow boundary condition.

Figure 13 gives a view of the tracer concentration field at different cross sections before and after the mixing element.

Starting with a circular induction profile the concentration field spreads smoothly in the laminar flow area before the mixer and is highly inhomogeneous after passing through the mixer.

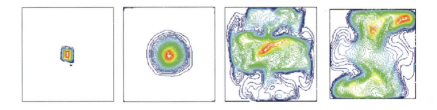

Fig. 13. Snapshots of the tracer concentration field at x_0, x_1, x_2, x_3 (from left to right)

Figure 14 illustrates the influence of the mesh resolution on the numerical solution. There, the time averaged axial velocity is compared at 60.5 after the mixer for levels 2, 3, and 5. Quite remarkable is the high degree of agreement between the results on levels 2 and 4 as opposed to those of level 3. We still don't have a satisfactory explanation for this phenomenon.

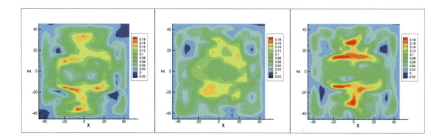

Fig. 14. Time averaged axial velocity at x_2 on levels 2, 3, and 5 (from left to right)

4.2.5 Comparison to Experimental Data
In this section, the simulation results that utilize the Finite Volume and Discontinous Galerkin solvers are compared to the experimental measurements provided by our project collaborators from Magdeburg.

For the DG scheme, we took $B=1$, $C=200/h$, and $D = \left(4 \cdot 10^{-4} + 2 \cdot 10^{-5} \mid \underline{u}_h \mid\right)h$.

We start with the area before the mixer and compare in Figure 15 the time averaged axial velocity profiles at x_1. As opposed to the simulation, the experimental

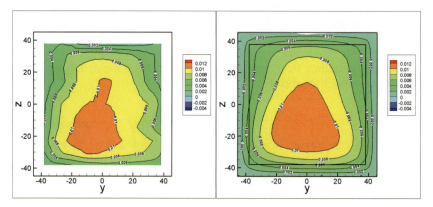

Fig. 15. Time averaged axial velocity before the mixer. Experiment (left), Finite Volume simulation on level 5 (right).

velocity profile seems to be distorted, and, moreover, one can detect significant differences between experimental velocity profiles at x_0 and x_1 although the geometry is uniform. This might be an indication of certain problems with the experimental setup.

The next plot (Figure 16) displays the time averaged axial velocity components after the mixer (cross section at x_2). We see rather good agreement in shape as well as in magnitude for the Finite Volume solution. The DG results are rather poor which might be due – at least to some extent – to low grid resolution and missing turbulence model.

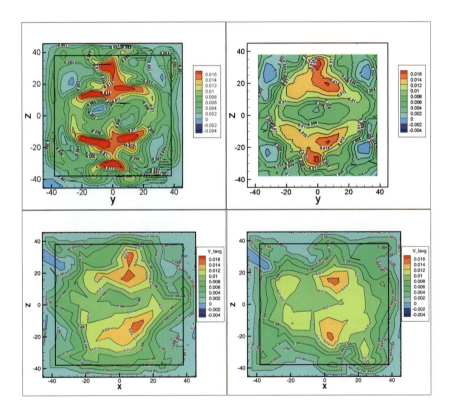

Fig. 16. Time averaged axial velocity after the mixer. Experiment (top left), FV on level 5 (top right), Linear DG on level 3 (bottom left), Quadratic DG on level 2(bottom right).

One sees less agreement when comparing the horizontal (Figure 17) and vertical (Figure 18) velocity components. The extrema seem to be close, but there is generally less agreement in the overall structure of the flow.

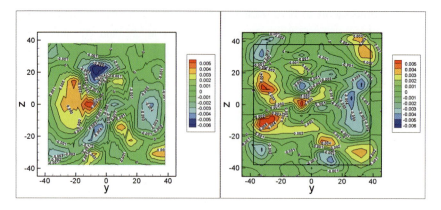

Fig. 17. Time averaged horizontal velocity after the mixer. Experiment (left), FV on level 5 (right).

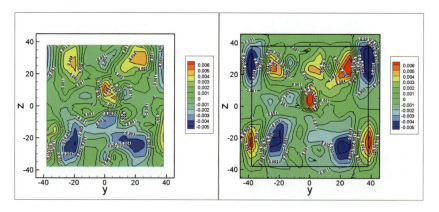

Fig. 18. Time averaged vertical velocity after the mixer. Experiment (left), FV on level 5 (right).

The next plot (Figure 19) compares the power spectra of the axial velocity component at point (x_3 , 0, 0) on the main axis of the mixer. The experimental measurements (peak at ~0.36 Hz) agree rather well with the Finite Volume results, but both DG frequencies are a bit off: ~ 0.4 for the piecewise linear approximation on level 3 and ~0.43 for the piecewise quadratic approximation on level 2. One possible explanation is the absence of an LES model in our DG implementation resulting in an unrealistically low value of the viscosity term.

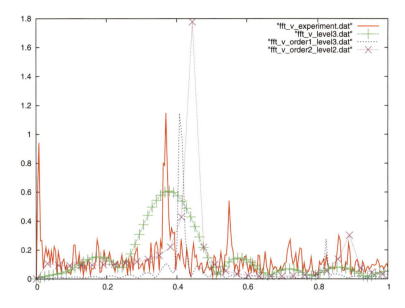

Fig. 19. Power spectra of the axial velocity at x_2. Experiment, FV on level 3, linear DG on level 3, quadratic DG on level 2 (from top to bottom).

5 Concluding Remarks

In the course of our work on SPP 1141 we implemented and tested adaptive mesh refinement strategies and applied them to simulations of flow and mixing in static mixers with complex geometry. We also developed a Discontinuous Galerkin scheme for the incompressible Navier-Stokes equations. The results of our simulations suggest that the higher order methods (such as the DG method) have advantages when applied to smooth flows, however, they tend to be more expensive (in terms of the number of degrees of freedom) than the node centered Finite Volume method if we try to resolve small flow features and, thus, have a problem imposed upper limit on the element size. A promising approach could be an adaptive higher order scheme based on the same principles that we applied to the Finite Volume method.

We plan to continue the work on the higher order schemes for the incompressible Navier-Stokes equations, with a special focus on adaptive methods.

Acknowledgments. The authors would like to thank our colleagues Dominique Thévenin and Andreas Lehwald of the University of Magdeburg for fruitful collaboration.

References

[1] Arnold, E.F., Brezzi, F., Cockburn, B., Marini, L.D.: Unified Analysis of Discontinuous Galerkin Methods for Elliptic Problems. SIAM J. Num. Anal. 39, 1749–1779 (2002)

[2] Bastian, P., et al.: UG - A Flexible Software Toolbox for Solving Partial Differential Equations. Computing and Visualization in Science, 27–40 (1997)

[3] Cockburn, B., Kanschat, G., Schötzau, D.: A Note on Discontinuous Galerkin Divergence-free Solutions of the Navier-Stokes Equations. Journal of Scientific Computing 31(1-2), 61–73 (2007)

[4] Dubiner, M.: Spectral methods on triangles and other domains. Journal of Scientific Computing, 345–390 (1991)

[5] Germano, M., Piomelli, U., Moin, P., Cabot, W.: A dynamic subgrid-scale eddy viscosity model. Phys. Fluids A, 1760–1765 (1991)

[6] Ghia, U., Ghia, K.N., Shin, C.T.: High-Resolutions for incompressible flow using the Navier-Stokes equations and a multigrid method. J. Comput. Phys., 387–411 (1982)

[7] Hauser, A., Sterz, O.: UG-interface for CAD geometries. Technical Report. Forschungsverbund WiR Baden-Wurttemberg, Heidelberg (2004)

[8] Nägele, S.: Mehrgitterverfahren für die inkompressiblen Navier-Stokes Gleichungen im laminaren und turbulenten Regime unter Berücksichtigung verschiedener Stabilisierungsmethoden. Universität Heidelberg, Heidelberg, PhD Thesis (2003)

[9] Oden, J.T., Babuška, I., Baumann, C.E.: A discontinuous HP finite element method for diffusion problems. Journal of Computational Physics 146(2), 491–519 (1998)

Part 4: Macro- and Micro-Mixing in Micro Channel Flow

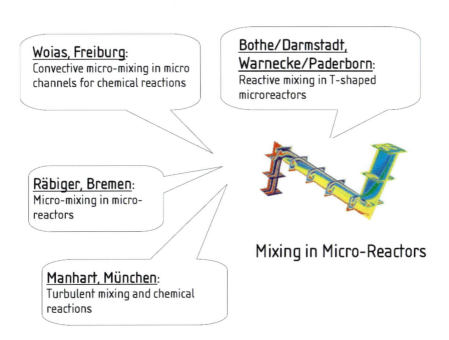

Woias, Freiburg:
Convective micro-mixing in micro channels for chemical reactions

Bothe/Darmstadt, Warnecke/Paderborn:
Reactive mixing in T-shaped microreactors

Räbiger, Bremen:
Micro-mixing in micro-reactors

Manhart, München:
Turbulent mixing and chemical reactions

Mixing in Micro-Reactors

Micro channels are well suitable to study the effects of micro mixing on mass transport and reaction phenomena. Due to the small scales and laminar flow conditions the micro mixing process is not masked by complex macroscopic mixing problems such as turbulence and back mixing. Despite the fact that flow fields under laminar flow conditions are predictable by numerical calculations, the behavior of reacting flows in micro mixers is still difficult to understand. For this purpose different measurement methods have been developed, like micro particle image velocimetry and confocal laser scanning microscopy. The influence of the flow regime on the yield and selectivity has been investigated experimentally by means of a parallel consecutive reaction system. In the course of the theoretical work a combined approach for the numerical simulation of turbulent mixing with chemical reaction at high Schmidt-numbers was developed for the specific case of a precipitation process in a T-mixer.

Computational Analysis of Reactive Mixing in T-Microreactors

Dieter Bothe[1], Alexander Lojewski[2], and Hans-Joachim Warnecke[3]

[1] Center of Smart Interfaces, TU Darmstadt,
  Petersenstraße 32, 64287 Darmstadt
[2] Chair for Mathematics (CES), RWTH-Aachen University,
  Schinkelstraße 2, 52062 Aachen, Germany
[3] Chair for Chemical Engineering, University of Paderborn,
  Warburger Straße 100, 33098 Paderborn, Germany

Abstract. Reactive mixing in T-shaped microreactors is studied based on numerical simulations. The flow conditions under consideration are laminar and stationary, but with a rather complex secondary flow which promotes mixing. A hybrid numerical approach is developed which allows for full resolution of all relevant scales and thereby enables a computational analysis of how chemical reactions interact with convective and diffusive transport. The approach is extended to include instantaneous reactions. Results of the numerical calculations are in excellent agreement with local data from experimental measurements. Simulations under different operating conditions help to understand the scaling behaviour of reactive mixing in microreactors.

1 Introduction

Most applications of microreactors aim at an intensification of chemical transformations or at the design of new classes of chemical processes, especially fast or highly exothermic chemistry. The large area-to-volume ratio of microreactors gives prospect of, e.g., better yield and selectivity than for conventional designs, since diffusive fluxes of mass and heat in micro-devices scale with the area, while the rate of changes corresponding to sources and sinks are proportional to the volume. Since the mechanisms of chemical reactions act on an Ångström length scale, several orders of magnitude have to be bridged by transport processes even in microreactors in order to homogenize the initially segregated inlet streams on the molecular level. This mixing has to happen fast enough such that the chemical transformations are not masked by transport processes. The fact that the local state of mixing can massively affect for instance selectivities of chemical reactions is well-known for macro-systems (Rys 1992, Bourne 2003) but equally applies to micro-chemical processes. Consequently, transport processes can still limit the process performance even if the dimensions of the system are small and, hence, need to be studied by both experimental and theoretical methods.

The present study aims at a computational analysis of reactive mixing in T-shaped microreactors based on first principles. For this purpose, a hybrid approach for numerical simulations is developed which allows for the resolution of all relevant time and especially length scales, thereby enabling so-called Direct Numerical Simulations of reactive mixing in T-microreactors.

2   Computational Approach

Basic to our approach is the knowledge of the velocity field within the microreactor or a sufficiently accurate approximation thereof. Employing today's CFD-tools, this assumption is valid for the stationary, complex laminar flow conditions under consideration. Simulation of reactive mixing processes then requires the numerical solution of additional species equations which account for convective and diffusive transport of chemical components as well as their chemical transformation. The main problem here is the occurrence of very small length scales in the concentration fields especially for species transport in liquids with large typical Schmidt numbers Sc. Indeed, since these small scale structures can only be smeared out by diffusion, the resulting length scale for convective-diffusive mixing is the Batchelor length scale (cf. Ottino 1994, Bothe et al. 2008)

$$\lambda_{conc} = \frac{\lambda_{vel}}{\sqrt{Sc}},$$

where λ_{vel} denotes the characteristic length scale of the velocity field. Although the *integral* length scale decreases more slowly with increasing Sc, the problem of resolving all relevant length scales remains: local steep concentration gradients are smeared by numerical diffusion, leading to artificial mixing and to significant deviations in the computed reaction rates. In case of fast or (almost) instantaneous chemical reactions, the educts coexist only in a thin reaction zone and the speed of reaction is determined by the gradients in the educt concentrations. In all cases, intensification of chemical reactions in microreactors necessarily requires use of the full reactor volume for the chemical transformations. Therefore numerical simulations need to span the full volume, while at the same time resolving small length scales. With standard techniques this is not possible today in a 3D simulation. Instead, we employ a hybrid approach using a decomposition of the reactor into a *mixing zone* and a *reaction zone*. The mixing zone is characterized by a strong secondary flow which promotes convective mixing in cross directions. Due to no-slip at the channel walls and viscous energy dissipation, the secondary flow diminishes such that the velocity field becomes more and more oriented towards the axial direction. This is the entry into the reaction zone, the largest part of the T-microreactor, in which mixing of educts is only due to diffusion. Inside this zone, there is no back-flow. Exploiting this fact and neglecting diffusion in axial direction, the stationary reaction-convection-diffusion equations are rewritten as parabolic equations with the axial direction as the new progress variable. This allows for a successive computation of the concentration profiles on cross sections along the channel. The advantage lies in the sufficiency of a two-dimensional spatial grid with the possibility of significantly finer resolution.

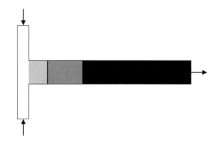

Fig. 1. Different zones in the hybrid numerical model

Figure 1 illustrates the different zones within this hybrid numerical model. The light grey one is the mixing zone with backflow regions. Within this part, full 3D computations for both the velocity and the species fields are performed. This zone ends when no more backflow appears. It is followed by a region in which the secondary flow is still significant but with the axial component pointing into down-stream direction. In this grey zone, the velocity field comes from a 3D computation with moderate resolution, while the species equations in parabolised form are solved on a fine 2D grid. In the next part (dark grey) of the micro-channel, the flow field is close to a fully developed Poiseuille-type flow and, hence, the velocity is either set to this profile or an exponential decay of the secondary flow is assumed. Consequently, only the species equations are solved numerically there which is done on the same 2D grid as before.

The coupling between the different zones is straight forward: data is passed from left to right, using interpolation if required. The zone in light grey of full 3D simulation is somewhat longer in order to suppress back effects of the down-flow boundary condition. In addition, adaptive grid refinements need to be extended over the interface into the middle zone.

With this overall approach, full resolution of all relevant length scales is possible in case of complex stationary flow conditions and allows for a computational analysis of reactive mixing. The detailed mathematical and numerical treatment of the different zones is explained in the subsequent sections.

3 Computational Analysis of the Mixing Zone

The numerical investigations presented here are performed for T-shaped micro-mixers with rectangular cross sections. The geometry of the basic microreactor design (see Figure 2) consists of two inlet channels, each with a length of 8 mm and a depth of 100 µm. The mixing channel is 20 mm long with the same depth as the inlet channels. The width of the inlet channels are 100 µm each, the width of the mixing-channel is 200 µm; we use the abbreviation "T-200" for this geometry. For this reactor under stationary flow conditions, the mixing zone extends for several 100 µm into the main channel. To minimize back-effects of the boundary conditions at the artificial outflow boundary, the mixing zone is defined to have a length

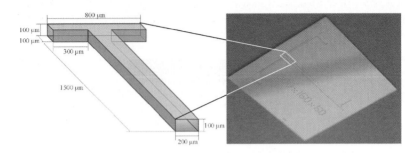

Fig. 2. Mixing zone of the T-shaped microreactor

of 1500 µm even for strictly laminar flow conditions at very small Reynolds numbers. Preliminary calculations (cf. Bothe et al. 2004) showed that this length is sufficient to avoid non-physical flow behaviour. For a reduction of the computational domain, shortened inlet channels with fully developed duct flow as inlet conditions are used.

For channels of these dimensions and for liquid flow, the transport of mass and momentum is adequately described (cf. Guo and Li 2003) by the incompressible Navier-Stokes equations, which in non-dimensional form read as

$$\nabla \cdot \mathbf{u} = 0, \qquad \partial_t \mathbf{u} + \mathbf{u} \cdot \nabla \mathbf{u} = -\nabla p' + \frac{1}{\text{Re}} \Delta \mathbf{u} \tag{1}$$

with $\mathbf{u}$ the velocity field and p' the reduced pressure. The Reynolds number

$$\text{Re} = \frac{U d_H}{\nu} \tag{2}$$

is based on the mean velocity U and the hydraulic diameter d_H of the channel. Equation (1) is complemented by appropriate boundary conditions which are the inflow conditions mentioned above, no-slip conditions at the lateral walls and an outlet pressure condition which includes a homogeneous Neumann condition for the velocity.

In such a fluid flow, the transport of an ideally diluted chemical species that undergoes a chemical reaction of order n, say, is governed by the species equation

$$\partial_t c_i + \mathbf{u} \cdot \nabla c_i = \frac{1}{\text{Re Sc}_i} \Delta c_i + \frac{d_H}{L} \text{Da}_\text{I} c_i^n \tag{3}$$

where c_i denotes the dimensionless concentration of species i, the quantity L denotes the channel length and

$$\text{Sc}_i = \frac{D_i}{\nu}, \qquad \text{Da}_\text{I} = \frac{L k c_0^{n-1}}{U} \tag{4}$$

are the Schmidt number for species i and the Damköhler(I) number, respectively; here D_i is the species diffusivity, k the rate constant and c_0 a reference concentration. Note that the factor d_H / L appears in (3) since the Reynolds number is based on the diameter of the channel, while the Damköhler number contains the hydrodynamic residence time which is based on the length of the channel.

Employing equations (1) and (3), numerical simulations of reactive mixing are performed on 3D-grids with rectangular cells. Using the commercial CFD-Tool ANSYS® FLUENT® all transport equations are discretised by the Finite Volume Method. In order to resolve the smallest length scales, adaptive grid refinements are done for all areas where steep gradients of the concentration fields are present.

3.1 Hydrodynamics

Due to the short residence time in microreactors, mixing has to be fast in order to enlarge the contact area between chemically reacting species. While turbulent flow provides fast convective mixing, it requires high energy input resulting in large pressure drops especially for micro-systems. Laminar velocity fields with secondary flow show lower pressure drops but can still promote mixing significantly. In addition, such flows provide defined flow conditions. Therefore, stationary laminar flows with secondary flow patterns are important for applications in microreactors. For the specific geometric design considered here, three different stationary flow regimes can be observed up to a Reynolds number of about 240, where time-periodic flow phenomena set in (cf. Bothe et al. 2006, Dreher et al. 2009). At Reynolds numbers below 40, strictly laminar flow behaviour (Re < 40) occurs without secondary flow. Then for 40 < Re < 138 the so-called vortex flow regime shows secondary flow in form of a double vortex pair which is formed due to centrifugal forces. This symmetry is destroyed if the Reynolds number exceeds Re = 138. In this case fluid elements reach the opposite side of the mixing channel (see Fig. 3) resulting in good cross-directional mixing. Among these stationary flow conditions, the engulfment flow is the most effective one and leads to a rapid increase of the contact area between the two inlet flows.

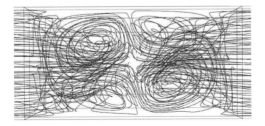

Fig. 3. Streamlines at the beginning of the mixing zone (Re = 149)

3.2 Energy Dissipation

Laminar flows in straight ducts show a linear pressure drop along the axial direction, whereas in a T-shaped micromixer a modified behaviour is observed due to deflection of the inflowing streams.

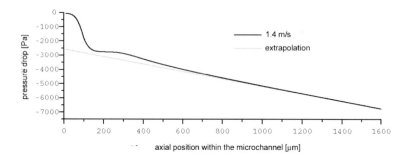

Fig. 4. Pressure drop along the mixing channel for a mean velocity of 1.4 m/s

This is displayed in Figure 4 (cf. Bothe et al. 2006) for a flow condition (Re=186) in the engulfment flow regime. In this case, a relatively large energy input is needed for the deflection of the flow into the mixing channel and to build up the intertwining vortex structure. This results in the high extra pressure drop over the first approximately 120 μm. This pressure decline is followed by a plateau region which roughly corresponds to the zone of strong cross-sectional mixing. Further down-stream, the axial pressure profile becomes linearly decreasing as is expected for fully developed duct flow. The varying behaviour of the pressure along the mixing channel allows for an automatic detection of the mixing zone within numerical simulations. For mean velocities of 0.1–1.4 m/s the zone of convective cross-sectional mixing ends at a distance of 50–200 μm from the entrance into the mixing channel.

A quantitative analysis of cross-directional mixing can be based on the mass specific rate of energy dissipation in cross directions which is determined by the 2×2 block in the rate-of-deformation tensor which corresponds to the cross directions; see (Bothe et al. 2006) for more details. Normalised with respect to the full mass-specific rate of energy dissipation, this yields the fraction of dissipated energy which causes relative fluid motion perpendicular to the axial direction. From the viewpoint of chemically reacting flows, complete mixing in cross directions without axial mixing would be the optimal case. Therefore, larger values of this normalised cross-sectional energy dissipation are preferable. Figure 5 shows the evolution of this quantity along the channel for the different flow types. In all cases the values abruptly increase in the deflection zone, followed by a longer decrease. If no more flow in cross direction occurs, the rate of energy dissipation in cross direction reaches zero, while the total dissipation rate reaches the value for a fully developed duct flow.

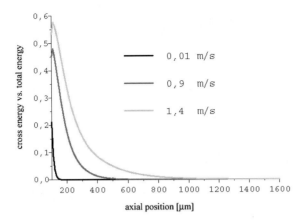

Fig. 5. Ratio between total and cross sectional energy dissipation for different mean velocities

Increasing the mean velocity and, hence, the secondary flow leads to a larger fraction of the dissipated energy that is employed for cross sectional mixing. Furthermore, an increase of the length of the mixing zone can be seen. The effective value of the dissipated energy in cross direction determines the shrinkage of unmixed structures and the increase of additional contact area across which diffusive mixing takes place. An acceleration of the mean flow velocity thus leads to an intensification of convective mixing. This mixing in cross direction dies out along the mixing channel. Therefore, successive changes of the flow direction as realized in zigzag-type structures provide a means to increase the mixing efficiency.

3.3 Passive Mixing

Tracer experiments with the fluorescent dye Rhodamin B were performed to evaluate the mixing performance quantitatively. For this purpose a defined amount of the dye was continuously given to one of the inlet flows. In the simulations, the dye was treated as a passive diluted component with a given maximum concentration. The distribution of Rhodamin B was used to quantify the mixing state on subsequent cross sections. The obtained results show good accordance with the µ-LIF experiments performed by our cooperation partner within this priority research program 1141; see, e.g., (Hoffmann et al. 2006).

Figure 6(a) – (f) displays the change of contact area within the same cross section for increasing mean velocities. At low flow rates both inlet streams run parallel through the mixing channel, either with or without formation of vortices, and the planar contact area remains unchanged (cf. Fig. 6(a)). At higher velocities the two vortex pairs get intertwined, which leads to a roll-up of regions with different concentrations (Fig. 6(b)–(d)). As a consequence, the specific contact area is enlarged, which is characteristic for the engulfment regime and an essential requirement for efficient diffusive mixing. In Fig. 6(e)–(f), smearing of the contact

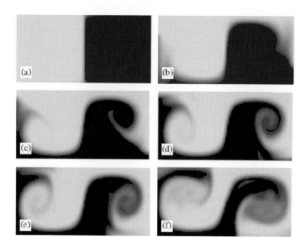

Fig. 6. Tracer profiles on the cross section of the mixing channel 300 μm behind its entrance for mean velocities of (a) 0.9 m/s, (b) 1.05 m/s, (c) 1.1 m/s (d) 1.15 m/s, (e) 1.2 m/s and (f) 1.4 m/s

area becomes visible, which results from reduced diffusion paths due to smaller segregation scales and, hence, reduced time needed for diffusive dissipation of gradients.

3.4 Intensity of Mixing and Specific Contact Area

As a quantitative measure for the mixing condition the intensity of mixing, i.e.

$$I_M = 1 - \sqrt{I_s} \, , \tag{5}$$

is used which is derived from the intensity of segregation defined by Danckwerts in 1952 as

$$I_S = \frac{\sigma^2}{\sigma^2_{max}} \, . \tag{6}$$

Here

$$\sigma^2 = \frac{1}{|V|} \int_V (c - \bar{c})^2 \, dV \tag{7}$$

and σ^2_{max} is the variance of a totally segregated system with the same amount of substance and the same maximum concentration c_{max}. This leads to

$$\sigma^2_{max} = \bar{c}(c_{max} - \bar{c}) \tag{8}$$

which allows assessing a species distribution by itself. On the other hand, since
c_{max} will usually change with time or space, this cannot be used for defining a
mixing time or a mixing length. In this case, the variance at the initial time or at
the entrance is used instead. Since I_s is normalised, it reaches a value of 0 for a
completely segregated system and a value of 1 for the homogeneously mixed case.

For a meaningful characterisation of mixing quality, the intensity of segregation
has to be supplemented with a scale of segregation. For this purpose we employ
the measure

$$\phi(V) = \frac{1}{|V|} \int_V \|\nabla f\| dV \qquad \text{with} \qquad f = \frac{c}{c_{max}}, \qquad (9)$$

where $|V|$ denotes the volume of the spatial region V and ∇f is the (Euclidean)
length of the gradient of the normalised concentration. Mathematically, this quantity is the total variation of the scalar field f. For a segregated species distribution
it exactly gives the specific contact area. The reciprocal of ϕ has the dimension of
a length and can be interpreted as an average distance between regions of high and
low species concentration. For further information see (Bothe, Warnecke 2007),
(Bothe et al. 2008).

To evaluate the mixing behaviour within tracer experiments, the state of mixing
is investigated on the cross section 300 µm behind the entrance into the main
channel. It is found that the intensity of mixing is approximately zero in the first
two flow regimes (laminar and vortex). For mean velocities exceeding 1–1.1 m/s,
the intensity of mixing increases significantly; this is shown in Figure 7. Simultaneously, the scale of segregation decreases from about 200 µm, which corresponds
to the channel width of the considered T-200, to approximately 50 µm. This effect
is caused by the intertwinement of the two fluid streams which generates additional contact area.

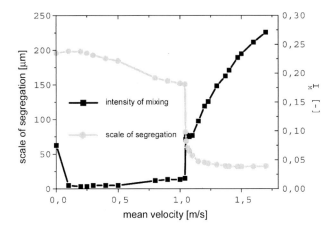

Fig. 7. Intensity of mixing (grey) and scale of segregation (black) vs. mean velocity

4 Computational Analysis of the Reaction Zone

Due to their small dimensions and the resulting efficient mass transport, micro-mixers are well suited for fast chemical reactions which are normally limited by the diffusive transport step, i.e. by micro-mixing. A computational analysis of the interactions of transport and fast chemical reactions requires the resolution of all involved time and length scales for the full microreactor since a fast reaction im-mediately takes place if the educts are brought into contact. Therefore highly resolved computational grids are necessary already in the mixing zone of the mi-cro-channel. This is especially important because fully resolved concentration pro-files are required as input for the 2D model of the reaction zone.

As a basic example the formation of $[CaFluo4]^{3-}$ was studied. This chemical re-action is relevant for experimental investigations of reactive mixing since the product is a fluorescent dye which can be detected. The reaction reads as

$$Ca^{2+} + Fluo4^{5-} \xrightleftharpoons[k_d]{k} [CaFluo4]^{3-}. \tag{10}$$

The fluorescent complex is formed from Ca^{2+}- and $Fluo4^{5-}$-ions in a fast second order reaction. In the literature the rate constant for the forward reaction is re-ported as $k = 1{\times}10^6 \ m^3/(mol \ s)$ (cf. Arakawa, Wada 1983), (Eberhard, Erne 1989), while the dissociation of the complex is so slow $(k_d = 370 \ s^{-1})$ that it can be ne-glected. Note that these values are only estimates with possibly large error bounds. The intrinsic reaction rates, i.e. the purely chemical rates not masked by transport processes, are usually unavailable. Nevertheless, the latter are needed for a detailed numerical simulation if the reaction is the rate determining step.

Numerical simulations of reactive mixing using the present hybrid approach re-quire a combination of 3D- and 2D-simulations, since data from 3D-simulations on highly refined grids is used as input for the 2D-model. Therefore, first of all, the Navier-Stokes equations (1) are solved on a uniform 3D-grid which represents the mixing zone. These simulations provide the full 3D velocity field which is then stored separately for each cross section. Afterwards, 3D-simulations of (1) and all species equations are performed on adaptively refined grids. Thereby the species concentrations are computed with sufficient resolution to be used as input data for the 2D-simulation of the reaction zone. As mentioned above, the reaction (10) is considered as an irreversible reaction of second order between educts A and B to a product P. If c_A, c_B and c_P denote the molar concentrations (normalized to a common reference concentration c_0) of A, B and P, respectively, then the evolution of these quantities is governed by the species equations

$$\partial_t c_A + \mathbf{u} \cdot \nabla c_A = \frac{1}{Re\,Sc_A}\Delta c_A - \frac{d_H}{L}Da_I\,c_A\,c_B, \tag{11}$$

$$\partial_t c_B + \mathbf{u} \cdot \nabla c_B = \frac{1}{Re\,Sc_B}\Delta c_B - \frac{d_H}{L}Da_I\,c_A\,c_B, \tag{12}$$

$$\partial_t c_P + \mathbf{u} \cdot \nabla c_P = \frac{1}{\mathrm{Re}\,\mathrm{Sc}_P} \Delta c_P + \frac{d_H}{L} \mathrm{Da}_{\mathrm{I}}\, c_A\, c_B. \tag{13}$$

In (11)-(13), the Damköhler(I) number is given by

$$\mathrm{Da}_{\mathrm{I}} = \frac{k\, c_0\, L}{U}. \tag{14}$$

In the computation for the mixing zone, these species equations are solved in 3D in their stationary form. In the Finite Volume discretisation, the reaction rate is included as a volumetric source term which is implemented into each species equation. An important detail of the implementation within FLUENT® is that all individual species concentrations are additionally stored before a new time step is started. This allows for reloading of the data required for the calculation of the source term. This way, the same source term is calculated to the same result in the different species equations, resulting in mass conservation.

In order to check whether sufficient resolution can be achieved in the 3D simulations of the mixing zone, we performed test simulations with increasing resolution. Figure 8 shows the distribution of the product on the cross section at $z = 250\ \mu m$ for two different spatial resolutions of the grid. These series of computations reveals that grid independence is attained for an adaptively refined grid with a minimal cell size of $h = 0.3\ \mu m$. Such a fine resolution can even not be realised for the entire mixing zone. This underlines the need of the hybrid approach described above. Fortunately, sufficiently high resolution can be achieved for the part of the mixing zone in which backflow occurs. Therefore, we are able to compute the fully resolved concentration field which is required as the inlet condition for the parabolised species equation to be described below.

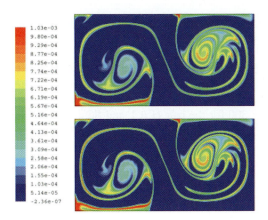

Fig. 8. Distribution of the product [CaFluo4]$^{3-}$ at $z = 260\ \mu m$ (T-200 mixer) on grids with cell size h = 0.3 μm (top) and h = 0.075 μm (bottom)

4.1 Reactive Mixing with Instantaneous Reactions

The simulation of reactive mixing with finite rate chemistry according to (11)-(13) is only possible if the Damköhler numbers are not too large since this would lead to extremely stiff differential equations. Therefore the numerical simulation of reactive mixing with instantaneous chemical reactions requires special solution methods or extremely small time steps. In the limiting case of infinitely fast chemical reactions like neutralizations or radical reactions, the stiffness can sometimes be removed by a method already used for theoretical purposes by (Toor and Chiang 1959). Indeed, subtracting (12) from (11) yields the convection-diffusion equation

$$\partial_t \phi + \mathbf{u} \cdot \nabla \phi = \nabla \cdot \left(\frac{1}{Pe(\phi)} \nabla \phi \right) \tag{15}$$

where

$$\phi := c_A - c_B \tag{16}$$

and

$$Pe(\phi) = U \, d_H \, / \, D(\phi). \tag{17}$$

Now, because of the infinite speed of reaction, the species A and B *cannot coexist*. Consequently, solving (15) for ϕ eventually yields c_A and c_B by means of the relations

$$c_A = \max\{\phi, 0\} \quad \text{and} \quad c_B = \max\{-\phi, 0\}. \tag{18}$$

Elimination of the reaction term as explained above yields a pure transport problem which can be solved numerically with resolution of all length scales by means of adapted grids and parallel computing. In the stationary flow case under consideration, the grid adaptation can even be done manually in a few successive steps.

This approach to very fast reactions has been applied to the neutralization of hydrochloric acid and sodium hydroxide. Since the diffusivities of the reactants are not equal (Sc = 300 for the pair H^+/Cl^- and Sc = 470 for Na^+/OH^-), the diffusion coefficient for the scalar ϕ depends on the value of ϕ according to

$$D(\phi) = \begin{cases} D_{HCl} & \text{for } \phi \geq 0 \\ D_{NaOH} & \text{for } \phi < 0 \end{cases}. \tag{19}$$

For experimental visualization purpose the neutralization is combined with the deprotonation of a *fluorescent* dye depending on the pH-value which leads to a parallel reaction of the form

$$H^+ + OH^- \rightarrow H_2O$$

$$HFl + OH^- \underset{\longleftarrow}{\overset{\longrightarrow}{\rightleftharpoons}} H_2O + Fl^-$$

with Fl^- denoting the fluresceine. Note that the dissociation of water can be neglected here. Fluresceine shows fluorescence at low pH-values.

The experimental measurements (Hoffmann et al. 2006) were performed by feeding a mixture of fluresceine with HCl in one inlet of the T-micromixer and an aqueous NaOH solution in the other; cf. Fig. 9. The timescale of the reaction, which takes place at the contact area between both solutions, is in the range of a few nanoseconds, hence quasi-instantaneous.

The solution of the transport equation of fluresceine allows for the calculation of the fluorescent signal intensity. The calculated intensity is compared with the experimental results as depicted in Fig. 10 and shows good agreement with experimental measurements of our cooperation partner (Räbiger et al., IUV Bremen). The lighter spots inside the dark regions in the left part correspond to slightly negative concentration values due to numerical errors. Furthermore, the geometry used in the experiment is slightly different compared with the standard one: the aspect ratio is decreased from 0.5 to 0.45 with a T-microchannel of 600 μm width and a height of 270 μm. The Reynolds-number was 250 and the signal of the fluorescence dye is detected on a cross section 250 μm down the mixing channel.

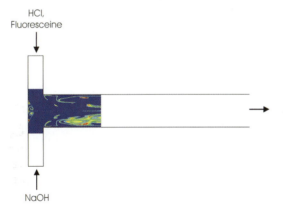

Fig. 9. Visualization of the neutralization of hydrochloric acid and sodium hydroxide with fluresceine - principal setup

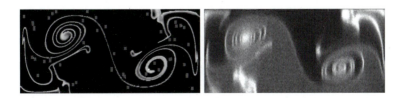

Fig. 10. Fluorescent signal obtained from simulation (left) and experiment (right)

While the approach given above is well suited for computing the reactive mixing of two educts in an instantaneous reaction, it will not give the product concentration. Furthermore, local mass conservation of the reaction cannot be used to calculate the local product distribution due to diffusive and convective transport processes. Consequently, since the product concentration needs to be known in case of consecutive reactions, it remains to solve the stationary reaction-convection-diffusion equation corresponding to (13) but the stiffness of the differential equation complicates the direct numerical solution for fast reactions. Observe that the kinetic term is localized at a small layer around the interface between the educts; it is even only active *on* this interface in case of an instantaneous reaction. To circumvent this problem, we exploit (11) or (12) to replace the kinetic term by the net transport of one of the educts into the grid cells. Under stationary conditions, the concentration field of the product is therefore determined from

$$\mathbf{u} \cdot \nabla c_P - \frac{1}{\mathrm{Re}\,\mathrm{Sc}_P} \Delta c_p = -\left(\mathbf{u} \cdot \nabla c_A - \frac{1}{\mathrm{Re}\,\mathrm{Sc}_A} \Delta c_A \right), \tag{20}$$

where the concentration c_A of species A is obtained from the scalar variable ϕ by means of (18). The right-hand side of (20) is calculated for each cell and implemented as a source term. Since FLUENT® cannot handle hanging nodes for the manual calculation of the right-hand side in (20), this approach can only be used on uniform grids. Finite Differences are used to calculate the gradients in (20).

4.2 Parabolised Species Equation

Despite of possible secondary flow existing for the chosen flow regime, the axial velocity component always points downwards along the axial direction already shortly behind the entry into the mixing channel. This reorientation allows handling the axial flow direction as a pseudo-time variable, so that the evolution of the concentration profile can be computed on successive cross sections, thereby following the main axial flow direction. Considering stationary flow conditions and neglecting species diffusion in axial direction, the parabolised form of the species equation is derived from the stationary 3D species equation (3). This leads to

$$\partial_z c_i + \begin{bmatrix} u/w \\ v/w \end{bmatrix} \cdot \nabla_{x,y} c_i = \frac{1}{w} \nabla_{x,y} \cdot \left(D_i \nabla_{x,y} c_i \right) + \frac{r(c_1,\dots,c_m)}{w}, \tag{21}$$

where c_i is the molar species concentration, x, y denote the cross sectional and z the axial coordinate and u, v, w are the corresponding velocity components. Finally, D_i is the species' diffusivity and r is the rate function of the chemical reaction. The model is complemented by homogeneous Neumann boundary conditions concerning the species distribution. Let us note that although the velocity satisfies the no-slip condition at the lateral boundaries, we observe in 3D simulations that the ratios u/w and v/w attain finite limits at these walls.

The implementation of the parabolised species equation into the commercial CFD-solver FLUENT® 6.2 requires the conversion of (21) into an equation with divergence structure. Replacing also the pseudo-time variable z by the symbol t, this leads to

$$\partial_t c_i + \nabla \cdot \left(\begin{bmatrix} u/w \\ v/w \\ 1/w \end{bmatrix} c_i - \frac{D_i}{w} \nabla c_i \right) = \frac{r}{w} + c_i \nabla \cdot \begin{bmatrix} u/w \\ v/w \end{bmatrix} - \nabla \left(\frac{D_i}{w} \right) \cdot \nabla c_i \tag{22}$$

which is discretised by the Finite Volume Method within the segregated solver of Ansys® Fluent®, employing a second order spatial discretisation and a first order time discretisation. Inside the mixing zone, velocity vectors in (22) are interpolated from 3D-velocity data stored for all cross sections in this zone.

4.3 Validation of the Parabolised Species Equation

To evaluate the new approach and its implementation into the subroutines of FLUENT®, the results taken from a simulation on a 3D-grid were compared with the results obtained from a simulation with the parabolised species equation. For this validation purpose, a simplified velocity field given by the superposition of the Poiseuille velocity profile for a rectangular duct and a cross sectional flow that represents two counter-clockwise spinning vortices is used. For the initial species distribution a simple diagonal profile was chosen. The initial concentration profile of educt B corresponds to 1-A. Both species react in an irreversible second order reaction according to

$$A + B \rightarrow P \tag{23}$$

with the rate function

$$r = k\, c_A c_B. \tag{24}$$

As the rate constant we used the one of the formation of Ca-Green (Baroud et al. 2003), i.e. $k = (1.0 \pm 0.4) \cdot 10^6\, \mathrm{m^3(mol\, s)^{-1}}$. The species distribution of species A obtained from a 3D-simulation with the full model is compared to the outcome of the 2D-simulation of the parabolised species equations. The result is depicted in Figure 11 and shows excellent agreement between the two solutions.

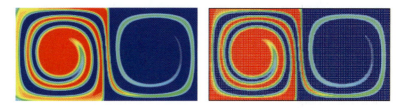

Fig. 11. Comparison between the concentration of species A on a cross section at 1mm of a 3D-simulation (full model, left) and a 2D-simulation (parabolised species equations, right)

For the next level of validation, the artificial velocity field was replaced by the velocity field of an engulfment flow at Re = 186. As reaction system the formation of [CaFluo4]$^{3-}$ was chosen. For the numerical set-up, the diffusion coefficient for Ca(II)-ions was taken from (Song et al. 2003) and the diffusivity of Fluo4 from (Kling 2004). For the complex [CaFluo4]$^{3-}$ the same diffusion coefficient as for Fluo4 was used.

The comparison between the 3D- and the 2D-approach for the mixing zone shows good accordance comparing the evolution of the mean concentration of the educt Fluo4^{5-}. In case of the 3D-model the necessary resolution was only possible for the T-200 microreactor up to the cross section at $z = 500$ µm; therefore the comparison could only be made for a distance of 250 µm in axial flow direction.

The numerical simulation of the reaction zone based on the parabolised species equation has also been validated experimentally, considering the formation of [CaFluo4]$^{3-}$ again. Both in the experiment and in the simulation the inlet concentrations were $[Ca^{2+}] = 1$ mol/m^3 and $[Fluo4^{5-}] = 0.001$ mol/m^3 in water. The reaction was performed under isothermal conditions at room temperature. In the simulations only the educts and the complex are considered, i.e. further species which are present in the experiments like co-ions and the buffer (cf. Hoffmann et al., 2009) are neglected.

For a comparison between the numerically computed concentrations of [CaFluo4]$^{3-}$ and the experimentally measured fluorescence signal, a conversion from concentrations to intensities is required. Here a linear correlation has been employed which was developed by our cooperation partners (Räbiger et al., IUV, Bremen) and is valid for low concentrations. Fig. 12 displays both the computed and the measured fluorescence intensities. The agreement between measured and computed fluorescence signals is very good.

Quantitative comparisons between cross sectional averages of the complex concentration still show deviations between simulation and experiment. There are many possible reasons. Regarding the *modelling* these reasons include the omittance of additional chemical species and the neglection of electrical charges. Concerning the *numerical simulations*, sources for deviations are discretisation errors and non zero divergence of the velocity field. On the *experimental* side, errors result from, e.g., deviations of material, reactor and process parameters. Current simulation runs especially with reduced diffusivities show improved agreement. This issue will be further investigated by the authors.

Fig. 12. Fluorescent signal intensity: Comparison between simulation (left) and experiments (right) at the axial position $z = 10320$ µm. Re = 186, T-400 microreactor

All in all, taking into account the complexity of reactive mixing processes on the different time and length scales as well as the enormous challenges on the experimental side to perform quantitatively reliable measurements in micro-system, the level of agreement between experiment and simulation as illustrated by Figures 10 and 12 is promising. Hence the combination of 3D-simulations of the mixing zone with 2D-computations of the reaction zone using the parabolised species equations is an appropriate instrument for further investigations of reactive mixing especially in T-microreactors.

4.4 Scale Effects

For given geometric similarity, the flow behaviour is solely determined by the Reynolds number, regardless of the dimensions of the device, if the in- and outlet conditions are scaled correspondingly. For the complete system of model equations (1) and (3), hydrodynamical and physico-chemical similarity can not be achieved simultaneously for reactors of different size. This is due to the fact that the Reynolds number depends on the product $U d$ and the Damköhler(I) number on the quotient L/U. Hence, for fixed ratio d/L, not both numbers can be kept fixed while changing the size. To quantify the effect of scaling, a geometrical scaling factor λ is introduced. For this purpose $L' = \lambda L$, $d' = \lambda d$ and $U' = U/\lambda$ with $\lambda \ll 1$. The resulting dimensionless numbers for the micro- and macro-scale system satisfy

$$\mathrm{Re}' = \mathrm{Re}, \qquad \mathrm{Sc}' = \mathrm{Sc}, \qquad \mathrm{Da}'_I = \lambda^2 \mathrm{Da}_I. \qquad (25)$$

This phenomenon is due to changes in the involved time scales of convection, diffusion and reaction. Under the same scaling as above, they behave as

$$\tau'_H = \lambda^2 \tau_H, \quad \tau'_D = \lambda^2 \tau_D, \quad \tau'_R = \tau_R. \qquad (26)$$

Hence, compared to the time scale of the chemical reaction, the characteristic times of the transport processes are massively reduced. This leads to a shift in the range of accessible reaction times, i.e. certain reactions that are too fast to be performed in a macro-scale reactor can be accomplished in micro-scale systems. Further details on the scaling behaviour can be found in (Bothe et al. 2006).

Besides such general considerations, it is now interesting to investigate how the size of the microreactor influences more involved quantities like space-time yield and selectivities. For this purpose, numerical simulations of the scaling behaviour are performed with the hybrid approach explained above. To be more specific, three T-microreactors with geometric similarity have been studied in the engulfment flow regime (Re = 186). The smallest one is denoted T-200 with dimensions 200x100x100 (cf. Figure 2) as described in section 3. In addition, two scaled versions of this microreactor are used, abbreviated as T-400 and T-600, the dimensions of which are two, respectively three times larger. The type of chemical reaction studied is the same as the formation of $[\mathrm{CaFluo4}]^{3-}$. In order to vary the time scale of the chemical reaction, the following (artificial) rate constants were used: $k_{RI} = 1000 \ m^3 mol^{-1} s^{-1}$, $k_{RII} = 1{\times}10^5 \ m^3 mol^{-1} s^{-1}$ and $k_{RIII}=\infty$. Again, the back-ward reaction (dissociation of the complex) was neglected in the simulations.

The analysis of the concentration fields reveal that the chemical reaction is limited by the transport processes since the reaction is more than three orders of magnitude faster than the diffusive transport:

The diffusion time was calculated using the smallest length scale of segregation on cross sections. This yields the diffusion times

$$\tau_D = \frac{(1/\phi)^2}{2D} \tag{27}$$

with ϕ from (9) computed on cross sections. The results in Table 1 show that the theoretically predicted scaling behaviour for the diffusion time was also found in the numerical simulations. Due to the small diffusion time, the T-200 microreactor provides the highest conversion as can be seen in Fig.13. Here the conversion of educt A at an axial position z is defined as

$$1-\dot{n}_A\left(z\right)/\dot{n}_A^{\,in} \tag{28}$$

with $\dot{n}_A^{\,out}(z)$ being the flux of species A in outflow direction through the cross section at given axial position z.

In all three cases the highest conversion was found for the smallest reactor. If the scaling behaviour of the dimensions is taken into account, i.e. if the conversion is plotted against a scaled axial position as shown in Fig. 14, it turns out that all three reactors provide the same conversion. This is reasonable since the reaction is limited by the diffusive transport. In fact this is true for the three different rate constants; cf. Table 1.

In the limiting case of an instantaneous reaction, the conversion is completely determined by the solution of (15). Hence, since (15) shows the same invariance under re-scaling of the dimensions as (1) and (3), the conversion as a function of the scaled axial coordinate is not influenced by the reactor dimensions if the

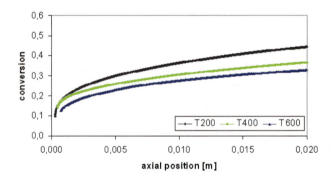

Fig. 13. Conversion of Fluo4^{5-} for a rate constant of $k_{RII} = 1{\times}10^5 \ m^3/(mol \ s)$

Table 1. Comparision between reaction and diffusion time for different rate constants and reactor sizes

	τ_R	τ_D (T-200)	τ_D (T-400)	τ_D (T-600)
k_{RI}	1×10^{-3} s	0,57 s	2,56 s	4,70 s
k_{RII}	1×10^{-5} s	0,60 s	2,55 s	5,25 s
k_{RIII}	0 s	0,64 s	2,45 s	6,05 s

geometric design as well as the Reynolds and Schmidt number are kept constant. For fast reactions this still holds approximately which explains the outcome displayed in Fig. 14.

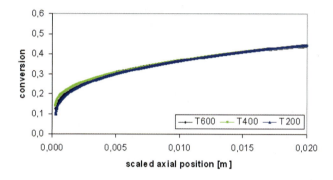

Fig. 14. Conversion of Fluo4^{5-} for a rate constant of $k_{RII} = 1\times10^5$ m^3/(mol s)

Table 2 shows the space-time yield for the three different reactors and the different reaction speeds. Here the space-time yield (STY, for short) is defined as $\dot{N}_P/V$ with V the reactor volume and $\dot{N}_P$ the rate of outflow of moles of product P. From the simulation results, a scaling behaviour proportional to λ^{-2} is found for the part of the reactor volume which extends up to an equivalent cross section placed at $z=20$ mm (T-200), $z=40$ mm (T-400) or $z=60$ mm (T-600), respectively. This is to be expected also theoretically. Note that the rate of outflow of moles of product is given as

$$\dot{N}_P = \int_A c_P \mathbf{u} \cdot \mathbf{n} \, dA . \tag{29}$$

Now with the concentration being scale independent, this flux scales as

$$\dot{N}'_P = \lambda \dot{N}_P \tag{30}$$

which leads to the result above since the volume up to the fixed cross section scales with λ^3.

Table 2. Space-time yield of [CaFluo4]$^{3-}$ up to an equivalent cross section

	STY (T-200) [mol/m^{3}s]	STY (T-400) [mol/m^{3}s]	STY (T-600) [mol/m^{3}s]
k_{RI}	14,5	3,8	1,7
k_{RII}	15,5	3,9	1,7
k_{RIII}	15,2	3,7	1,7

4.5 Variation of the Reynolds Number

A previous analysis of the non-reactive mixing behavior has shown that an acceleration of the flow velocity promotes the mixing quality; see (Bothe et al. 2006). It is therefore interesting to study the case of reactive mixing. For this reason, the formation of CaFluo4 has been simulated under variation of the Reynolds number between Re = 120 and Re = 220. The resulting conversion is shown in Fig. 15 and, as presumed, a better mixing in cross directions results in a higher conversion.

For Re = 120, the flow regime is the vortex flow without intertwinement of the ingoing streams. Mixing between the feeds can therefore be only achieved by diffusion and, hence, the conversion is significant smaller than in the engulfment regime. Fig. 15 shows that in case of enhanced convective mixing in the engulfment flow regime, the reaction is nearly finished at the beginning of the mixing zone. This is due to the smaller length scales of segregation which enhance diffusion.

In case of lower Reynolds number (Re = 160, engulfment flow regime) the convective mixing leads to length scales which are larger than those found for Re = 240. Because of the differing residence time, which is higher in case of Re = 160 than for Re = 240, the same conversion is obtained at the outlet of the reactor.

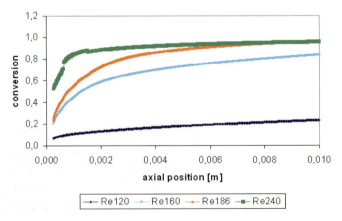

Fig. 15. Conversion of Fluo4^{5-} vs. axial position for different Reynolds numbers

5 Conclusions

Microreactors exploit diffusion as an efficient mixing mechanism alternatively to the turbulent mixing typically used on the macro-scale. Nevertheless, convective mixing may still be required to decrease the initial scale of segregation down to a micro-scale on which diffusion is fast enough. This may be achieved by symmetry breaking secondary flows. Under such conditions, microreactors allow shifting the window of accessible chemical reactions, for which sufficiently high selectivities can be combined with high conversion, to faster ones. Still, depending on the concrete time scales involved, the millimetre scale may also be of interest – especially if large throughput and small pressure drops are required.

In contrast to macro-scale reactors, the transport processes in a microreactor are often well-defined and reproducible. This makes microreactors an ideal means for research on local transport phenomena and their interplay with chemical reactions. Combined with numerical simulations based on first principles, a new and deeper understanding of the effective mechanisms can be developed. Since this requires sound mathematical models and detailed knowledge of model parameters, this in return pushes new experimental techniques – like methods to determine intrinsic kinetics of fast chemical reactions.

References

Arakawa, Y., Wada, O.: Extraction and fluorimeric determination of Organotin compound with Morin. Anal. Chem. 55(12), 1902–1904 (1983)

Baroud, C.N., Okkels, F., Ménétier, L., Tabeling, P.: Reaction-diffusion dynamics: Confrontation between theory and experiment in a microfluidic reactor. Physical Review E 67, 60104 (2003)

Bothe, D., Stemich, C., Warnecke, H.-J.: Theoretische und experimentelle Untersuchungen der Mischvorgänge in T-förmigen Mikroreaktoren - Teil I: Numerische Simulation und Beurteilung des Strömungsmischens. Chem. Ing. Tech. 76, 1480–1484 (2004)

Bothe, D., Stemich, C., Warnecke, H.-J.: Fluid mixing in a T-shaped micro-mixer. Chem. Eng. Sci. 61, 2950–2958 (2006)

Bothe, D., Stemich, C., Warnecke, H.-J.: Computation of scales and quality of mixing in a T-shaped microreactor. Computers & Chemical Engineering 32, 108–114 (2008)

Bothe, D., Warnecke, H.-J.: Berechnung und Beurteilung strömungsbasierter komplexlaminarer Mischprozesse. Chem. Ing. Tech. 79, 1001–1014 (2007)

Bourne, J.R.: Mixing and the selectivity of chemical reactions. Organic Process Research & Development 7, 471–508 (2003)

Dreher, S., Kockmann, N., Woias, P.: Characterization of laminar transient flow regimes and mixing in T-shaped micromixers. Heat Transfer Engineering 30, 91–100 (2009)

Eberhard, M., Erne, P.: Kinetics of Calcium binding to fluo3 determined by the stopped flow fluorescence. Biochem. Biophys. Res. Commun. 163, 309–314 (1989)

Guo, Z.Y., Li, Z.-X.: Size effect in microscale single-phase flow and heat transfer. Int. J. Heat and Mass Transfer 46, 149–159 (2003)

Hoffmann, M., Schlüter, M., Räbiger, N.: Experimental investigation of liquid–liquid mixing in T-shaped micro-mixers using μ-LIF and μ-PIV. Chem. Eng. Sci. 61, 2968–2976 (2006)

Hoffmann, M., Schlüter, M., Räbiger, N.: Microscale Flow Visualisation. In: Hessel, V., Renken, A., Schouten, J.C., Yoshida, J. (eds.) Micro Process Engineering: A Comprehensive Handbook, Fundamentals Operations and Catalysts, vol. A (3 volume Set) (2009)

Kling, K.: Visualisieren des Mikro- und Makromischens mit Hilfe zweier fluoreszierender und chemisch reagierender Farbstoffe. PhD-thesis, University of Hannover (2004)

Ottino, J.M.: Mixing and chemical reactions: a tutorial. Chem. Eng. Sci. 49, 4005–4027 (1994)

Rys, P.: The mixing-sensitive product distribution of chemical reactions. Chimia 46, 469–476 (1992)

Song, H., Bringer, M.R., Tice, J.D., Gerdts, C.J., Ismagilov, R.F.: Experimental test of scaling of mixing by chaotic advection in droplets moving through microfluidic channels. Applied Physics Letters 83, 4664–4666 (2003)

Toor, H., Chiang, S.: Diffusion-controlled chemical Reactors. AIChE Journal 5, 339–344 (1959)

Experimental Analysis and Modeling of Micromixing in Microreactors

Marko Hoffmann, Michael Schlüter, and Norbert Räbiger

Institute of Environmental Process Engineering, University of Bremen
Leobener Str. UFT, 28359 Bremen, Germany

Abstract. For the investigation of chemical reactions in micro- and minichannels the threedimensional velocity field and three dimensional concentration field of an inert and reactive tracer was measured in a T-shaped micromixer. For this purpose a measurement method for micro particle image velocimetry and confocal laser-scanning microscopy has been developed. By solving the continuity equation the calculation of three dimensional streamlines, residence time distribution and local energy dissipation is possible. It becomes out that a regime transition to engulfment flow occurs at a certain Reynolds number dependent on reactor dimensions that can be predicted by the modified model of Soleymani. This transition leads to a different efficiency of micromixing measurable by the reactionproduct of a fast chemical reaction by confocal laserscanning microscopy. The influence of the flow regime on the yield and selectivity has been investigated by means of a parallel consecutive reaction system.

1 Introduction

Micromixing is essential for all biochemical and chemical reactions and therefore a major optimization goal for industrial applications such as chlorination, alkylation, polymerization, crystallization and precipitation. The effect of micromixing on heat transfer plays a dominant role for example in chlorination and alkylation and affects strongly the yield and product quality. In polymerization, micromixing may control the molecular weight distribution and in crystallization the crystal size distribution and the average size of crystals depends on it. This examples shows that the knowledge of micromixing and it's control is an important tool to enhance product quality, yield and selectivity of reactions.

Is was shown early within the DFG priority program 1141 that microreactors are well suitable to study the effects of micromixing on masstransfer and reaction (Schlüter et al., 2004). Due to the small scales and laminar flow conditions the micromixing process in microreactors is not masked by complex macroscopic mixing problems such as turbulence and backmixing. Despite the fact that flow fields under laminar flow conditions are calculable precisely by numerical simulations the behaviour of reacting flows in micromixers is still difficult to predict. One reason is the influence of manufacturing tolerances that are difficult predictable in their effect on hydrodynamics, mass- and heat-transfer. Another reason is the lack in adequate models for interface effects like surface tension or zeta potential. Especially if chemical reactions are initiated by micromixing a complex interaction between different transport phenomena takes place that is only understandable by local measurements and numerical simulation.

For this purpose a close cooperation between manufacturing of microfluidic devices (project IMTEK, Freiburg, Prof. Woias), numerical simulation (project CES, Aachen, Prof. Bothe; TC, Paderborn, Prof. Warnecke) and experimental investigation (project IUV, Bremen, Prof. Räbiger) was installed within the DFG priority program 1141. This report focus on the experimental results obtained within the DFG project at Institute of Environmental Process Engineering, University of Bremen (Prof. Räbiger) between 2002 and 2008.

2   Experimental Methods

To study mass transfer and chemical reactions in dependency of micromixing a noninvasive measurement method with a very high spatial resolution is necessary. A typical experimental set-up for measurements in microfluidic devices is given in Fig. 1.

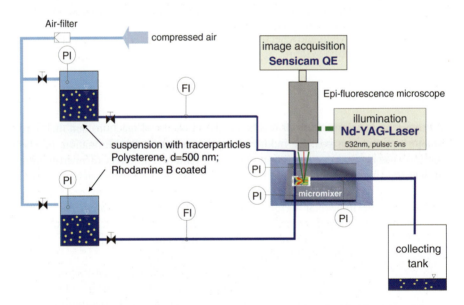

Fig. 1. Experimental set up for optical measurements in microfluidic devices

To prevent any pulsation the fluid is fed out of pressure containers into the microfluidic device. The mass flow is measured by two Bronkhorst High-Tech BV Flow indicators. Both fluids are mixed in the micro fluidic device (e.g. T-mixer) visible in a Microscope to enable a high spatial resolution.

As microfluidic devices different T-shaped and zigzag-shaped micromixers have been provided by IMTEK, Freiburg (project DFG WO 883/8-1) and IMSAS, Bremen manufactured by silicon etching (Fig. 2).

Fig. 2. Microchannel fabricated in silicon with a bonded glass lid

An overview of investigated geometries is given in table 1.

Table 1. Ranges of the geometrical parameters of the T-shaped micro-mixer used in the experimental investigation

Width of the mixing channel A / μm	Width of the inlet channels B / μm	Depth of all channels C / μm	Hydraulic diameter of the mixing channel / μm	Hydraulic diameter of the inlet channel / μm
200 - 600	100 - 300	100 - 300	133 - 400	100 - 300

For measuring flow fields, the non-intrusive particle image velocimetry (PIV) has been adapted to microscale applications. For μ-PIV nano particles with low inertia are added to the flow (polystyrene particles $d_P = 500\ nm$ coated with Rhodamine B, (micro Particles GmbH Berlin) and its displacement within the flow field is detected by illumination with two laser pulses (New Wave Research Inc. Laser, wavelength *532 nm*, pulse width *5 ns*, pulse distance *1 μs*). The defined pulse distance enables the detection of particle velocities and directions according to the local flow field. For μ-PIV a double frame CCD-Camera (PCO Sensicam QE) mounted on a Epi-Fluorescence Microscope (Olympus BX51WI Microscope; Plan Achromat C Objective 20X/0.4) is used. The data evaluation is performed with a standard cross-correlation scheme based on a FFT (VidPIV-Software, ILA GmbH Jülich). A typical result of correlation is given in Fig. 3 that shows the 2D+2C velocity field inside the entrance region of a T-shaped micromixer as a horizontal slice in the middle of the channel i.e. at half the channel depth. Due to sufficient particle seeding and high signal/noise ratio the cross-correlation was used with a interrogation size of *32 x 32* pixel^2 and half-overlapping.

With this system a lateral resolution of *7 x 7 μm^2* in a measurement depth of approximately *13 μm* is achievable.

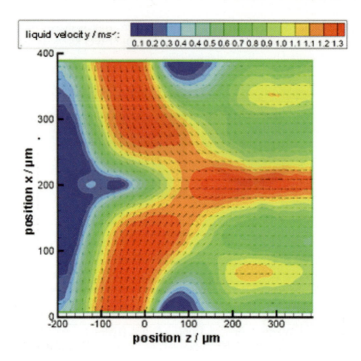

Fig. 3. Velocity field - entrance region of a T-shaped micromixer

For the measurement of the local distribution of reactants and products inside the micro device the Laser Induced Fluorescence (LIF) is used. In all forms of LIF, a laser is used to excite a fluorescent species within the flow. Typically, the tracer is an organic fluorescent dye such as fluorescein or rhodamine. The dye absorbs a portion of the excitation energy and spontaneously re-emits a portion of the absorbed energy as fluorescence. The fluorescence is measured optically and used to infer the local concentration of the dye (Crimaldi et al. 2008). For three dimensional concentration measurements a Confocal Laser Scanning Microscope (CLSM) is used. The major advantage of CLSM is the possibility of collecting emitted light only from a focus plane. For the following experimental investigations a Carl Zeiss LSM 410 was used with a Helium Neon Laser *(543 nm, 1mW)* as excitation source and a *20x/0.5* Plan Neofluar objective. The optical slice thickness is *8.0 μm*, the axial resolution *5.0 μm* and the lateral resolution *0.6 μm*. A typical experimental set-up is given in Fig. 4. A buffer solution (deionised water, *pH=8.2, T=20°C*) is feed out of pressure containers into the micro device to prevent any pulsation while the mass flow is measured by two mass flow controllers FI (Bronkhorst High-Tech BV).

One inlet stream is enriched with the fluorescent dye (e.g. Rhodamine B) in a very low concentration thus the fluorescence intensity of the fluorochrome can be assumed as proportional to the concentration of the dye c (Hoffmann et al., 2006).

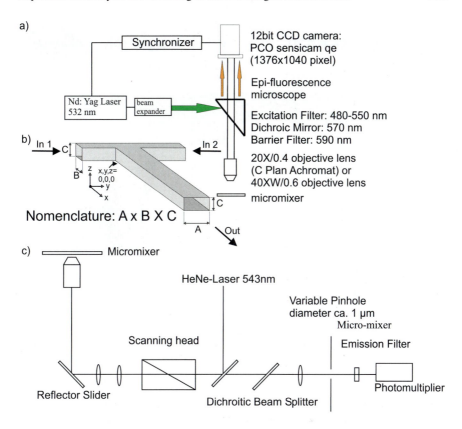

Fig. 4. Experimental set-up for concentration field measurements

Both streams entered into the micro device (e.g. T-shaped micromixer) were the mixing and/or reaction process takes place under the confocal laserscanning microscope. In this microscope the illumination by a laser beam induces the fluorescence which is detected by a photomultiplier. In front of the detector a pinhole is arranged on a plane conjugate to the focal plane of the objective. When light strikes the pinhole that comes from planes above or below the focal plane it is out of focus and does not contribute to the image (Wilhelm et al., 2003). The laser scans rapidly along a single focal plane in order to complete a full-field image on the detector unit. The scan is repeated for multiple focal planes to reconstruct a three dimensional image. Grey values corresponding to the local dye concentration and were calibrated previously.

An indispensable condition for using the CLSM is the time-invariance of the flow field because the temporal resolution of the scanning technique is in the range of several milliseconds depending on the spatial resolution. It has been shown that the flow field in micro-devices is stationary and reproducible even though very complex structures occurs (Fig. 5, bottom).

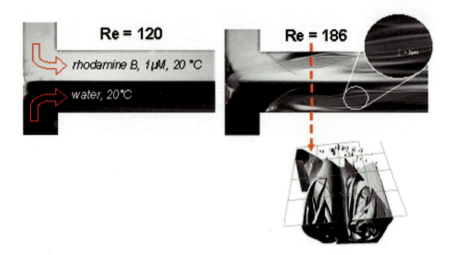

Fig. 5. Plane and spatial distribution of dye in a T-shaped micro-mixer for two different flow regimes (measured by CLSM)

By using the continuity equation, Eq. (1) for incompressible flow it is possible to calculate the out-of-plane velocity component w from the two in-plane components u and v, as shown by (Sousa et al., 2002) for a turbulent flow around a surface-mounted obstacle.

$$\frac{\partial u_i}{\partial x_i} = 0 \quad with \quad x_i = (x,y,z); u_i = (u,v,w) \tag{1}$$

Integration of Eq. (1) in the z-direction (out-of-plane) yields the expression to gain the w-component Eq. (2).

$$w(x_i, y_i, z_k) = w(x_i, y_i, z_{k-1}) - \int_{z_{k-1}}^{z_k} \left[\frac{\partial u}{\partial x}(x_i, y_i, z) + \frac{\partial v}{\partial y}(x_i, y_i, z) \right] dz \tag{2}$$

discretized in the nodes i, j, k of a three-dimensional (measurement) grid. The derivatives in the integral of Eq. (2) were computed in the x, y-plane, employing (second-order) central differences.

This technique for calculating the out-of-plane component was recently applied to microfluidic flows (Hoffmann et al., 2007; Bown et al., 2007 and Kinoshita et al., 2007). With the knowledge of the third velocity component w it is possible to visualize the 3D-structure in the entrance region of a T-shaped micromixer, as shown in Fig. 6. The out-of-plane component w is unequal zero, indicating that the flow structure is three dimensional in the entrance region of this type of mixer. Furthermore, six streamlines (3D) are chosen to indicate different residence times due to the three dimensional flow structure (Hoffmann et al., 2007).

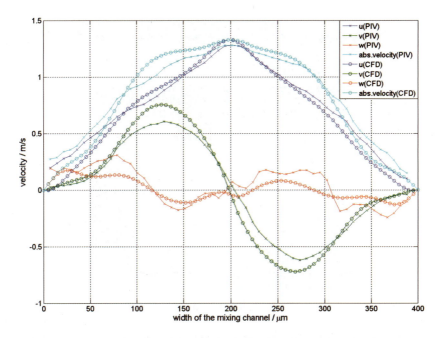

Fig. 6. Components u,v,w, entrance region of a T-shaped micromixer, x=167 µm; out-of-plane component w is calculated by means of the continuity equation; comparison with CFD results

By using the third velocity component *w* it is also possible to calculate the local energy dissipation ε with the deformation tensor **D** due to

$$D = \frac{1}{2}\left(\nabla u + (\nabla u)^T\right) \tag{3}$$

$$D = \begin{bmatrix} \dfrac{\partial u}{\partial x} & \dfrac{1}{2}\left(\dfrac{\partial u}{\partial y}+\dfrac{\partial v}{\partial x}\right) & \dfrac{1}{2}\left(\dfrac{\partial u}{\partial z}+\dfrac{\partial w}{\partial x}\right) \\[3mm] \dfrac{1}{2}\left(\dfrac{\partial u}{\partial y}+\dfrac{\partial v}{\partial x}\right) & \dfrac{\partial v}{\partial x} & \dfrac{1}{2}\left(\dfrac{\partial v}{\partial z}+\dfrac{\partial w}{\partial y}\right) \\[3mm] \dfrac{1}{2}\left(\dfrac{\partial u}{\partial z}+\dfrac{\partial w}{\partial x}\right) & \dfrac{1}{2}\left(\dfrac{\partial v}{\partial z}+\dfrac{\partial w}{\partial y}\right) & \dfrac{\partial w}{\partial x} \end{bmatrix} \tag{4}$$

$$\varepsilon = 2v \cdot \varepsilon = D : D \tag{5}$$

An example for the local energy dissipation in dependance of the position within the mixing device is given in figure 7.

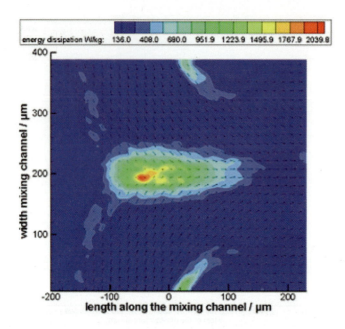

Fig. 7. Field of local energy dissipation in a T-shaped micromixer, T400x200x200

By using the image processing software IMARIS (Bitplane AG, Switzerland) the generation of a volume, rendered 3-D image is possible out of a PIV-data set (e.g. *50* slices for a channel *200μm* in depth). By using this commercial image processing software the reconstruction of a 3D-concentration field is possible (Fig. 5).

By exploiting the dependency between fluorescence intensity and dye concentration, it is possible to get the distribution of dye as three-dimensional concentration field inside the micro device.

For small concentrations, i.e. *1 μmol/l*, the intensity of the emitted light by the fluorescent dye Rhodamine B is linearly dependent on the concentration of the dye and can be simplified by a series expansion so that I_f is proportional to the concentration c of the dye (Fig. 8).

The calibration of greyscales with dye concentrations allows the quantitative analyses of mixing quality as well as potential for diffusive mixing. For a quantitative analysis of Danckwerts' intensity of segregation the mixing quality M

$$M = 1 - \sqrt{I_s} = 1 - \frac{\sigma}{\sigma_{max}} \tag{6}$$

can be calculated by means of the gray values of the cross section areas along the mixing channel length. $M=0$ corresponds to a totally segregated system whereas a value of $M=1$ corresponds to a homogeneous mixture.

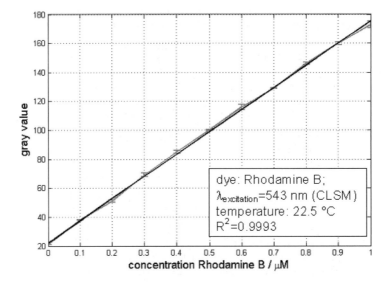

Fig. 8. Calibration grayvalue vs. fluorescence dye concentration (Rhodamine B); for each measuring point: mean grayvalue and standard deviation of 50 vertical slices (Hoffmann et al., 2007).

As mentioned e.g. by Bothe et al. (Bothe et al., 2006) the mixing quality is not sensitive to the length scales on which the segregation takes place which is very important for describing micromixing in laminar flows. Bothe is using the potential for diffusive mixing

$$\Phi(V) = \frac{1}{|V|} \int_V \|\nabla f\| dV \quad \text{with} \quad f = \frac{c}{c_{max}} \tag{7}$$

as a significant parameter for the total driving force within the concentration field for diffusive dissipation of concentration gradients (the concentration of the species is c). The potential for diffusive mixing is also calculable by means of the gray values of the cross section areas along the mixing channel length with a special MATLAB routine.

3   Results and Discussion

Fig. 5 makes clear that even though mostly laminar mixing occurs in micro-devices very fine structures are achievable if the flow regime given by the Reynolds number Re

$$Re = \frac{u \cdot d_h}{\nu} \tag{8}$$

has been chosen correctly. In Equation (8) u denotes the average velocity in the mixing channel, d_h the hydraulic diameter and v the kinematic viscosity. Figure 5, below shows for example fine structures with short diffusion lengths down to $3\mu m$ in a T-shaped micro-mixer. On the other hand, only a reduction of the flow rate to one third causes a tremendous decrease of contact area between two educts and mixing performance (Fig. 5, top).

The development of mixing is shown visually in Fig. 5. It becomes obvious that in engulfment flow the two inlet streams are forced together into a spiral structure with increasing tortuosity. This induces an increasing contact area between both streams for mass transfer by diffusion (micromixing) and reaction.

To study the mechanism of micromixing the mixing quality and potential for diffusive mixing are calculated for several cross sections downstream the mixing device (Fig. 9 and Fig. 10). Approximately after five times the width of the mixing channel no significant increase of mixing quality and potential for diffusive mixing is detectable. The roll up of concentration profiles in Fig. 9 demonstrates an increasing contact area downstream the micromixer but Fig. 10 makes clear that the effect on potential for diffusive mixing and mixing quality is only marginal. Obviously, the flow has to be reflecting or bend again to enhance the mixing quality

The characterization of the flow regime is possible by taking into account the Reynolds number as well as the aspect ratio and ratio of inlet velocities. In Fig. 11 the contact area is shown in dependency of different ratios of inlet flows. $\dot{V}_1$ denotes the inlet flow marked by the passive tracer Rhodamin B. In $\dot{V}_2$ no fluorescent dye is soluted. The measurement shows a decreasing potential for diffusive mixing Φ with decreasing ratio of $\dfrac{\dot{V}_1}{\dot{V}_2}$: (a) $\Phi = 10767 m^{-1}$, (b) $\Phi = 8690 m^{-1}$ and (c) $\Phi = 7533 m^{-1}$.

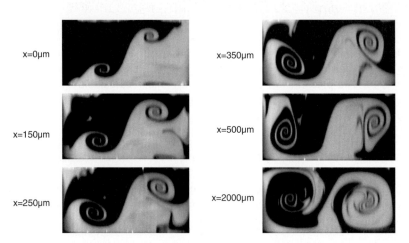

Fig. 9. Concentration distribution at cross sections of the mixing channel at six different positions, ranging from 0 μm to 2000 μm for a mean velocity of 0.5 m/s (Re=207), T-shaped micromixer T600x300x300; system Rhodamine B/DI water.

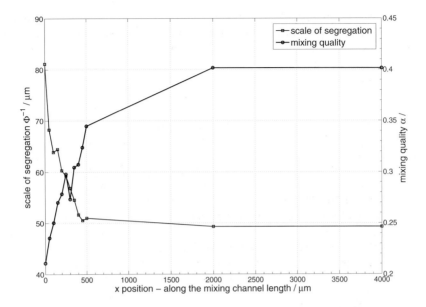

Fig. 10. Scale of segregation and mixing quality along the mixing channel for a mean velocity of 0.5 m/s (Re=207), T-shaped micromixer T600x300x300; system Rhodamine B/DI water.

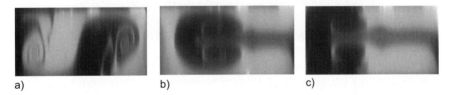

a)   b)   c)

Fig. 11. Concentration distribution along the cross section, T-shaped micromixer 400x200x290, x=387μm, Re=160, $a\,)\,\dot{V}_1 = \dot{V}_2$, $b\,)\,\dot{V}_1 = \frac{1}{2}\dot{V}_2$, $c\,)\,\dot{V}_1 = \frac{1}{3}\dot{V}_2$

System: Rhodamine B/DI water

The influence of geometry and ratio of inlet flows was modeled by Solimany by using numerical simulation (Soleymani et al., 2008) based on some experimental data from literature. A validation of the model by Solimany with the experimental data gained within this project and by our cooperation partner (Woias et al., IMTEK, Freiburg (Kockmann, 2008) shows a good agreement and high reliability of the model.

More precisely the numerical simulation performed by Bothe et. al (2008) shows the flow regime and velocity field within the micromixer. A qualitative validation conducted with the CFD results is given in figure 12, where the tracer distribution of a passive scalar (i.e. Rhodamine B) is compared with numerical simulation.

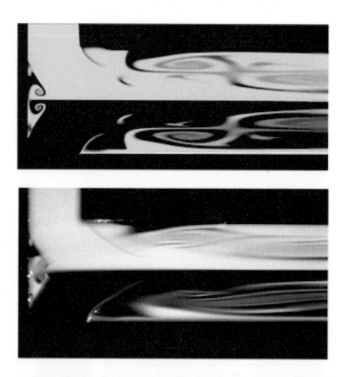

Fig. 12. Experimental (bottom) and numerical calculated concentration field (top), micromixer geometry 200x100x100 μm and Re=186, middle of the channel depth

In spite the fact that the CFD simulations were performed with about 9 million grid cells, the extremely fine structures cannot be resolved with the numerical simulation. With a Schmidt number of approx. *3600* (Rhodamine B, large molecule, diffusion coefficient $2.8*10^{-10}$ m^2/s) the numerical diffusion dominates the diffusive mass transfer on too coarse computational grids. This is one of the main limiting factors in the usage of CFD in the investigation of the mixing performance in static micromixers. In a numerical simulation an ideal geometry is used (no geometry tolerances, no surface roughness) with a huge number of grids. These facts show the necessity of the experimental investigation as a basis for the modelling of mixing in microreactors.

As shown in figure 10 in a simple T-mixer only a mixing quality of about *M=0.4* can be achieved under laminar flow conditions. Dreher et al. (2009) reaches higher mixing qualities in T-mixers under turbulent flow conditions but shows also that the mixing quality alternates due to velocity fluctuations downstream. A more reliable method to reach higher mixing qualities is the usage of multiple changes in flow direction by rectangular bends. Figure 13 shows the

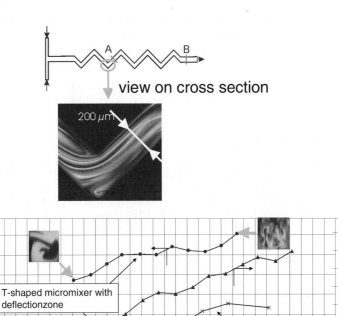

Fig. 13. Potential for diffusive mixing and mixing quality - zig-zag-shaped micromixer; mixing quality compared with a T-shaped micromixer (T400X200x200); system Rhodamine B/DI water

potential for diffusive mixing as well as the mixing quality for a zig-zag-shaped micromixer and makes clear that complete mixing is achievable even at low Reynolds-numbers.

Reactive Mixing

For chemical and biochemical reaction all reactants have to get in direct contact by micromixing. Due to this fact many macroscale application uses reactive mixing for investigating the phenomena of micromixing (e.g. Faes et al., 2008). Due to the high spatial resolution within a microchannel the micromixing is observable very detailed in dependency of both, three-dimensional velocity- and concentration field. Several problems have to be solved to investigate reactive mixing in a microchannel:

- the time constant of reaction must be in the range of the residence time within the channel
- the reaction product must be markable by a fluorescence dye
- the experimental setup must be adaptable to the excitation and emission frequency of the fluorescence dye
- the operation conditions must be acceptable (pressure, temperature, toxicity, ...)

These conditions are difficult to reach especially because not even the kinetic constants are known under micromixing conditions.

A well known system to study micromixing in macroscales is Fluo^{-4} that reacts with Ca $^{2+}$ (Kling et al., 2004). Fluo^{-4} is a fluorescent dye with Rhodamine moiety and fluoresces by chelating with Ca $^{2+}$ ion. Hence, the produced complex ion can be visualized as a chemical product by the LIF. The reaction rate constant between both reactants is in the order of 10^6 m^{3}mol^{-1}s^{-1}, thus the reaction can be regarded as a rapid reaction. The chemical product is visualized in a T-shaped micromixer with a cross-section of *600x300 µm²* (Fig. 14).

Fig. 14. Reaction Ca$_2^{++}$ Fluo-4 (fluorescence tracer): visualization of the chemical product; T-shaped micromixer - cross-section 600x300 µm², Re=186; product distribution - horizontal slice (left), product distribution along the cross-section area - 4mm downstream (right).

The product distribution along the vortex pair in both images also depicts the three-dimensional flow structure for higher *Re* in T-shaped micromixers.

Due to the differences in residence time distribution and spacial concentration of reactants the hydrodynamic conditions (transition to engulfment flow) should influence the yield and selectivity of parallel consecutive reactions remarkably. To check this influence the well known forth Bourne reaction was used in a T-shaped micromixer. The fourth Bourne reaction is a parallel reaction system where the acid-catalyzed hydrolysis of 2,2-dimethoxypropane (DMP) is conducted in parallel with the neutralization of HCl with NaOH. With very rapid mixing, the HCl catalyst is neutralized by NaOH and minimal hydrolysis of DMP occurs. The DMP concentration is measured by gas chromatography at the outlet of the T-shaped micromixer. Figure 15 shows the potential for diffusive mixing with the typical step to higher Φ in the range of $Re=180$ due to the transition to engulfment flow. Despite the fact of intensified mixing at $Re>180$ no significant decrease in DMP concentration occurs. This result makes clear, that local micromixing-effects are not detectable by overall measurements at the channel outlet. Obviously the final mixing at the outlet connection is more dominant than the micromixing within the channel. Therefore local concentration measurements should be done.

For a comparison of numerical simulations given by our cooperation partners (Bothe et al., CCS, Aachen) with experimental data the above mentioned $Fluo^{-4}$, Ca^{2+} reaction has been used with the development of the fluorescence product in dependency of the x-position within the reactor (Fig. 16). It becomes obvious that even if a reasonable agreement between the slopes of numerical and experimental data is achievable the quantitative values are still deviating in the order of 100%. From the experimental point of view this can be explained mainly by the low quality of the pictures taken with the Confocal Laser Scanning Microscope (see Fig. 14). Another reason might be the complexity of the calibration procedure as well as small differences in the numerical and real reactor geometry.

Finally it has to be emphasized that even though the quantitative analyses is improvable the numerical simulation as well as the experimental local measurement of the product concentration was carried out successfully for a chemical reaction within a microchannel.

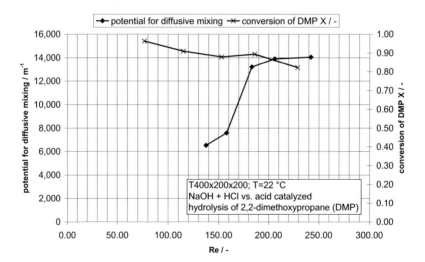

Fig. 15. Potential for diffusive mixing Φ and conversion X of DMP, OH^- + H^+ vs. acid catalyzed hydrolysis of 2,2-dimethoxypropane for different Re

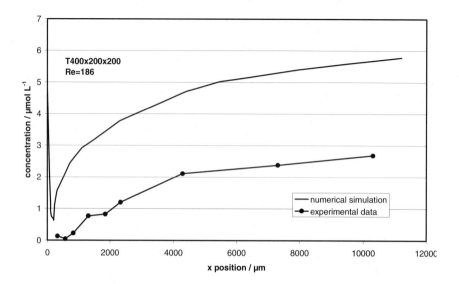

Fig. 16. Comparison between numerical simulations and experimental data for the development of the [CaFluo4]3⁻-concentration within a T-shaped micromixer (400x200x200) at Re = 186

4 Conclusion

For the investigation of micromixing the threedimensional velocity field and three-dimensional concentration field of an inert and reactive tracer was measured in a T-shaped micromixer. For this purpose a measurement method for micro particle image velocimetry and confocal laser scanning microscopy has been developed. By solving the continuity equation the calculation of three dimensional streamlines, residence time distribution and local energy dissipation is possible. It becomes out that a regime transition to engulfment flow occurs at a certain Reynolds number dependent on reactor dimensions that can be predicted by the modified model of Soleymani et al. (2008). This transition leads to a different efficiency of micromixing measurable by the reaction product of a fast chemical reaction by confocal laserscanning microscopy. The influence of the flow regime on the yield and selectivity has been investigated by means of a parallel consecutive reaction system. It has been shown that the measurement of streamlines, local energy dissipation rates and local concentration fields of single reaction products is possible by means of micro particle image velocimetry and confocal laser scanning microscopy even if a reaction takes place. For using this technique in process design the fluorescence dye has to be selected very carefully.

The results show that the prediction of yield and selectivity will be possible for chemical reactions in the near future to intensify processes by better reactor designs and operating strategies. Nevertheless more research is necessary to come to more reliable results for this and other reaction systems and reactor geometries.

References

Arakawa, Y., Wada, O.: Extraction and fluorimeric determination of Organotin compound with Morin. Anal. Chem. 55(12), 1902–1904 (1983)

Baroud, C.N., Okkels, F., Ménétier, L., Tabeling, P.: Reaction-diffusion dynamics: Confrontation between theory and experiment in a microfluidic reactor. Physical Review E 67, 60104 (2003)

Bothe, D., Stemich, C., Warnecke, H.-J.: Theoretische und experimentelle Untersuchungen der Mischvorgänge in T-förmigen Mikroreaktoren - Teil I: Numerische Simulation und Beurteilung des Strömungsmischens. Chem. Ing. Tech. 76, 1480–1484 (2004)

Bothe, D., Stemich, C., Warnecke, H.-J.: Fluid mixing in a T-shaped micro-mixer. Chem. Eng. Sci. 61, 2950–2958 (2006)

Bothe, D., Stemich, C., Warnecke, H.-J.: Computation of scales and quality of mixing in a T-shaped microreactor. Computers & Chemical Engineering 32, 108–114 (2008)

Bothe, D., Warnecke, H.-J.: Berechnung und Beurteilung strömungsbasierter komplex-laminarer Mischprozesse. Chem. Ing. Tech. 79, 1001–1014 (2007)

Bourne, J.R.: Mixing and the selectivity of chemical reactions. Organic Process Research & Development 7, 471–508 (2003)

Dreher, S., Kockmann, N., Woias, P.: Characterization of laminar transient flow regimes and mixing in T-shaped micromixers. Heat Transfer Engineering 30, 91–100 (2009)

Eberhard, M., Erne, P.: Kinetics of Calcium binding to fluo3 determined by the stopped flow fluorescence. Biochem. Biophys. Res. Commun. 163, 309–314 (1989)

Guo, Z.Y., Li, Z.-X.: Size effect in microscale single-phase flow and heat transfer. Int. J. Heat and Mass Transfer 46, 149–159 (2003)

Hoffmann, M., Schlüter, M., Räbiger, N.: Experimental investigation of liquid–liquid mixing in T-shaped micro-mixers using μ-LIF and μ-PIV. Chem. Eng. Sci. 61, 2968–2976 (2006)

Hoffmann, M., Schlüter, M., Räbiger, N.: Microscale Flow Visualisation. In: Hessel, V., Renken, A., Schouten, J.C., Yoshida, J. (eds.) Micro Process Engineering: A Comprehensive Handbook, Fundamentals Operations and Catalysts, vol. A (3 volume Set) (2009)

Kling, K.: Visualisieren des Mikro- und Makromischens mit Hilfe zweier fluoreszierender und chemisch reagierender Farbstoffe. PhD-thesis, University of Hannover (2004)

Ottino, J.M.: Mixing and chemical reactions: a tutorial. Chem. Eng. Sci. 49, 4005–4027 (1994)

Rys, P.: The mixing-sensitive product distribution of chemical reactions. Chimia 46, 469–476 (1992)

Song, H., Bringer, M.R., Tice, J.D., Gerdts, C.J., Ismagilov, R.F.: Experimental test of scaling of mixing by chaotic advection in droplets moving through microfluidic channels. Applied Physics Letters 83, 4664–4666 (2003)

Toor, H., Chiang, S.: Diffusion-controlled chemical Reactors. AIChE Journal 5, 339–344 (1959)

A Numerical Approach for Simulation of Turbulent Mixing and Chemical Reaction at High Schmidt Numbers

Florian Schwertfirm and Michael Manhart

Fachgebiet Hydromechanik, Arcisstrasse 21, 80339 München
e-mail: f.schwertfirm@bv.tum.de, m.manhart@bv.tum.de

Abstract. Many chemical engineering processes take place in aqueous solutions at high Reynolds numbers. An accurate numerical prediction of such processes is very challenging as the reaction rate depends on large scale features, such as stirring motions, geometry of the mixing device, etc., and on the distribution of the mixture fraction on the smallest scales. The spacing between the largest and the smallest scales can become enormous with increasing Reynolds and Schmidt number so one is not able to compute all scales directly with todays computer power. In this contribution we present a complete numerical approach for the description of a precipitation process in an aqueous solution in a confined impinging jet reactor, also called T-mixer. We first present a detailed analysis of the flow and passive scalar mixing in such a device at $Sc = 1$ by direct numerical simulation. The large scale mixing at high Sc is analysed for the same flow with the SEMI DNS method and the results are validated with LIF measurements. With both results, a description of the precipitation is possible along Lagrangian particle paths, resulting in two consecutive simulations for describing the precipitation process. For an integrate approach we combine a DNS of the flow field with a filtered density function simulation of the scalar field where the chemical source term is closed. We develop a model for the micro-mixing term in the FDF transport equation and show first results of the micro-mixing and precipitation process in a T-mixer at $Sc = 1000$.

1 Introduction

Turbulent mixing is of great importance in process engineering because it can considerably influence product properties [1, 6].

In the context of chemical reaction engineering, mixing is the process of bringing two or more species together on a molecular scale, allowing for chemical reaction. Usually the reaction rate and other processes like nucleation etc. are a function of the oversaturation and, by acting as a sink, reduce the oversaturation. The time scale with which the oversaturation is built up by the mixing process is the mixing time t_M and the time scale by which the oversaturation is reduced is the time scale of the chemical reaction t_R. Especially when t_R is smaller or in a similar range than t_M, the turbulent mixing is the rate determining step and has a dominating influence on all subsequent steps such as e.g. nucleation, growth, agglomeration and aggregation in a precipitation process. The prediction of the mixing time is therefore a crucial step for developing numerical tools for simulating a wide class of chemical engineering processes.

From a fluid mechanic point of view this is a challenging task, as for a correct prediction of the mixing time a wide range of scales has to be accounted for. In a general concept of mixing, three phases can be distinguished, namely (i) large-scale-mixing, (ii) meso-mixing and (iii) micro-mixing. While large-scale-mixing is determined by the large scale vortical motions of the turbulent flow field and the scalar integral scale L_Φ (e.g. the initial conditions), meso-mixing is the mechanical reduction of the scalar segregation scale down to the Batchelor scale by the turbulent cascade. Molecular mixing or the so-called micro-mixing is defined as the destruction of concentration fluctuations by molecular diffusion [1, 6] and is significant only at the Batchelor scale. The Batchelor length scale $\eta_B = \eta_K/\sqrt{Sc}$ [2] is dependent on the Schmidt number Sc and as Sc can easily reach ≈ 1000 in aqueous solutions, it can become much smaller than the Kolmogorov length scale η_K. For an accurate description of the mixing time, the simulation approach must be able to describe all three subsequent mixing steps, ranging from the large scale structures, which are often inhomogeneous in real applications, down to the molecular mixing at the Batchelor scale.

The aim of this project was to develop a numerical simulation technique for the accurate prediction of precipitation processes in aqueous solutions (e.g. high Sc). As the time scales of precipitation processes are small, often so called T-Mixers are used as with these continuous flow mixers small mixing times can be achieved. To produce the most accurate results possible the flow is described by direct numerical simulation (DNS) and the scalar field is described by a filtered density function simulation (FDF). In this way the chemical source term is closed. To develop this method the flow in the specific mixing device has to be studied and validated and a micro-mixing model has to be developed. This contribution is structured as follows. In the following chapter DNS is used to study the flow and mixing inside a T-Mixer at low Reynolds numbers and at $Sc = 1$. The results are validated with experimental data and a first approach for a simulation of the precipitation process is proposed. In the next chapter, a modelling approach for high Sc number flows in a Eulerian framework is introduced and results of the mixing in the T-Mixer with this subgrid scale model are presented. As with this model only large scale scalar structures can be predicted, it is not adequate to simulate the molecular mixing process at high Sc. To close the description of the mixing process even at high Sc a so called filtered density function (FDF) approach in conjunction with a DNS of the flow field is introduced in the fourth chapter and first results of the mixing process inside a T-Mixer are shown. As the chemical source term is closed in this description, a direct simulation of the precipitation process was incorporated in the FDF-DNS simulation and also, first results for the particle size distributions are presented.

2 Direct Numerical Simulation of the Flow and Mixing Inside a T-Mixer at Low *Sc*

The basic geometry used throughout this study is shown in figure 1. It consist of two opposing feeding pipes with diameter D which open into a main duct at an angle of 90, with quadratic cross- section and a side length of $0.5H = D$. The main duct is closed at one side and the feeding pipes are flush with this wall. The flow enters the mixer symmetrically via the feeding pipes, i. e. with the same Reynolds number, and leaves the mixer at the open side of the main mixing duct. The origin of the coordinate system is set at the intersection point of the symmetry axis of the feeding pipes with the symmetry axis of the main duct. The positive x direction is defined to follow the flow down the main duct. The y direction follows along the axis of the feeding pipes. The z axis is defined to close the right-hand side coordinate system. The flow Reynolds number is defined with the main duct width H and the bulk velocity u_b in the main duct $Re = u_b H / v$.

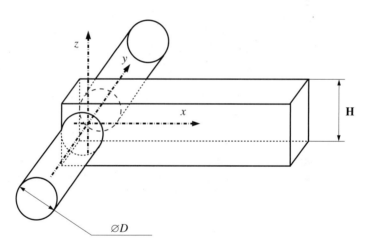

Fig. 1. Geometry of a T-Mixer

For the DNS of the flow field the code MGLET is used [12, 15]. The incompressible Navier-Stokes equations, namely the conservation of mass,

$$\frac{\partial u_i}{\partial x_i} = 0 \tag{1}$$

and momentum,

$$\frac{\partial u_i}{\partial t} + u_j \frac{\partial u_i}{\partial x_j} = -\frac{1}{\rho} \frac{\partial p}{\partial x_i} + \frac{\partial}{\partial x_j} \left(v \left(\frac{\partial u_i}{\partial x_j} + \frac{\partial u_j}{\partial x_i} \right) \right), \tag{2}$$

together with the passive scalar transport equation,

$$\frac{\partial \Phi_\alpha}{\partial t} + u_i \frac{\partial \Phi_\alpha}{\partial x_i} = \Gamma \frac{\partial^2 \Phi_\alpha}{\partial x_j^2} \tag{3}$$

are solved in the Finite Volume (FV) formulation on a Cartesian grid, using a staggered variable arrangement. The discretization in space and the approximation of the derivatives and the interpolation is accomplished by a second order central scheme. For the integration in time a third order Runge-Kutta method is used. The incompressibility constraint is satisfied by solving the Poisson-equation for the pressure with an Incomplete Lower-Upper (ILU) decomposition and applying a correction step for the velocities and the pressure.

The geometry of the T-mixer is represented in the numerical simulation by a main duct with square cross section with a side length H and a length of $L_D = 4H$. Two circular feeding pipes intersect perpendicular and flush with the main duct at the bottom wall. The diameter of the feeding pipes is $0.5H$ and its length is $L_T = 2H$ (see Figure 1). Compared to former simulations (see [26]) the geometry was extended to include the feeding pipes as it turned out that setting the inflow boundary condition directly at the entry plane of the feeding pipes in the main duct leads to a completely different flow field in the main duct.

This geometry is completely embedded in a regular Cartesian computational domain with the box size $(L_x = 4H, L_y = 5H, L_z = H)$. The no-slip condition for the velocities and zero-gradient condition for the scalar at the walls of the geometry were modelled by using the Immersed Boundary (IB) technique. In the IB technique, the node values of cells which are intersected by the geometry are not computed by solving the Navier-Stokes equations. Instead the values are interpolated with a high order scheme, so that the boundary condition is exactly fulfilled at the surface of the geometry. Cell values lying outside the flow field and inside the geometry are not computed. This method has been successfully used in various flow problems, for further details see [15].

At the outflow a zero-gradient and at the inflow-planes of the feeding pipes laminar parabolic inflow profiles with peak inflow velocity V_0 are prescribed, as the Reynolds number in the feeding pipes is in the laminar regime. The scalar is set to $c(y/H = -1.5) = 1.0$ and $c(y/H = 1.5) = 0.0$ at the inflow planes of the feeding pipes, leading to a maximal difference in mixture fraction of $\Delta c = 1.0$, whereas the zero-gradient condition is set at the outflow.

The grid is chosen so that it resolves the wall friction and the Kolmogorov length scale in the flow-field. In the region of intense mixing ($x/H \leq 1.51$) an equidistant grid with spacings of $(\Delta x, \Delta y, \Delta z) = (0.011H, 0.005H, 0.005H)$ is used. In this area, the grid spacing in x direction is larger, because the main velocity gradients are in the y and z direction. The grid spacing in the x direction results in a distance of the first pressure point to the bottom wall of $x^+ = 1.55$. According to Johannson and Andersson [10], the turbulence of the impinging jets in the stagnation point area is not in equilibrium, thus the Kolmogorov length scale is the lower limit of the resolution requirement. The grid is stretched from $1.51 < x/H < 3.75$ in the x

direction towards the outflow and from $y/H < -0.5$ and $y/H > 0.5$ in the y direction towards the inflow planes. This results in a grid consisting of 17.2×10^6 grid points of which 10.24×10^6 lie within the main duct of the T-mixer.

2.1 Flow Field Inside the T-Mixer

The most important feature of the flow inside this T-mixer configuration is the break of symmetry. This is best visualized by a vector plot of the mean velocities in planes normal to the x-direction. As can be seen in figure 2, the two inflow jets pass each other and impinge on the opposite wall. This 'passing' of the jets gives rise to a large helical vortex which fills the whole mixer.

Four additional features are visible at $x/H = 0.0$: (1) a strong shear layer forms between the to inflow jets, (2) a secondary vortex evolves in the corner where the jets impinge on the wall, (3) there is back flow into the feeding pipes, and (4) there are large regions of relatively small mean velocities where also secondary vortices evolve. When following the x axis, the main vortex develops towards axisymmetry with respect to the x axis, giving rise to secondary vortices at $x/H = 0.75$ in each corner of the main duct and finally being almost axisymmetric towards the end of the main duct. The rotational energy is strong enough to give rise to one rotation around the x axis within the computational domain. The streamwise velocity is small in the vortex core and even slightly negative between $0.25 \le x/H \le 1.25$.

The structure of the passing inflow jets also dominates other important turbulent quantities. In the strong shear layer between and at the edges of the jets the production of turbulent kinetic energy (TKE) has its peak values. At the impingement zones of the jets at the opposite walls, the production of TKE becomes negative, whereas the dissipation of TKE has its maximum there.

Figure 3 shows the cross-sectional averages of production and dissipation along the centerline of the mixer for $Re = 500$. This figure clearly shows that the turbulence around the impingement zone of the two inflow jets is far from equilibrium.

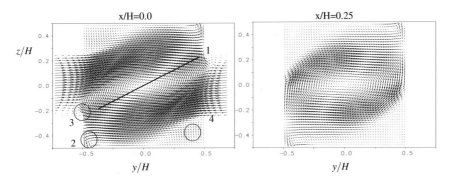

Fig. 2. Vector plots of the mean flow field at $x = 0H$, $x = 0.25H$ at $Re = 500$

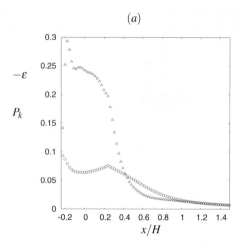

Fig. 3. (*a*): Production P_k (triangle) and dissipation ε (circle) of turbulent kinetic energy averaged over cross sections, normalized with V_o^3/H

The production exceeds the dissipation of TKE for $x/H < 0.5$ and from $x/H > 0.5$ on, it is vice versa. This demonstrates that for the flow inside this configuration the production and dissipation of TKE is locally not in balance, the turbulence is inhomogeneous in space and not in local equilibrium. Turbulence models and also models for passive scalar mixing which are based on equilibrium assumptions are therefore likely to fail. A more detailed analyses of the flow can be found in [22].

2.2 Turbulent Mixing at $Sc = 1$

The break of symmetry, the structures of the jets and the main vortex are also visible in the scalar statistics. As can be seen in the mean scalar field (figure 4) the 'passing' of the jets causes relatively unmixed fluid to enter the opposite feeding pipe.

When moving down the main duct, the rotational character of the flow causes the structure of the mean concentration field to rotate clockwise, while slowly approaching a state of homogeneous mean concentration (figure 4, $x/H = 0.25$, and 1.25 respectively). In this process one can define a large scale variance (LSV) as the deviation of the local mean concentration from the fully mixed state $\xi'^2 = (\langle \Phi \rangle - 0.5)^2$ and a small scale variance (SSV) as the local fluctuation around its local mean $\langle \Phi'^2 \rangle$ [11, 6]. The LSV is reduced via the term $P_\Phi = 2 \langle \Phi' u'_j \rangle \frac{\partial \langle \Phi \rangle}{\partial x_j}$, which is often refered to as macro-mixing, and appears with opposite sign as a production term in the SSV transport equation. The SSV is destroyed via the scalar dissipation rate ε_Φ, which is responsible for micro-mixing. The production of SSV shows negative values in the impingement zone, indicating counter gradient turbulent transport which can not be captured by simple mixing models. To compare the importance of large

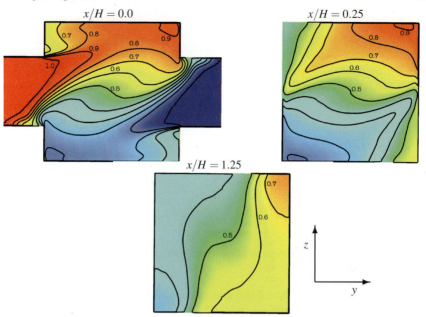

Fig. 4. Isolines of the mean scalar field $\langle\Phi\rangle/\Delta\Phi$, normalized by initial scalar difference

and small scale mixing, two time scales have been introduced [6]. The large scale mixing time is

$$t_{LS} = \frac{\langle\xi'^2\rangle}{-2\langle u'_j\Phi'\rangle\frac{\partial\langle\Phi\rangle}{\partial x_j}} = \frac{\langle\xi'^2\rangle}{P_\Phi},$$ (4)

and the small scale mixing time is

$$t_{SS} = \frac{\langle\Phi'^2\rangle}{2\Gamma\left\langle\left(\frac{\partial\Phi'}{\partial x_j}\right)^2\right\rangle} = \frac{\langle\Phi'^2\rangle}{\varepsilon_\Phi}.$$ (5)

Both time scales are defined by the ratio of the respective quantity to its sink term, e. g. SSV to its dissipation rate. While t_{SS} is well defined everywhere in this flow, t_{LS} would become negative and even infinite where $P_\Phi \to 0$. It is obvious that a definition of large scale mixing time on a local basis is not suitable for this complex flow. However, on an integral scale the LSV must decrease when following the main stream direction. For this reason, we compare the mixing times t_{LS} and t_{SS} calculated from the cross sectional averages of the quantities along the main duct in Figure 5.

Both mixing times have their lowest values in the impingement zones and increase along the course of the main mixing duct. From $x/H > 3$ on, t_{LS} decreases again towards the outflow. Generally, t_{SS} is lower than t_{LS}, and thus the large scale mixing is the dominant process controlling the mixing performance of this

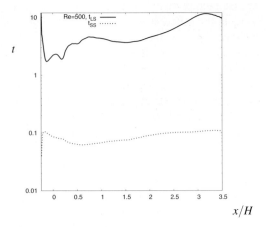

Fig. 5. Mixing time scales calculated from cross sectional averages

configuration. It should be noted that this statement holds only for the Schmidt number of unity considered in this section.

2.3 Precipitation along Lagrange Paths

The DNS described above provides detailed data of all flow and scalar quantities at the specific *Re* and *Sc*. With this information it is possible to avoid some simplifications used by traditional mixing models and develop a first complete numerical description of precipitation processes. In this approach the population balance equation (PBE) describing the dynamics of the precipitation is formulated for a small volume of fluid. This volume of fluid is represented in the DNS simulations by a fluid particle which follows the simple Lagenvin differential equation describing a Lagrangian path:

$$\frac{dX_i^+(t,Y)}{dt} = u_i^+(X^+,t) \tag{6}$$

where $X_i^+(t,Y)$ is the position of the particle which originated from Y at the time t and u_i^+ is the fluid velocity at this position. Following this Lagrange path in the T-mixer, all flow and scalar quantities which are necessary as boundary conditions for the PBE are extracted from the DNS simulations. In this way, quantities required for modelling the PBE are provided at every time instant and locally at the position of the fluid particle, avoiding dubious assumptions about distributions of these quantities inside the T-mixer and allowing for a wide and realistic spread of these quantities. The simulation of the precipitation process is conducted in two steps: first the DNS is performed and a large number of fluid particles are tracked. The fluid particles are initialized equally on the inflow plane of the feeding pipes and once a particle exits the T-mixer it is initialized at its starting position. In this way it is guaranteed that each particle has at least three complete passing through the T-mixer and owing to the different starting positions a wide range of different path histories covering the whole range of possible flow conditions in the T-mixer are generated. In a second step the PBE is solved on each Lagrange path, producing a particle size distribution for this path. Depending on the solution method used for the PBE, this size distribution is ranging from a complete distribution when applying the global model with a finite element Galerkin h-p representation of the PSD to only the mean diameter and mean mass of the PSD when applying the method of moments for solving the PBE. However, by computing statistics over the large number of PSD from each fluid particle path from the DNS, a wide PSD is generated in any case. The results of this approach are published in [20, 9] and yield good agreement with experimental data.

The quantities sampled from the DNS are the local and time dependent dissipation of turbulent kinetic energy $\varepsilon = 2\nu s'_{ij}s'_{ij}$ and the mixture fraction $\Phi^+(X^+,t)$. Since precipitation happens at $Sc = 1000$ the modified Engulfment model [1, 20] is used to describe the micro-mixing between the Kolmogorov and the Batchelor scale. In this model the mixture fraction $\Phi^+(X^+,t)$ is interpreted as the meso-mixed composition of the fluid particle, which can change in time and which is micro-mixed by the action of the dissipative structures of the flow field.

The wide range of different histories of ε, which is shown exemplary in figure 6 for three different paths, leads to a wide range of micro-mixing times and therefore to a wide range of nucleation and growth rates, resulting in broad PSDs [20, 9]. Although this approach already leads to good predictions for PSD of precipitation processes some shortcomings need to be addressed. First, the background concentration interpolated on the particle position is calculated by solving the passive scalar transport equation (3) at $Sc = 1$ and therefore does not resemble the filtered scalar field at $Sc = 1000$ with the DNS grid of the flow field as a filter width. Second, the precipitation process does not have a backlash on the passive scalar field. The two points are addressed in the next chapters.

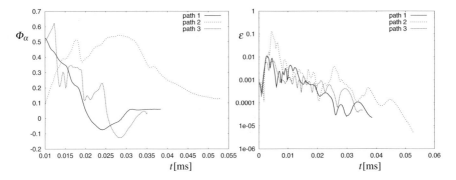

Fig. 6. Mixture fraction and dissipation of TKE (normalized with U_b^4/ν) along Lagrange paths from the same starting positions

3 Semi Direct Numerical Simulation of the Mixing Inside a T-Mixer at High Sc

As mentioned before, precipitation processes in aqueous solutions take place at high Sc. The Schmidt number of barium chloride and sulfuric acid considered in this project and used for experiments in the research group of Prof. Peukert is $Sc = 1000$. In this case, structures in the scalar field can become much smaller than the Kolmogorov length scale η_K. According to Batchelor [2], the smallest scalar length scale is related to the Kolmogorov length by $\eta_B = \eta_K/\sqrt{Sc}$. It is apparent that the Batchelor length scale η_B can not be computed by a DNS of the scalar field as the numerical resolution requirements far outweigh the available computational resources even at low Re. In the preceding chapter the meso- and micro-mixing between the Kolmogorov and the Batchelor scale was accounted for with the modified Engulfment model [1, 20]. However, in the preceding approach the background concentration which determines the composition of the fluid particle was taken from the solution of the passive scalar transport equation which was solved at $Sc = 1$. To account for the high Sc in the passive scalar transport equation a Large Eddy Simulation (LES) model for the scalar transport equation at high Sc in combination with a DNS of the flow field was developed.

3.1 ADM Model for the Turbulent Mass Flux

Filtering the passive scalar transport equation spatially yields for high Sc

$$\frac{\partial \overline{\Phi}}{\partial t} + \frac{\partial \overline{u}_j \overline{\Phi}}{\partial x_j} = \frac{\nu}{Sc}\frac{\partial^2 \overline{\Phi}}{\partial x_j^2} - \frac{\partial}{\partial x_j}M_i \quad . \tag{7}$$

The subgrid scalar mass flux is defined as

$$M_i = \overline{u_i \Phi} - \overline{u}_i \overline{\Phi} \quad . \tag{8}$$

The concept of SEMI DNS exploits the different length scales in the flow and scalar field at high Sc. Using a filter width which is in the same order of magnitude as the Kolmogorov length scale $\Delta \approx \eta_k$ yields the velocity field unchanged $u_i = \overline{u}_i$ and only separates the scalar field into resolved and unresolved scales $\Phi = \overline{\Phi} + \Phi'$. Applying this specific filter operation to the governing equations results in a DNS of the flow combined with a LES of the scalar field. This combination is termed SEMI DNS. In this context only the subgrid scale mass flux has to be modelled. This is done by constructing an approximate $\tilde{\Phi}$ to the unfiltered scalar field by an approximate deconvolution relation and using this approximate for computing the subgrid scale mass flux:

$$M_i = \overline{u_i \tilde{\Phi}} - \overline{u_i} \overline{\Phi} \quad . \tag{9}$$

With this method the filtered scalar fields in turbulent channel flow could be computed up to $Sc = 1000$ and yielded good agreement with DNS results up to $Sc = 10$ and experimental results up to $Sc = 36000$. Further details can be found in [21].

3.2 *Mixing at High Sc Inside a T-Mixer*

A SEMI DNS was used to compute the turbulent mixing inside a T-mixer at $Sc = 1930$ for validation with LIV measurements and at $Sc = 1000$ for computing Lagrangian paths with a filtered background concentration which were used again for determining time dependent boundary conditions for the precipitation simulation.

Shown in figure 7 are the time averaged scalar fields at $x/H = 0.0$ at $Sc = 1930$ from SEMI DNS (left) and at $Sc = 1$ from DNS (right). The concentration distributions are almost identical as the large scale scalar field is determined by the flow field and the scalar boundary conditions, which are the same in both cases. The intensity of the concentration fluctuations increase as the molecular diffusivity decreases with increasing Sc.

Figure 8 (left) shows the cross sectional averages of the concentration fluctuations at two different Sc and results from LIF measurements. The increases in intensity of the *rms* values is clearly visible and the match between the SEMI DNS method and the experimental results is excellent. The increasing fluctuations also effect the concentration along a Lagrangian path, as shown in figure 8 (right). As the filtered concentration field is interpolated on the Lagrange path, the concentration seen by the fluid particle changes rapidly. The Lagrange path resembles a fluid particle which is significantly larger than the Batchelor scale. The data extracted along a large number of Lagrangian paths from the SEMI DNS was again used as boundary conditions for the solution of the PBE.

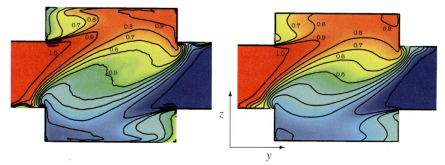

Fig. 7. $\langle \Phi \rangle / \Delta \Phi$; left: SEMI DNS $Sc = 1930$, right: DNS $Sc = 1$; $x/H = 0.0$

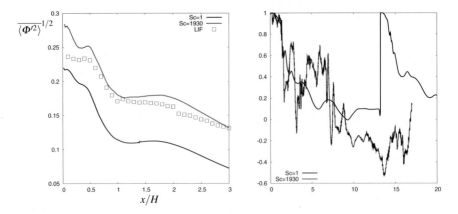

Fig. 8. left: Cross sectional average of concentration fluctuations along main duct of the T-mixer; right: concentration along Lagrange path

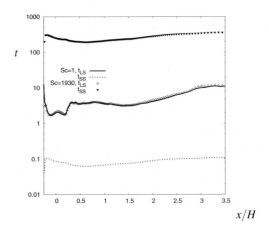

Fig. 9. Mixing time scales calculated from cross sectional averages at high Sc

Figure 9 shows the mixing times calculated from the cross sectional averages at $Sc = 1930$ from the SEMI DNS. As it could be expected, the large scale mixing time does not change with Sc, as it is dominated only by the large scale structures and both, the large scale variance as well as the production of small scale variance P_Φ are identical. The small scale mixing time however is overestimated, due to the fact that the SEMI DNS does not resolve the small scales and therefore the dissipation of scalar variance is under resolved. As the dissipation by the ADM model is not taken into account the calculated t_{SS} is too high.

4 Direct Numerical and Filtered Density Function Simulation of the Flow and Mixing Inside a T-Mixer at High *Sc*

In the previous approach for computing high *Sc* number flows, the scalar transport equation was filtered and the effect of the subgrid scale fluctuations onto the filtered field was modelled with the ADM model. As it could be shown this approach is capable of predicting the large scale scalar field as well as the resolved scalar fluctuations quite accurately. However, when dealing with reactive flows, the smallest structures of the scalar field around the Batchelor scale are most important as chemical reaction takes place at these scales[1]. Additionally the previous approaches for numerical description of the precipitation processes were based on two consecutive simulations with no possible integration of backlash of the precipitation process on the passive scalar field. To alleviate these shortcomings, a new approach based on the filtered density function formulation is presented in this section, the micro-mixing model is developed here and first results on the mixing and precipitation process inside the T-mixer are presented.

4.1 The FDF-DNS Method

In the DNS-FDF approach a spatial filter operation $G(\mathbf{x})$ is applied to the field variables with a filter width in the same order of magnitude as the Kolmogorov scale $\Delta \approx \eta_K$. Therefore the velocity fields are unchanged by the filter operation and for high *Sc* flows only the small scale fluctuations of the scalar field are removed $\Phi = \overline{\Phi} + \Phi'$. Using this filter operation the FDF is defined as:

$$P_L(\Psi;x_i,t) = \int_{-\infty}^{+\infty} \rho \left[\Psi, \Phi(x_i',t) \right] G(x_i' - x_i) dx_i', \tag{10}$$

where ρ is the "fine-grained" density and represents a N dimensional delta function

$$\rho \left[\Psi, \Phi(\mathbf{x},t) \right] = \prod_{\alpha=1}^{N_s} \delta \left[\Psi_\alpha - \Phi_\alpha(x,t) \right]. \tag{11}$$

Φ_α is the underlying scalar array and Ψ_α the corresponding composition variable. In the following discussion we only focus on a single scalar and set $N = 1$. The FDF defined in equation (10) describes the local and instantaneous PDF of the scalar field in the at x_i with which every statistical quantity of the scalar field can be computed. Is $Q(\Phi(x_i,t))$ a function of the scalar field, it's filtered value can be computed by integrating over the composition space

---

[1] In the scalar transport equation this is recognizable in the non-linearity of the chemical source term.

$$\overline{Q(x_i,t)} = \int_{-\infty}^{+\infty} Q(\Psi)P_L(\Psi;x_i,t)d\Psi. \tag{12}$$

An useful quantity is the "conditional filtered value" of the function Q which is defined as

$$\overline{Q(x_i,t)|\Psi} = \frac{\int_{-\infty}^{+\infty} Q(x_i',t)\rho\left[\Psi,\Phi(x_i',t)\right]G(x_i'-x_i)dx_i'}{P_L(\Psi;x_i,t)}. \tag{13}$$

It represents the filtered value of $Q(\Phi(x_i,t))$ under the condition that $\Phi = \Psi$. It therefore resembles the filtered value of all events of $Q(\Phi(x_i,t))$ which happen at Ψ within the filter width.

Using the time derivative of definition (10) the transport equation for the FDF $P_L(\Psi;x_i,t)$ can be derived from the transport equation of the passive scalar $\Phi(x_i,t)$:

$$\frac{\partial P_L(\Psi;x_i,t)}{\partial t} =$$

$$\frac{\partial}{\partial\Psi}\left\{\left[\overline{\left(\frac{\partial u_i\Phi}{\partial x_i}|\Psi\right)} - \overline{\left(\frac{\partial}{\partial x_i}\left(\Gamma\frac{\partial\Phi}{\partial x_i}\right)|\Psi\right)} - \overline{(\omega(\Phi)|\Psi)}\right]P_L(\Psi;x_i,t)\right\} \tag{14}$$

Under the assumption that the velocity distribution is in the worst case linear within the filter width the convective term in the FDF transport equation can, with a small error [21], be simplified to:

$$\frac{\partial}{\partial\Psi}\left\{\left(\overline{\frac{\partial u_i\Phi}{\partial x_i}|\Psi}\right)P_L\right\} = -u_i\frac{\partial P_L}{\partial x_i}. \tag{15}$$

The whole procedure results in the FDF transport equation

$$\frac{\partial P_L}{\partial t} + u_i\frac{\partial P_L}{\partial x_i} = -\frac{\partial}{\partial\Psi}\left\{\left(\overline{\Gamma\frac{\partial^2\Phi}{\partial x_i^2}|\Psi} + \omega(\Psi)\right)P_L\right\}. \tag{16}$$

The left hand side is the material derivative of the FDF and is closed as the flow field is represented by a DNS. The last term on the RHS is the sink or source term due to chemical reaction and is closed in the FDF formulation, which is one reason for the attractiveness of the FDF methods. The first term on the right hand side is the conditional filtered diffusive term and as the one time, one point Eulerian FDF gives no information on the spatial derivative of the scalar this term is not closed and must be modelled. The conditional filtered diffusive term can be reformulated in several ways [3, 19] and we focus here on the splitting into diffusion by the resolved scalar field and conditional diffusion by the sub filter scalar fluctuations

$$\frac{\partial P_L}{\partial t} + u_i\frac{\partial P_L}{\partial x_i} = -\frac{\partial}{\partial\Psi}\left\{\left(\Gamma\frac{\partial^2\overline{\Phi}}{\partial x_i^2} + \overline{\Gamma\frac{\partial^2\Phi'}{\partial x_i^2}|\Psi} + \omega(\Psi)\right)P_L\right\}. \tag{17}$$

The diffusion due to the resolved scalar field can be computed with the filtered scalar field and therefore only the second term on the RHS, namely the so called micro-mixing term is unclosed and must be modelled.

4.2 Formulation of a Model for the Micro Mixing

As the FDF resembles a local and instantaneous PDF it is natural to adapt models developed for the PDF-RANS context for the DNS-FDF method. The most common model is the "linear mean square estimation" (LMSE) model [5, 14, 16, 27]. Although this model is known to have some deficiencies [4, 27] it is known that the performance of the model improves when a more local description of the flow is used, e.g. LES over RANS. As the description of the flow field with a DNS is more local in time and space as in a LES, the LMSE model can be expected to perform well and was therefore a natural choice.

With the LMSE model, the conditional filtered subgrid diffusion is replaced by a linear relaxation of the particle composition to the filtered composition:

$$\frac{\partial}{\partial \Psi} \left[\overline{-\Gamma \frac{\partial^2 \Phi'}{\partial x_i^2} | \Psi} \, P_L \right] = \frac{\partial}{\partial \Psi} \left[\Omega_M \left(\Psi - \overline{\Phi} \right) P_L \right], \tag{18}$$

where Ω_M is the relaxation constant and is called the mixing frequency. When neglecting the diffusion of subgrid scalar variance the mixing frequency is defined as:

$$\Omega_M = \frac{\varepsilon_\Phi}{2 \overline{\Phi'^2}} \quad \text{with} \quad \varepsilon_\Phi = 2\Gamma \overline{\frac{\partial \Phi'}{\partial x_i} \frac{\partial \Phi'}{\partial x_i}}. \tag{19}$$

ε_Φ is the dissipation of subgrid scalar variance and $\overline{\Phi'^2}$ the subgrid scalar variance itself. By analysis of DNS data of mixing up to $Sc = 49$ [24] it could be shown [24, 25] that the LMSE model (18) along with definition (19) is well suited for modelling the conditional subgrid diffusion in the DNS-FDF approach. The dissipation of the subgrid scalar variance is not closed. By analyzing the transport equation of the subgrid scalar dissipation rate following model equation for the subgrid scalar dissipation rate could be developed [24, 25]:

$$\frac{D\varepsilon_\Phi}{Dt} = \frac{C_D}{\ln(Sc)} \left(\frac{\varepsilon}{\nu} \right) \Phi^{*2} + C_s \left(\frac{\varepsilon}{\nu} \right)^{1/2} \varepsilon_\Phi - C_d \frac{\varepsilon_\Phi}{\overline{\Phi'^2}} \varepsilon_\Phi. \tag{20}$$

The material derivative of ε_Φ is modelled by three processes: (i) the production due to flux of scalar energy into the scalar dissipation range, (ii) the production due to vortex stretching by the smallest turbulent eddies, and (iii) the destruction of subgrid scalar dissipation by molecular diffusion, which are represented by the first, second and last term on the RHS respectively. The model constants are $C_D = 0.02, C_s = 1.0$ and $C_d = 3.0$, Φ^{*2} is the scalar variance present at the Kolmogorov length scale and ε the dissipation of turbulent kinetic energy. It could be shown that this model equation gives a good prediction of the scalar dissipation rate in turbulent channel flow [24, 25].

4.3 Small Scale Mixing at High *Sc* Inside a T-Mixer

A convenient way of solving the FDF transport equation (17) along with the model for the conditional subgrid scalar diffusion (18) is the Monte-Carlo method. The Monte-Carlo method uses the statistical equivalence of the solution of the FDF transport equation with the solution of Ito stochastic differential equations (Ito-SDE) when the coefficients are chosen accordingly [16, 7]. As there are no mixed derivatives in the FDF transport equation (16) there are two separate stochastic processes, one for the position in physical space x_i and one for the position in compositional space Ψ, resulting in two separate Ito-SDEs. When the LMSE model is used to describe the conditional diffusion by the sub filter fluctuations the Ito-SDEs for the stochastic particles are

$$dX_i^+(t) = u_i(X_i^+,t)dt \tag{21}$$

$$d\Psi^+(t) = \left(\Gamma \frac{\partial^2 \overline{\Phi}}{\partial x_i^2} + \omega(\Psi^+) \right) dt - \Omega_M \left(\Psi - \overline{\Phi} \right) dt, \tag{22}$$

where X_i^+ is the position of the particle in physical space and Ψ^+ it's position in composition space. The local fluid velocity acts as a drift term in physical space where the diffusion by the filtered scalar field as well as the sink term act as drift terms in compositional space. Together with the model equation for the subgrid scalar dissipation rate (20) this system of equations is closed and can be solved for a large number of stochastic particles. The stochastic particles follow the transport equation (21) in physical space and the particle composition follows equation (22). The transport equation for the scalar dissipation rate (20) is solved on each particle where large scale quantities like dissipation of TKE, filtered concentration field and fluid velocity are interpolated from the DNS grid of the flow computation on the particle position. This system of equations was solved for approximately 25 particles per grid cell (smallest cell) which resulted in approximately 365×10^6 particles in the main duct of the T-mixer. The particles where initialized in a stochastic manner [8] in the inflow plane of the feeding pipes guaranteeing a constant particle density.

Analogue to the mixing times defined in chapter 2 the the mixing frequency (19) resembles the inverse time scale with which the subgrid scalar variance is reduced $t_M = 1/(2\Omega_M)$. In figure 10 this time scale, calculated from the cross sectional averages drawn from the properties of the stochastic particles, is shown along with the time scales computed from the DNS at $Sc = 1$. As can be seen with increasing Sc the micro-mixing time scale increases dramatically, although it stays well lower than the micro-mixing time scale as predicted by the SEMI DNS method. In the impingement zone, where the highest turbulence intensities are, t_M is in the same order of magnitude as t_{LS}. Towards the exit of the mixer, as the flow relaminarizes, the micro-mixing time increases.

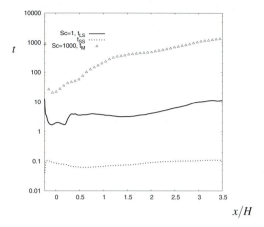

Fig. 10. Comparison of mixing time scales calculated from cross sectional averages at $Sc = 1$ (DNS) and $Sc = 1000$ (DNS-FDF)

4.4 Precipitation Simulation at High *Sc* Inside a T-Mixer

In the transport equation of the stochastic particle the chemical reaction term is closed and therefore no further modelling is required. For a numerical prediction of the PSD, the PBE was solved with the method of moments (MM) [20] directly on each stochastic particle.

Figure 11 shows the PSD of precipitation of $0.5m$ $BaCl_2$ and $0.33m$ H_2SO_4 inside a T-mixer at $Re = 1100$ from experimental measurements [20] and the nu-

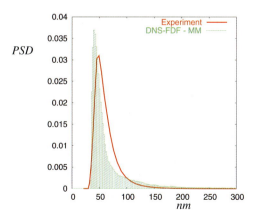

Fig. 11. PSD of precipitation process $0.5m$ $BaCl_2 + 0.33m$ H_2SO_4 at $Re = 1100$ inside a T-mixer

merical result for the PSD from the combined DNS-FDF approach with the method of moments. The numerically generated PSD was taken from the stochastic particles which left the T-mixer at the outflow at one time instant. Due to the strong decay of the turbulence towards the exit of the mixer the process of nucleation was regarded as finished and the remaining oversaturation was assumed to decay by pure growth of the nuclei. Without further adjustment of the parameters the match between simulation and experiment is very good. Even with the MM, a wide spread of the PSD is achieved, which points out once more the importance of the different mixing histories of the stochastic particles.

5 Conclusions

In the course of this work a combined approach for the numerical simulation of turbulent mixing with chemical reaction, in our specific case a precipitation process in a T-mixer, at high Sc was developed. To achieve the most accurate method possible with todays computer resources a DNS-FDF method was developed. The DNS-FDF method is based on a DNS of the flow field. For this reason DNS of the flow (and mixing at $Sc = 1$) in a T-mixer configuration was performed and validated with experimental results. It could be shown that the flow field can be accurately reproduced by DNS. The detailed information of local and time dependent turbulence statistics from the DNS results were used to develop a method of predicting the PSD along Lagrangian paths. To account for the high Sc a SEMI DNS method was developed and applied to the flow and mixing in the T-mixer. In the SEMI DNS method the flow field is fully resolved and the scalar field is filtered on the Kolmogorov scale. With this method it could be shown that the large scale mixing is independent of Sc and that the filtered scalar fields as well as the scalar variance accurately match experimental LIF measurements. For integrating the PBE in the flow solver and allow for backlash of the precipitation process on the passive scalar field a DNS-FDF method was developed. With the help of DNS of turbulent mixing in a channel flow up to $Sc = 49$ it could be shown that the LMSE model is suited for the DNS-FDF approach and that the traditional formulation of the mixing frequency is adequate. A model for the subgrid dissipation of scalar variance could be developed and showed a good agreement with data from DNS of the turbulent channel flow. The DNS-FDF approach was applied to the flow and mixing in a T-mixer at $Sc = 1000$ and was coupled with a PBE on each stochastic particle. First results from these simulations for the micro-mixing and the PSD were shown and are promising. As micro-mixing at very high Sc itself is hard to study directly by DNS and as the PSD of the precipitation process is very sensitive to the turbulent mixing process it can be used as an indicator for the quality of the micro-mixing model. Therefore the Reynolds number effects on the mixing time and PSD have to be carefully studied and compared with experimental results. Additionally future work includes the expansion of the DNS-FDF approach to a LES-FDF approach by including a stochastic term accounting for the unresolved convective term in the FDF transport equation and adding an intermediate stage in the model for the subgrid scalar dissipation rate to model the flux of spectral scalar energy in the unresolved scales of the inertial-convective subrange.

References

1. Baldyga, J., Bourne, J.R.: Turbulent Mixing and Chemical Reactions. Wiley, Chichester (1999)
2. Batchelor, G.K.: Small-scale variation of convected quantities like temperature in turbulent fluid. Part 1. general discussion and the case of small conductivity. J. Fluid Mech. 5, 113–133 (1959)
3. Colucci, P.J., Jaberi, F.A., Givi, P., Pope, S.B.: Filtered density function for large edddy simulation of turbulent reacting flows. Phys. Fluids 10(2), 499–515 (1998)
4. Dopazo, C., O'Brien, E.E.: Isochoric turbulent mixing of two rapidly reacting chemical species with chemical heat release. Phys. Fluids 16, 2057 (1973)
5. Dopazo, C., O'Brien, E.E.: Statistical treatment of non-isothermal chemical reactions in turbulence. Combust. Sci. Technol. 13, 99 (1976)
6. Fox, R.O.: Computational Models for Turbulent Reacting Flows. Cambridge University Press, Cambridge (2003)
7. Gardiner, C.W.: Handbook of Stochastic Methods for Physics, Chemistry and Natural Sciences. Springer, Heidelberg (1990)
8. Gobert, C.: Lagrangesche Monte Carlo Simulation zur Vermischung bei hohen Schmidt Zahlen, TU-München, Diplomarbeit (2006)
9. Gradl, J., Schwarzer, H.C., Schwertfirm, F., Manhart, M., Peukert, W.: Precipitation of nanoparticles in a T-mixer: Coupling the particle population dynamics with hydrodynamics through direct numerical simulation. Chemical Eng. and Proc. 45, 908–916 (2006)
10. Johansson, P.S., Andersson, H.I.: Direct numerical simulation of two opposing wall jets. Phys. Fluids 17 (2005)
11. Liu, Y., Fox, R.O.: CFD predictions for chemical processing in a confined impinging-jets reactor. AIChE Journal 52, 731–744 (2006)
12. Manhart, M.: A zonal grid algorithm for DNS of turbulent boundary layers. Computers and Fluids 33(3), 435–461 (2004)
13. Mitarai, S., Riley, J.J., Kosaly, G.: Testing of mixing models for Monte Carlo probability density function simulations. Phys. Fluids 17(4) (2005)
14. O'Brian, E.E.: The Probability Density Function (pdf) Approach to Reacting Turbulent Flows, pp. 185–218. Springer, Heidelberg (1980)
15. Peller, N., Le Duc, A., Tremblay, F., Manhart, M.: High-order stable interpolations for immersed boundary methods. Int. J. for Numerical Methods in Fluids 52, 1179–1193 (2006)
16. Pope, S.B.: PDF methods for turbulent reactive flows. Prog. Energy Combust. Sci. 11, 119 (1985)
17. Pope, S.B.: Lagrangian PDF Methods for Turbulent Flows. Annu. Rev. Fluid Mech. 26, 23–63 (1994)
18. Pope, S.B.: Turbulent Flows. Cambridge University Press, Cambridge (2000)

19. Raman, V., Pitsch, H., Fox, R.O.: Hybrid large-eddy simulation/Lagrangian filtered-density-function approach for simulating turbulent combustion. Combustion and Flame 143, 56–78 (2005)

20. Schwarzer, H.C., Schwertfirm, F., Manhart, M., Schmid, H.J., Peukert, W.: Predictive Simulation of Nanoparticle Precipitation based on the Population Balance Equation. Chem. Eng. Sci. 61, 167–181 (2006)

21. Schwertfirm, F., Manhart, M.: ADM Modelling for Semi-Direct Numerical Simulation of Turbulent Mixing and Mass Transport. In: Humphrey, J., Gatski, T., Eaton, J., Friedrich, R., Kasagi, N., Leschziner, M. (eds.) Turbulence and Shear Flow Phenomena, Williamsburg, USA, pp. 823–828 (2005)

22. Schwertfirm, F., Gradl, J., Schwarzer, H.C., Manhart, M., Peukert, W.: The low Reynolds number turbulent flow and mixing in a Confined Impinging Jet Reactor. Int. J. Heat and Fluid Flow 28, 1429–1442 (2007)

23. Schwertfirm, F., Manhart, M.: DNS of passive scalar transport in turbulent channel flow at high Schmidt numbers. Int. J. of Heat and Fluid Flow (2007), doi:10.1016/j.ijheatfluidflow.2007.05.012

24. Schwertfirm, F.: Direkte Simulation und Modellierung des Mikromischens bei hohen Schmidt Zahlen, TU-München, Dissertation (2008)

25. Schwertfirm, F., Manhart, M.: Developement of a DNS-FDF approach to non homogeneous non-equilibrium mixing for high Schmidt number flows, Direct and Large Eddy Simulation 7, Trieste, Italy (September 2008)

26. Telib, H., Iollo, A., Manhart, M.: Analysis and low order modeling of the inhomogeneous transitional flow inside a T-mixer. Phys. Fluids 16, 2717–2731 (2004)

27. Villermaux, J.: Micromixing phenomena in stirred reactors. Encyclopedia of Fluid Mechanics (1986)

Theoretical and Experimental Investigations of Convective Micromixers and Microreactors for Chemical Reactions

Simon Dreher[1], Michael Engler[2], Norbert Kockmann[3], and Peter Woias[1]

[1] Laboratory for Design of Microsystems, Department of Microsystems Engineering (IMTEK), University of Freiburg, Germany
[2] Sick AG, Waldkirch, Germany
[3] Lonza AG, 3930 Visp, Switzerland

Abstract. Convective micromixers allow fast mixing with characteristic times below 1 ms by creating vortices in bends or junctions of microchannels. Their robustness and comparatively high throughput make them suitable for process intensification purposes. In this contribution, T-shaped micromixers with rectangular cross section and meandering structures as basic elements are numerically investigated for Reynolds numbers from 0.01 to 1000 in the mixing channel. Static and transient mixing regimes in T-mixers are described up to the transition to turbulence, where a worsening of mixing is found. Different experimental results confirm the simulations and lead to the convective lamination model. This model links the energy dissipation to mixing similar to a modeling of turbulent mixing. Additional to mixing of two fluids, also wall contact of fluids in channel flow is investigated with a catalyzed chemiluminescence reaction. With this method, dead zones can be made visible.

1 Introduction

Precise and fast mixing is a key concept in process intensification. A common method to accelerate mixing is the miniaturization of mixing devices. In interdigital mixers and split-and-recombine mixers, fluid lamellae are formed and stretched by geometric focusing of the flow or in repeated flow guiding elements. The finer these lamellar structures get, the faster is diffusive mixing. Consequently, those mixing devices need complex channel geometries and narrow channels to achieve short mixing times. Convective micromixers in contrast use vortices induced in curved or bent flow for the creation of lamellar structures. Comparatively wide channels in the range of several 100 μm are used, making these devices robuster against clogging and fouling and keeping the pressure loss in an acceptable range. Nevertheless, such microchannels are small enough to provide well-defined mixing conditions and extremely fast mixing with characteristic times below 1 ms.

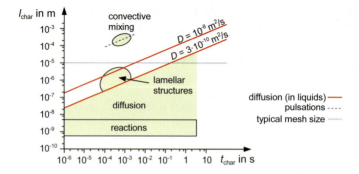

Fig. 1. Overview over time and length scales in mixing processes. For convective mixing, typical device sizes and mixing times are taken as characteristic length and time scales; for lamellar structures, lamellae thicknesses and corresponding time scales of diffusive mixing are marked.

In this contribution, mixing effects in the planar T-mixer and in meandering channels are investigated. These structures are widely used as very simple but effective mixing elements. Investigation of these mixers is difficult as processes involved cover several orders of magnitude in time and space, as displayed in Fig. 1. Length scales of the order of magnitude of 1 mm for the device geometry, 10 μm for fluid mechanics, less than 1 μm for lamellar sizes and less than 1 nm for reaction mechanisms cannot be simulated together, but require modeling of the single processes.

2 Characteristic Mixing Properties

To characterize the performance of convective micromixers, the mixing quality α and the mixing potential Φ are used. Based on these values, a method for describing the homogeneity of the mixing state is proposed.

2.1 The Mixing Quality α

The mixing quality α describes the mixing state in a cross section of the mixing channel based on the intensity of segregation introduced by Danckwerts (1952). It is formed by the variance σ_V of the normalized concentration c of one of the fluids normalized to the maximal variance $\sigma_{V,\max}$ according to

$$\alpha = 1 - \sqrt{\frac{\sigma_V^2}{\sigma_{V,\max}^2}} . \tag{1}$$

The totally mixed state means $\alpha = 1$, in the totally segregated state is $\alpha = 0$.

In contrast to the definitions given by Bothe et al. (2006) and Engler (2006), we calculate the variance of the concentration as a function of the volume flow $\dot{V}$ instead of a function of the cross sectional area, i. e.

$$\sigma_{\dot{V}}^2 = \frac{1}{\dot{V}_{\text{tot}}} \int_{\dot{V}_{\text{tot}}} (c - \bar{c})^2 \, d\dot{V} \,, \qquad \sigma_{\dot{V},\text{max}}^2 = \frac{\dot{V}_1 \dot{V}_2}{(\dot{V}_1 + \dot{V}_2)^2} c_{\text{max}}^2 \,. \tag{2}$$

Here, $\bar{c}$ is the normalized concentration of the fully mixed state, c_{max} the normalized concentration of the pure component for which the concentrations are given, and $\dot{V}_1$, $\dot{V}_2$ the volume flows of both components. This definition accounts for the different contribution of regions with low and high flow velocities to the outflow of the mixer.

2.2 The Mixing Potential Φ

The mixing quality α only describes the state of diffusive mixing on the micro scale. It does not distinguish between segregated states with different distribution of the fluid structures, like large unmixed fluid volumes and fine lamellar structures. This convective distribution plays a significant role, specially for convective mixing of fluids with high Schmidt numbers Sc (occuring e. g. in liquids), which means that diffusive mass transport is slow compared to convective transport. For mainly convective mixing, the concentration distribution can be characterized by the ratio of the interfacial length between the fluids to the area of the mixing channel cross section. The mean concentration gradient, also called mixing potential, is a generalization of this interfacial length for partially diffusively mixed fluids. It takes into account the width of already diffusively mixed zones, which actually separate the unmixed zones. It can be interpreted as the potential for further diffusive mixing. Bothe et al. (2006) proposed the definition

$$\Phi = \frac{1}{A} \int_A \sqrt{(\nabla c)^2} \, dA \tag{3}$$

as the average of the absolute value of the concentration gradient ∇c over the cross sectional area A. Chao et al. (2005) use a quadratic mean value of the absolute value of the concentration gradient.

2.3 Stoichiometry of Mixing

Both the mixing quality and the mixing potential are integral values over the whole cross section of the mixing channel. At a partially mixed state, it is also necessary to identify if there are regions which are mixed very early and other regions with unmixed or non-stoichiometrically mixed fluids and how they participate in the mixing process.

For this consideration, a local mixing quality α_{local} is defined according to the integral mixing quality as

$$\alpha_{local} = 1 - \frac{|c - \bar{c}|}{\sqrt{\bar{c}(c_{max} - \bar{c})}} \; . \tag{4}$$

The local variant of the mixing potential Φ_{local} is defined as the absolute value of the concentration gradient

$$\Phi_{local} = |\nabla c| \; . \tag{5}$$

By plotting the local mixing potential Φ_{local} over the local mixing quality α_{local}, different ranges can be identified, see Fig. 2. Unmixed regions, which do not take part in the mixing process, are mapped to the lower left corner of the diagram. This is the starting configuration of the mixing process. The final configuration of the fully mixed, stable state is found at the lower right corner, i. e. at $\alpha_{local} = 1$ and $\Phi_{local} = 0$. For convective mixing, stretching and elongation of the interface leads to rising Φ_{local}. Simultaneously, diffusive mixing blurs the interfaces shifting the points to higher values of α_{local} in the diagram. In an effective mixing process, further diffusive mixing leads directly to the final state of $\alpha_{local} = 1$ and $\Phi_{local} = 0$. A mixing process towards stable, non-stoichiometric regions, like in single vortices, leads to long mixing times. In those regions, competitive chemical reactions can be favored leading to decreased yields in chemical production processes. A well-known example of such a competitive reaction system often used for mixing characterization is the iodide-iodate reaction as described in Sect. 5.3. As a special mixing process, back mixing of one pure component with already mixed fluid is reflected in a motion towards the non-stoichiometric stable region, see Fig. 2.

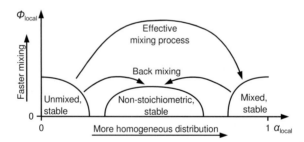

Fig. 2. Diagram of local mixing characteristics. During an effective mixing process, the position in the diagram for each point on the mixer cross section moves from the unmixed stable state over high values of Φ_{local} to the mixed stable state. The non-stoichiometric stable state should be avoided.

3 Flow and Mixing Regimes in T-Mixers

One of the basic elements of micromixers is the T-mixer. It is often used as a first contacting element providing fast mixing by the formation of vortices. As the channel dimensions are typically in the range of some $100\,\mu m$, the pressure loss is comparatively low and the mixers are tolerant against clogging and fouling, see Heim et al. (2006), Kockmann et al. (2007).

In this section, the flow and mixing behavior of a typical T-mixer with rectangular cross section is described in the Reynolds number range of $0.01 < Re < 1000$, i. e. from the strictly laminar regime to the transition to turbulence. In the following, the Reynolds number refers to the mixing channel. The geometrical setup of the investigated mixer is shown in Fig. 3, left. It is called $T\,600\times300\times300$, standing for a T-mixer with $600\,\mu m$ mixing channel width, $300\,\mu m$ inlet channel width, and $300\,\mu m$ channel depth. A mixer fabricated in silicon with a glass lid is shown in the right subfigure.

Fluid flow and mass transport are simulated in this benchmark geometry with CFD-ACE+ by the ESI group, a commercial computational fluid dynamics code based on the finite volume method. Flow regimes are investigated depending on the Reynolds number in the mixing channel. In this study, two fluids with properties equivalent to water at 20°C are mixed at a ratio of 1:1. The diffusion coefficient is set to $D = 2.8 \cdot 10^{-10}\,m^2/s$, which is typical for a large dye molecule like Rhodamine B in water, resulting in a Schmidt number of $Sc = \nu/D \approx 3600$. Other mixing ratios and the mixing of gases at different temperatures are examined in Dreher (2005), Kockmann et al. (2007). A grid size of 30×60 cells in the mixing cross section is chosen for most simulations, single simulations are also simulated with 60×120 cells. Grid refinement studies have shown that the fluid mechanics is resolved sufficiently with this choice of the grid size, whereas fine lamellar structures in the concentration

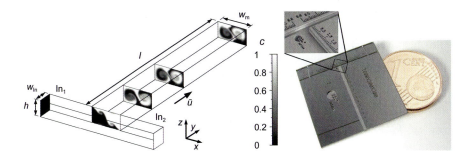

Fig. 3. Geometrical setup of the investigated T-mixer. *Left:* standard geometry of the $T\,600\times300\times300$ with $w_m = 600\,\mu m$, $w_{In} = 300\,\mu m$ and $h = 300\,\mu m$ and $l = 4000\,\mu m$. $\bar{u}$ denotes the mean velocity. *Right:* Silicon mixer chip with pyrex glass lid used for experiments. Near the inlets and the outlet, additional channels for pressure measurements are implemented.

field cannot be resolved resulting in high numerical diffusion, see also Gobert et al. (2006). The simulated concentration field thus highly overestimates mixing and can only be interpreted qualitatively. Nevertheless, large unmixed zones can be identified. Additionally, numerical results are verified by experiments described in Sect. 5 showing the same mixing behavior. Transient simulations of the instationary flow regimes are performed with timesteps of $10\,\mu s$ and $5\,\mu s$, where every 10th or 5th timestep is evaluated.

3.1 Static Flow Regimes

For low Reynolds numbers up to $Re = 240$, stationary flow patterns are observed. For $Re < 3$, the flow is strictly laminar, also referred to as creeping flow. The streamlines follow the $90°$ bend at the T-junction, no deflection in the z-direction can be seen, as shown in Fig. 4, left. Only diffusive mixing occurs, which is determined

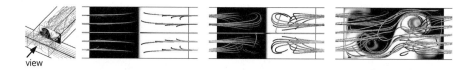

view

Fig. 4. Concentration profiles in a cross section and stream lines for the stationary regimes. *Left:* strictly laminar ($Re = 1$), *middle:* vortex regime ($Re = 100$), *right:* engulfment regime ($Re = 200$)

by the width of the mixing channel and the residence time in the mixing device. A simple estimation of the mixing time t_{mix} and the necessary mixing length l_{mix} is possible with the characteristic diffusion length $l_{diff} = \sqrt{2Dt}$ depending on the diffusion coefficient D and the time t. Choosing $l_{mix} \approx \bar{u}t = \frac{Re\,\nu}{d_h}t$ and $l_{diff} = 1/2\,d_h$ with the mean velocity $\bar{u}$, the kinematic viscosity ν and the hydraulic diameter d_h as a characteristic length, the mixing time and length get approximated by

$$t_{mix} \approx \frac{d_h^2}{8D}, \qquad\qquad l_{mix} \approx Re\frac{d_h\nu}{8D}. \qquad (6)$$

At $3 < Re < 140$, streamlines from the middle of the channel, where the highest velocities occur, are deflected towards the middle of the mixing channel by centrifugal forces and displace the slower fluid near the wall. This regime is called *vortex regime*. In both halves of the mixing channel, a counter-rotating vortex pair evolves, see Fig. 4, middle. In curved channels, this vortex pair is known as Dean vortex. The occurrence of these vortices is characterized by the Dean number

$$De = Re\sqrt{\frac{d_h}{R}} \qquad (7)$$

with the tube diameter d_h and the radius of curvature R. This number can be derived theoretically from the non-dimensionalized form of the Dean equations describing the secondary flow in slightly curved channels, see Berger et al. (1983). In the T-junction of the investigated mixer, the mean radius of curvature is in the order of magnitude of the diameter, and De is in the same order of magnitude as Re. The critical Dean number found for the transition to the vortex regime is $De_{crit} \approx 5$, which matches values found in literature, see e. g. Howell Jr et al. (2004) with $1.5 < De_{crit} < 5$ or Kockmann et al. (2006) with $De_{crit} \approx 10$. As the vortices are symmetrical to the middle plane of the mixing channel, the interface between the two fluids is not stretched and mixing is still dominated by diffusive mixing like in the straight laminar regime. In the theory of chaotic advection, the symmetry plane represents a separatrix between the Dean vortices inhibiting mass exchange and thus convective mixing.

At higher Reynolds numbers, the centrifugal force acting on the core flow dominates the motion of the fluid over viscous forces. For $Re > 140$, the symmetry to the middle plane of the mixing channel is broken and a diagonal shear flow forms in the cross section of the mixing channel forms. It is point symmetric to the mid point of the mixing channel cross sections. On the shear layer, two co-rotating vortices roll up forming lamellar fluid structures as shown in Figs. 4, right, and 5. The mechanism of this roll-up of the vortices is the Kelvin-Helmholtz instability. This regime is called *engulfment regime* reflecting the similarity to the structures found in turbulent mixing (Bałdyga and Bourne, 1999, Cybulski et al. , 2001).

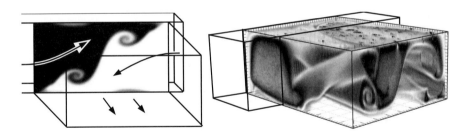

Fig. 5. Formation of the double vortex in the engulfment regime. *Left:* Formation of the shear layer and Kelvin-Helmholtz vortices, *right:* μLIF measurement of mixing in the T $400 \times 200 \times 200$ at $Re = 200$ showing the fine lamellar structure in the engulfment regime. By courtesy of M. Hoffmann, University of Bremen.

By the "turn-over" of one fluid to the opposite side of the mixing channel, the radius of curvature at the junction to the mixing channel is enlarged, lowering the friction loss due to the bend. Viscous friction forces counteract to the centrifugal force suppressing the elongation of the stream lines by the "turn-over". To reflect this behavior, the ratio of the centrifugal forces to the viscous forces is calculated with

$$\frac{F_c}{F_v} = \frac{2\dot{m}_{in}\bar{u}^2\Delta t/R}{\zeta_m\Delta t\eta(w_m+h)^2\bar{u}^2/8w_mh} = \frac{16}{\zeta_m}Re\frac{h/w_m}{1+h/w_m} \qquad (8)$$

where ζ_m is the Darcy friction factor and h/w_m the aspect ratio of the mixing channel. The radius for the centrifugal force is chosen to $R = \frac{w_m}{4}$. This expression can also be extended to describe the mixing of fluids with different properties and mixing ratios slightly differing from 1:1. The critical value of the ratio of centrifugal to viscous forces for the transition from vortex to engulfment regime found in the simulations as well as in mixing experiments is $F_c/F_v|_{crit} \approx 12$ (Engler, 2006) corresponding to $Re = 140$ at an aspect ratio of $h/w_m = 0.5$.

The roll-up of fluid lamellae greatly enhances mixing by stretching the interface and thinning the single fluid lamellae, see Fig. 5. Lamellae thicknesses of down to 3 µm were measured by M. Hoffmann with µLIF technique, see the contribution of our cooperation partners from IUV Bremen. Figure 6 shows the simulated mixing quality α at a mixing channel cross section 2 mm downstream from the T-junction plotted over the Reynolds number. While for the strictly laminar and vortex regimes (I and II) the mixing quality decreases with increasing Reynolds number due to the shorter residence time, at the engulfment regime (III) a steep increase of the mixing quality is observed. This enhancement of the mixing quality is the effect of the roll-up of fluid lamellae in the double vortex. With increasing Reynolds number, the vortices are intensified, which leads to a further rise of the mixing quality. For this regime, numerical diffusion blurs the simulated concentration field as the fine lamellar structures cannot be resolved by the grid of the simulation model. The simulated mixing quality is highly overestimated in this case. Nevertheless, the simulated trend of the mixing quality is also found experimentally by dye experiments as described in Sect. 5.2. The regimes IV–VI are described in detail in the following sections.

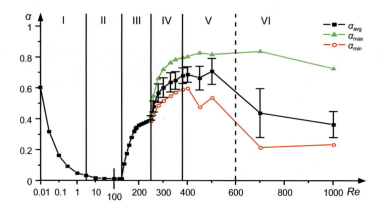

Fig. 6. Simulated mixing quality α at 2 mm downstream position in the mixing channel over the Reynolds number for the T $600\times300\times300$. The mixing regimes are marked with roman numerals. For the transient regimes, the averaged, maximal, and minimal mixing qualities α_{avg}, α_{max}, and α_{min} are given. The error bars for α_{avg} represent the standard deviation over time. For $Re > 140$, the values of the mixing quality are highly overestimated due to numerical diffusion, but follow the same trend as found in experimental measurements. Note the change from logarithmic to linear scaling for Re.

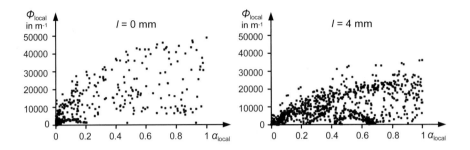

Fig. 7. Simulated local mixing characteristics in the T$600\times300\times300$ at $Re = 200$. Each point represents a grid point on the cross section. *Left:* values at the junction ($l = 0$mm), *right:* values at the end of the mixing channel ($l = 4$mm).

To evaluate the homogeneity and stoichiometry of the mixing process, plots of the local mixing potential Φ_{local} over the local mixing quality α_{local} are shown in Fig. 7 for the T$600\times300\times300$ at $Re = 200$. Each data point in the plot represents a grid point on the evaluated cross section of the mixing channel. The left subfigure shows the plot for the entrance into the mixing channel ($l = 0$mm) with mostly the unmixed stable state and some points on the interface with high Φ_{local}. In the right plot for the outlet ($l = 4$mm), points accumulate at $\alpha_{local} \approx 0.65$ for low values of Φ_{local}. These points could be identified as locations within the vortices. This means that in the vortices the actual mixing ratio is about 2:1 and 1:2 respectively instead of the desired ratio of 1:1. Another accumulation is found near the origin representing the middle of lamellae or less mixed zones in the corners, which can be seen in concentration plots of the cross section. These two zones are connected with an arced band of points showing the regions actually taking part at the mixing process. Numerical diffusion diminishes Φ_{local} acting as an additional diffusive term, but has less influence on large-scale inhomogeneities, like the non-stoichiometric mixing within the single vortices. Thus, the diagrams can still be evaluated qualitatively.

3.2 Periodic Vortex Shedding

With further increasing of the Reynolds number, the two vortices begin to fluctuate at $Re > 240$. Seen from a fixed cross section, the vortices move to the middle of the mixing channel and collapse to one single vortex. Immediately after this collapse, two new vortices form at the old positions of the vortices and the middle vortex vanishes, see Fig. 8. This fluid motion is repeated periodically. The point symmetry of the vortices to the center of the mixing channel cross sections is preserved at any time. Seen from the top of the mixing channel, the zone with the single vortex moves downstream with the fluid velocity. The frequency of the vortex pulsation

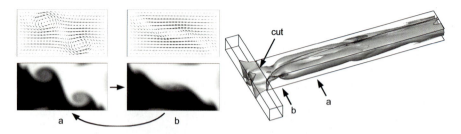

Fig. 8. Mechanism of the periodic vortex shedding at $Re = 300$. *Left:* in-plane velocity and concentration field at the beginning of the mixing channel for different points of time. The double vortex (a) conflates to one big vortex in the middle of the channel (b) and re-develops. *Right:* interface between the fluids at a fixed time. The location of the cross-sectional cuts shown in the left subfigure is marked. Wide regions of the double vortex (a) are interrupted by vortex breakdowns (b).

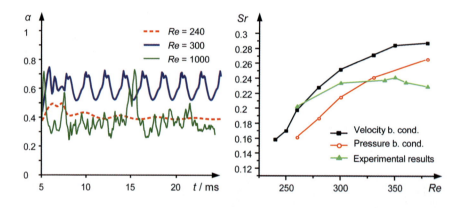

Fig. 9. *Left:* simulated mixing quality after 2 mm mixing channel over the time for different Reynolds numbers. For $Re = 240$, a disturbance at $t = 0$ leads to a damped oscillation leading to the static solution. At $Re = 300$, an oscillation with $f = 474$ Hz is found. For $Re = 1000$, the graph is non-periodic with low α disrupted by high peaks. *Right:* Strouhal numbers as dimensionless frequencies for simulations with different boundary conditions and measurements by stroboscopic imaging.

is determined by an evaluation of the mixing quality over time at a position 2 mm downstream in the mixing channel. In Fig. 9, left, the mixing quality is displayed over time for different Reynolds numbers. For $Re = 240$ in the T $600 \times 300 \times 300$, where a static solution still exists, a perturbation from the initial condition at $t = 0$ leads to a damped oscillation of the mixing quality with the frequency $f = 238$ Hz. It is damped to the static solution. For $Re = 300$, undamped periodic fluctuations are found with $f = 464$ Hz. These fluctuation frequencies can also be expressed in a dimensionless form by the Strouhal number Sr. It is defined as

$$Sr = \frac{f d_h}{\bar{u}} \qquad (9)$$

with the hydraulic diameter of the mixing channel d_h as characteristic length scale and the mean fluid velocity $\bar{u}$. In this dimensionless form, the effects of scaling the mixer geometries on the vortex shedding frequency can be predicted. In Fig. 9, right, the Strouhal number is plotted against the Reynolds number. For simulations with the velocity profile of fully developed channel flow as inlet boundary condition, the Strouhal number is in the range of 0.16 to 0.29. Simulations with constant pressure inlet condition show lower values. This is reasonable as the velocity boundary condition gives a stiffer system for the vortex formation. Frequencies obtained by experiments with stroboscopic imaging are in the same range as the simulated values, see Fig. 11. However, for higher Reynolds numbers ($Re > 300$), the Strouhal number is found to be nearly constant at $Sr = 0.23$ whereas the simulations predict a slight increase of the Strouhal number with increasing Reynolds number. The constant Strouhal number resembles the behavior of wake flow behind a cylinder or a sphere, where $Sr \approx 0.2$ is found over a wide range of the Reynolds number.

The fluctuations result in higher minimal and maximal values of the mixing quality than those found in the static engulfment regime, which means that the periodic vortex shedding enhances mixing. Looking at a vortex breakdown at the beginning of the mixing channel (see Fig. 8), this is remarkable: Next to the collapsed vortex, two zones of nearly unmixed fluids can be found. These zones appear near the wall of the mixing channel and opposite to the inlet of the corresponding fluid. In Fig. 11, a stroboscopic image of the mixing of pure water with a solution of fluoresceine shows this zone. The three-dimensional structure of these zones is depicted in Fig. 8, right. When these zones flow downstream, they get stretched very fast by the flow profile and mix into the double vortex. This leads to even better mixing than in the regions of undisturbed double vortices, explaining the enhancement of mixing. In fact, the position of the maxima of the mixing quality in Fig. 9, left, correspond to the time steps where the collapsed vortex passes the cross section where the mixing quality is evaluated. The mechanism of this mixing of alternating plugs is known as Taylor dispersion.

The mixing of the regions next to the vortex breakdown into the double vortex also enhances the stoichiometry of the mixing process. These zones consist of the fluid coming from the opposite inlet channel and therefore reduce the non-stoichiometry when mixed into the double vortices. For $Re = 300$, at every simulated time step the values at the grid points are plotted, showing an accumulation of points for low Φ_{local} at $\alpha_{local} \approx 0.8$, which corresponds to a mixing ratio of 3:2 and 2:3 respectively for large zones. The mixing ratios within the vortices at $Re = 200$ are 2:1 and 1:2.

3.3 Chaotic Regime with Transition to Turbulence

When the Reynolds number exceeds $Re = 380$, the vortex shedding gets irregular. Sometimes two vortex breakdowns follow each other in a short interval, while other intervals are longer. The main structure of the vortices and the breakdown remain the same. However, first chaotic fluctuations of the velocity field are superposed which are not symmetric to the center of the mixing channel showing turbulent characteristics. The mixing quality remains high, but the irregular vortex breakdown leads to larger fluctuations of the mixing quality. This transitional regime is called quasi-periodic vortex shedding emphasizing the similarity to the periodic vortex shedding.

Between $Re = 500$ and $Re = 700$, irregular asymmetric perturbations begin to dominate the flow. For $Re \gtrsim 500$, there are short pulses of well-mixed zones resembling the double vortex of the engulfment regime. Between them, in larger sections of the mixing channel, a wavy interface between the two components evolves. In these sections, multiple smaller vortices form. Mostly, they do not engulf each other but contain a large excess of one fluid. Additionally, asymmetric pressure distribution at the inlets leads to pulsations in the mixing ratio. They result in long fluid plugs of different composition. In consequence, the mixing quality drops for long times as shown in Fig. 9, left, disrupted by peaks of high mixing quality when engulfed vortices pass. The drop-down of the mean mixing quality is also shown in Fig. 6.

The asymmetry and the chaotic flow behavior in time and space indicate turbulence. However, it is supposed that for $Re < 2300$ the flow will relaminate further downstream. Thus, we prefer the classification as transition to turbulence.

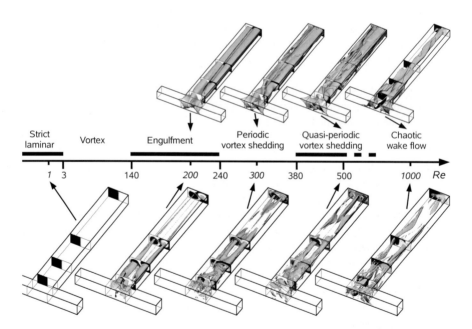

Fig. 10. Overview over flow regimes in the T-mixer. *Top pictures:* mixing interface ($c = 0.5$), *bottom pictures:* iso surfaces of helicity showing the vortices. On the cuts, the concentration is plotted.

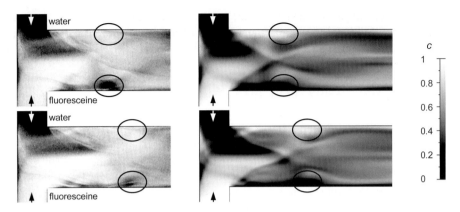

Fig. 11. Mixing of water without and with fluorescent dye at $Re = 300$ for two time steps with about 0.3 ms offset. *Left:* stroboscopic images of the experiment. *Right:* simulations of the depth-averaged normalized concentration c at corresponding time steps. The unmixed regions beside the vortex breakdown are marked.

3.4 Summary of the Mixing Regimes

The investigated flow regimes in the T-mixers are displayed in Fig. 10 for an overview. The top row of pictures shows the interface between the fluids, in the bottom row iso surfaces of the absolute value of the normalized helicity H are shown representing the location of the vortices. It is defined as

$$H = \frac{\mathbf{u} \cdot (\nabla \times \mathbf{u})}{|\mathbf{u}| \, |\nabla \times \mathbf{u}|} \tag{10}$$

with the velocity $\mathbf{u}$ and the vorticity $\nabla \times \mathbf{u}$ (Levy et al. , 1990). High absolute values of the normalized helicity indicate the core of vortices. In comparison to the vorticity, only vortices rotating around the main flow direction forming helical flow are marked, pure shear flow near the walls does not contribute to the value. Together with Fig. 6, it gives an overview over the flow and mixing regimes. In the strictly laminar and vortex regime, only diffusive mixing occurs. Due to the roll-up of fluid lamellae, mixing is strongly enhanced in the engulfment regime and at the periodic vortex shedding regime. However, mixing is non-stoichiometric, especially in the engulfment regime. The periodic vortex shedding further enhances mixing by Taylor dispersion, also leading to better stoichiometry. With the appearance of chaotic vortex shedding in the quasi-periodic regime and especially in the transition to the chaotic wake flow, the mixing quality drops, interrupted by bursts of high mixing quality. The optimal working conditions of the T-mixer are therefore in the engulfment and periodic vortex shedding regime.

As most mixing takes place in the first short part of the mixing channel, where the vortices form, a combination with additional mixing elements in the mixing channel can further enhance mixing. This is also favorable to minimize the non-stoichiometric effects found in the engulfment regime. We will treat such combinations in the following chapter.

4 Improved Mixing Structures

For further mixing of fluids after the the T-mixer as a contacting element, different meandering geometries are examined. The most simple element is a sharp 90° bend or "L". Similar to each halve of the T-mixer in the vortex regime, a Dean vortex pair evolves which provides mixing at low Reynolds numbers and Dean numbers. However, in the corner of the L-bend and, for higher Reynolds numbers, also at the inner wall behind the corner dead zones evolve. Such dead zones pose a risk for clogging and fouling and can collect bubbles during the filling of the structures. If possible, they are usually avoided. In round bends, also Dean vortices form, but dead zones are reduced. Putting these elements in series leads to meandering structures like those shown in Fig. 13.

Combinations of T-mixers with L-bends and S-shaped meanders made from round 90° bends show higher mixing quality than single mixing elements. Vortex formation in the T-mixer is not influenced by subsequent bends or meanders. For a rough estimation of the mixing quality at low Schmidt numbers, the mixing qualities for the single elements can be added. A combination of a T-mixing element with meandering channels also mixes fluid from the vortex pair in the engulfment regime. This levels the stoichiometry of mixing. Combining the T-mixer with an L-bend is also necessary for the parallelizing of mixers on multilayer chips as described by Engler (2006), Kockmann et al. (2006), and Kockmann (2008).

5 Experimental Characterization

In this section, different experimental methods are introduced to verify the simulations and to get further insight into the mixing process and its influence on chemical reactions. Mixing experiments with a non-reacting dye are used for visualization of the mixing regimes and the lamellar structures. Acid-base reactions with an indicator show the diffusively mixed zones and are used to determine the mixing quality. The mixing time is evaluated with the Iodide-Iodate reaction showing also the influence of mixing on chemical parallel competitive reactions.

5.1 Flow Visualization with Fluorescent Dye

Mixing in the convective mixers is visualized by dying one fluid with fluoresceine sodium. Illuminated with blue light, it fluoresces in yellow-green. This dye is also used for laser induced fluorescence described in the chapter of M. Hoffmann. With this method, three dimensional measurements of the concentration field can be performed, as shown in Fig. 5, right. However, the measurements takes several minutes where the concentration profile should not change. Measurements performed by M. Hoffmann from Bremen University were possible only for $Re < 240$, which matches well the transition to transient fluctuations also found by our simulations. Transient flow phenomena in the periodic flow regime can only be observed in a top view on the channel, i. e. a projection of the flow field in the chip plane.

The frequency of the fluctuations is in the order of magnitude of 500 Hz, which can be observed in experiments only with high-speed cameras or stroboscopes. We use a stroboscopic illumination of the fluidic chips by LEDs which replace the microscope lamp. Excitation and emission light are separated by optical interference filters with a cut-off wavelength of about 500 nm in the illumination and the image-forming light paths. With the filtered fluorescence signal, a high contrast between the two fluids can be achieved. The frequency of the stroboscopic light pulses is varied until sharp structures can be seen as slowly floating or even still pictures. As still pictures are also found at multiples and whole-numbered factors of the base frequency, the frequency separation has to be taken into account to determine the true base frequency of the fluctuations.

Figure 11, left, shows two pictures of the mixing zone for $Re = 300$ at different times in the fluctuation cycle. The right subfigures show simulated views on the T-mixer at corresponding time steps. The vortex breakdown can be identified by unmixed fluid packets from the opposite inlet as described in Fig. 3.1. Stroboscopic still images could be taken in the range of $260 < Re < 380$. For $240 < Re < 260$, the flow is too sensitive to external disturbances caused by mechanical shocks in the setup. At $Re > 380$, fluctuations get aperiodic and no frequencies with still images are found, which confirms the prediction of the CFD simulations.

5.2 Measurement of Mixing with Color Reactions

For the quantitative determination of the mixing state, a bromothymol blue solution is mixed with a basic buffer solution. In diffusively mixed regions, the color changes from yellowish green to dark blue. By image processing, the mixing quality α_A defined by the variance of the concentration as a function of the cross sectional area is evaluated. The measured values of α_A depending on the position l in the mixing channel are fitted with the function $\alpha_A = 1 - e^{-l/\lambda_m}$ with the fitting parameter λ_m. This parameter represents a characteristic mixing length where $\alpha_A = 63\%$ is reached.

For T-mixers, the mixing length decreases with an increasing Reynolds number up to $Re = 350$ showing the massive improvement of mixing with the vortex formation. For $Re > 350$, the mixing length is found to be constant at $\lambda_m \approx 4d_h$ resulting in a characteristic mixing time $\tau_m \propto 1/Re$. At $Re > 500$, the experimental curves are not well described by the exponential fit. After a steep increase of $\alpha_A(l)$, the curve gets flat and remains at lower values of α_A. This correlates with fast transient packages of less mixed fluid floating down the channels, as found in the simulations in the chaotic regime. The other convective mixing elements also show the constant mixing length at $Re > 350$, but here the mixing length remains constant for at least up to $Re = 1000$.

5.3 Iodide-Iodate Reaction

The iodide-iodate reaction introduced by Villermaux and Dushman, see e. g. Guichardon et al. (2000), is commonly used for the characterization of micromixers. An instantaneous neutralization and a fast reaction of a iodide iodate solution compete for an acid contained in the second solution. When the mixing time is in the order of magnitude of the reaction time for the second reaction, and regions with an exceed of acid exist for a while, iodine is formed. This is quantitatively characterized by the segregation coefficient X_s representing a normalized yield of the formation of iodine.

Figure 12, left, shows the dependence of the segregation index X_s on the second Damköhler number $Da_{II} = \frac{\tau_m}{\tau_r}$ for three scaled T-mixers. The mixing time τ_m is calculated by the experimentally found relation $\tau_m = \lambda_m/\bar{u} = 4d_h/\bar{u}$. The reaction time τ_r is estimated from the educt concentrations and the reaction rates (Engler, 2006, Guichardon et al. , 2000). The linear relationship between X_s and Da_{II} confirms that the iodide-iodate reaction can be used to determine the mixing time in convective micromixers.

In Fig. 12, right, the segregation index is plotted over the Reynolds number for differently scaled T-mixers. The graphs show better mixing for increasing Reynolds numbers up to $Re \approx 500$. For higher Reynolds numbers, the segregation index rises again showing the worsening of mixing, which is also observed in the simulations and the bromothymol blue color reaction. Particularly the T $200\times100\times100$ with its very low values of X_s shows this behavior. Other convective mixer structures as described in Fig. 4 do not show rising mixing times for increasing Reynolds numbers.

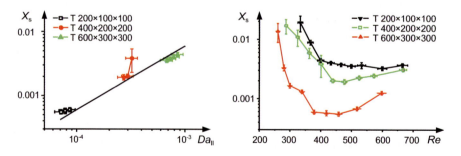

Fig. 12. Segregation coefficient X_s for different T-mixers. *Left:* dependency on the second Damköhler number Da_{II} for the Reynolds number range covered by the convective lamination model ($350 < Re < 500$). *Right:* dependency on the Reynolds number.

6 Convective Lamination Model

Numerical and experimental investigations of the convective micromixers are summarized in the convective lamination model first described by Engler (2006). It associates the energy dissipation with the mixing process. Convective mixing is superposed by diffusive mixing that homogenizes the fluids on the molecular scale. A characteristic diffusive mixing time of lamellae is given by $\tau_{\mathrm{m}} = \frac{d_1^2}{2D}$ with the lamellae thickness d_1. Comparing this diffusive mixing time with the convective mixing time $\lambda_{\mathrm{m}}/\bar{u}$, the final lamellae thickness gets $d_1 = \sqrt{2D\lambda_{\mathrm{m}}/\bar{u}}$. With the correlation of the mixing length and the hydraulic diameter of convective micromixers $\lambda_{\mathrm{m}} \approx 4d_{\mathrm{h}}$, the dimensionless lamellae thickness is calculated by

$$\frac{d_1}{d_{\mathrm{h}}} = \sqrt{\frac{8}{ReSc}} \ . \tag{11}$$

For $Re = 350$ and $Sc = 3600$, this means a final lamellae thickness of $d_1 \approx 1.5\,\mu\mathrm{m}$, which matches well the values measured by M. Hoffmann, see the contribution of our cooperation partners from IUV Bremen.

The pressure drop over convective mixers can be divided into a term linear to the mean fluid velocity and a quadratic term. The linear term represents the viscous pressure drop in channel flow according to Hagen-Poiseuille's law. A dimensionless representation of the pressure drop is the Euler number $Eu = \frac{\Delta p}{\rho \bar{u}^2}$. Considering only the quadratic part of the pressure drop caused by vortices, the Euler number becomes $Eu_{\mathrm{V}} = \frac{\Delta p_{\mathrm{V}}}{\rho \bar{u}^2} = \zeta/2$. From the pressure drop caused by vortices, the mean specific energy dissipation into the vortex formation can be calculated as

$$\bar{\varepsilon} = \frac{\Delta p_{\mathrm{V}} \cdot \dot{V}}{m_{\mathrm{eff}}} = \frac{\Delta p_{\mathrm{V}} \cdot \bar{u}}{\rho \lambda_{\mathrm{m}}} = \frac{Eu_{\mathrm{V}} \cdot \bar{u}^3}{\lambda_{\mathrm{m}}} \tag{12}$$

with the volume flow $\dot{V}$ and the effective mass m_{eff} in which the energy is dissipated. Assuming that vortices are created only in the active mixing zone, m_{eff} is determined by the mixing length λ_{m}. This assumption is true for T-mixers, for tangential mixers, which create one vortex by tangential inflow (Kockmann et al. , 2006), and for short meandering mixing elements. Solving for $\tau_{\mathrm{m}} = \lambda_{\mathrm{m}}/\bar{u}$, the mixing time gets

$$\tau_{\mathrm{m}} = \sqrt[3]{\frac{Eu_{\mathrm{V}} \lambda_{\mathrm{m}}^2}{\bar{\varepsilon}}} \propto \frac{1}{\sqrt[3]{\bar{\varepsilon}}} \ . \tag{13}$$

Inserting the correlation of the mixing length and the hydraulic diameter $\lambda_{\mathrm{m}} = 4d_{\mathrm{h}}$ found experimentally for $Re > 350$ and the Euler number $Eu_{\mathrm{V}} = 1.7$ as determined by simulations and experimentally verified, the mixing time calculates as

$$\tau_{\mathrm{m}} \approx 3\sqrt[3]{\frac{d_{\mathrm{h}}^2}{\bar{\varepsilon}}} \ . \tag{14}$$

This equation resembles the meso mixing time $\tau_{\mathrm{meso}} = 2\sqrt[3]{d_{\mathrm{h}}^2/\bar{\varepsilon}}$ derived from the disintegration of eddies in fully developed turbulent flow (Bałdyga and Bourne, 1999, Bourne, 2003). Similar equations are also valid for the other investigated convective mixing elements, where only the prefactor differs slightly. This shows the relationship between the lamellae formation in convective micromixers and the turbulent engulfment.

For comparison of different static mixers, the mixing motion is compared with the energy dissipation. Kockmann et al. (2006) and Engler (2006) introduce two dimensionless numbers, the mixing and mixer effectiveness, describing this benefit-cost ratio.

7 Wall Contact vs. Mixing in Microreactors

Besides the mixing of two fluids, also the fluid transport towards the reactor walls is an important parameter of microreactors. The convective transport of fluids towards the wall determines the heat transfer coefficient in reactors and heat exchangers. It is also necessary in reactors with catalyst-coated walls, where educts must be brought into contact with the catalyst and product must be removed from near-wall regions. Also quenching as the inverse process of catalysis depends on a good wall contact of the fluids. For thermal quenching, the heat of the reaction is immediately transferred to the wall and the reaction stops. The other mechanism is radical quenching where radicals are recombined and thus de-activated at the wall suppressing chain reactions. With quenching in micromixers, the mixing process and the reaction zone can be separated spacially ensuring equal reaction conditions for the whole reactants. One example of use is premixing of fuel gas with air.

7.1 Simulations and Optimization of Wall Contact

For the evaluation of wall contact in different meandering microreactors, flow and heat transfer from the wall into the fluid are simulated with CFD-ACE+. The standard geometry consists of a $300 \times 600\,\mu\mathrm{m}^2$ channel to match the mixing channels of the standard T-mixer, turning in $180°$ bends of $400\,\mu\mathrm{m}$ mean radius. This corresponds to a Dean number of $De \approx Re$, see also Sect. 3.1. Based on this meander, geometries with more and less changes of the sense of rotation are investigated and enhanced geometries are developed, see Fig. 13. To avoid an intersection of the channels, the geometry without any change of the sense of rotation had to be extended to the third dimension to form a helix. Like in the simulations of the T-mixers, water at $20°\mathrm{C}$ is used as liquid. At the channel walls of the meandering section, a constant wall temperature of $100°\mathrm{C}$ is assumed as boundary condition. Together with the simplification of temperature independent material properties, this allows a direct qualitative comparison to mass transfer for transport limited reactions at the

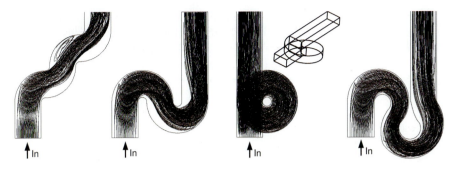

Fig. 13. Meandering reactor channels for the evaluation of wall contact with stream lines. *From left to right:* 90° bends, 180° bends, helix, optimized meander with 220° and 260° bends.

walls. Typical mesh sizes of 10 μm are used in the simulations, which resolves the hydromechanics and heat transfer sufficiently, as found in grid refinement studies.

For heat transfer, catalysis at the wall, and particularly quenching, it is necessary to get all parts of the fluid flow into contact with the wall. This can be described by a contact time distribution. It is defined as the residence time distribution within a region near the wall. In our geometries, it is chosen as the region up to 50 μm from each wall. This corresponds to the typical thermal diffusion length within the mean residence time in a 4.3 mm long channel.

In Fig. 14, the contact time distribution for the standard geometries with a total angle of 360° shows that the helix has the contact time distribution with the smallest part of core flow with contact times below 1 ms. The more often the sense of rotation

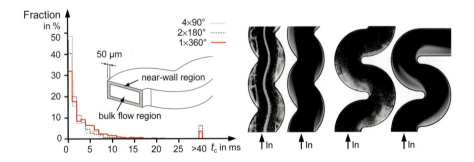

Fig. 14. Evaluation of the wall contact in meanders. *Left:* contact time distribution of the standard geometries. The bar marked with > 40 stands for all streamlines which end within the channel due to very low flow velocities near the wall. *Right:* Pictures of the luminol experiments. The left image of each pair is a photograph of the luminol experiment, the right one shows the simulated temperature distribution on a horizontal cut near the bottom wall.

Table 1. Basic solutions for the Luminol experiments. Both solutions are mixed 1:1 just before the experiments.

Solution 1	Solution 2
50 ml water	50 ml water
0.5 ml $NH_{3\,(aq)}$ 30%	0.5 ml $H_2O_{2\,(aq)}$ 30%
0.5 g Na_2CO_3	
0.05 g Luminol	

changes, the larger is the part of core flow in the meanders. This development of a core flow can also be seen in Fig. 13, where after every change of the rotation dead zones separate the main flow from the outer wall. The total heat flux from the wall into the fluid is maximal for the helix and decreases with the frequency of changes of the sense of rotation showing the effect of the dead zones. An improved meander structure with bends of 220° and 260°, where the dead zones are cut off the channels, shows a similar total heat transfer as the helix channel. Unfortunately, this improvement of planar meandering channels costs about the double pressure drop, as the channel width is reduced.

7.2 Quantification of Wall Contact with a Luminol Reaction

For comparison with the simulated temperature fields, a reaction system based on the oxidation of Luminol catalyzed by copper ions is used. For this experiment, the walls of the chips are plated with copper. In a first reaction step, copper is dissolved by a solution of hydrogen peroxide and ammonia and forms a complex:

$$Cu + H_2O_2 + 4NH_3 \longrightarrow [Cu(NH_3)_4]^{2-} + 2OH^- \tag{15}$$

This copper complex catalyzes the oxidation of Luminol with hydrogen peroxide, which is one of the strongest chemiluminescence reactions:

$$\tag{16}$$

The composition of the educt solutions is shown in Table 1. In regions where the fluid stays in contact with the wall for a long time, it begins to glow blue, which can be observed under the microscope. Due to the similarity of heat and mass transfer, the glowing fluid regions correspond to the hot regions in the simulations of heat transfer. Figure 14, right, shows a comparison of simulated and measured wall contact distributions. This experimental method provides an easy possibility to detect dead zones which lead to inefficient wall contact and hot spots in heat exchangers.

8 Conclusion

In this chapter, the mixing regimes in T-mixers between strictly laminar flow and the transition to turbulence were described. Good mixing is found in the engulfment regime and the periodic vortex shedding. Chaotic fluctuations at $Re > 500$ cause a drop of the mixing quality. A study of the local stoichiometry of the mixing process shows that subsequent meander structures are sensible. Besides simulations, also experiments with dye markers, color reactions and the iodide-iodate reaction system were performed. They confirm the simulated results, and give additional information on the diffusive micromixing, leading to the convective lamination model. This model links the energy dissipation and the mixing time for higher Reynolds numbers, showing the similarity to turbulent mixing. Finally, heat and mass transfer in meandering channels are characterized by a new reaction system. With this Luminol reaction, dead zones reducing the wall contact can be easily detected. This was shown in a design process of meandering microchannels improved for wall contact.

References

Bałdyga, J., Bourne, J.R.: Turbulent Mixing and Chemical Reactions, 1st edn. Wiley VCH, Weinheim (1999)

Berger, S.A., Talbot, L., Yao, L.S.: Flow in curved pipes. Annu. Rev. of Fluid Mech. 15, 461–512 (1983)

Bothe, D., Stemich, C., Warnecke, H.-J.: Fluid mixing in a T-shaped micro-mixer. Chem. Eng. Sci. 61, 2950–2958 (2006)

Bourne, J.R.: Mixing and the selectivity of chemical reactions. Org. Proc. Res. Dev. 7(4), 471–508 (2003)

Chao, S., Holl, M.R., Koschwanez, J.H., Seriburi, P., Meldrum, D.R.: Scaling for microfluidic mixing. In: Proc. of ICMM 2005, 3rd Int. Conf. on Microchannels and Minichannels, Toronto, Ontario, Canada (2005) 75236

Cybulski, A., Moulijn, J.A., Sharma, M.M., Sheldon, R.A. (eds.): Fine Chemicals Manufacture: Technology and Engineering. Elsevier, Amsterdam (2001)

Danckwerts, P.V.: The definition and measurement of some characteristics of mixtures. Appl. Sci. Res. A 3, 279–296 (1952)

Dreher, S.: Fertigung und Charakterisierung eines Mikroreaktors zur Aerosolbildung. Diploma thesis, Albert-Ludwigs-Universität Freiburg (2005)

Engler, M.: Simulation, Design, and Analytical Modelling of Passive Convective Micromixers for Chemical Production Purposes. Shaker Verlag, Aachen (2006)

Gobert, C., Schwertfirm, F., Manhart, M.: Lagrangian scalar tracking for laminar micromixing at high Schmidt numbers. In: Proceedings of FEDSM 2006, ASME, FEDSM 2006-98035, Miami, Florida, USA (2006)

Guichardon, P., Falk, L., Villermaux, J.: Characterization of micromixing efficiency by the iodide-iodate reaction system. Part I: Experimental procedure; Part II: Kinetic study. Chem. Eng. Sci. 55(19), 4233–4253 (2000)

Heim, M., Wengeler, R., Nirschl, H., Kasper, G.: Particle deposition from aerosol flow inside a T-shaped micro-mixer. J. Micromech. Microeng. 16, 70–76 (2006)

Howell Jr., P.B., Mott, D.R., Golden, J.P., Ligler, F.S.: Design and evaluation of a Dean vortex-based micromixer. Lab on a Chip 4, 663–669 (2004)

Kockmann, N.: Transport Phenomena in Micro Process Engineering. Heat and Mass Transfer. Springer, Berlin (2008)

Kockmann, N., Kiefer, T., Engler, M., Woias, P.: Convective mixing and chemical reactions in microchannels with high flow rates. Sensors and Actuators B 117(2), 495–508 (2006)

Kockmann, N., Dreher, S., Engler, M., Woias, P.: Simulation and characterization of microreactors for aerosol generation. Microfluidics and Nanofluidics 3(5), 581–589 (2007)

Levy, Y., Degani, D., Seginer, A.: Graphical visualization of vortical flows by means of helicity. AIAA Journal 28, 1347–1352 (1990)

Author Index

Printing and Binding: Stürtz GmbH, Würzburg